Numerical Methods for Simulation and Optimization of Piecewise Deterministic Markov Processes

Series Editor
Nikolaos Limnios

Numerical Methods for Simulation and Optimization of Piecewise Deterministic Markov Processes

Application to Reliability

Benoîte de Saporta
François Dufour
Huilong Zhang

iSTE

WILEY

First published 2016 in Great Britain and the United States by ISTE Ltd and John Wiley & Sons, Inc.

ISTE Ltd
27-37 St George's Road
London SW19 4EU
UK

www.iste.co.uk

John Wiley & Sons, Inc.
111 River Street
Hoboken, NJ 07030
USA

www.wiley.com

Library of Congress Control Number: 2015954427

British Library Cataloguing-in-Publication Data
A CIP record for this book is available from the British Library
ISBN 978-1-84821-839-0

Contents

Preface

The intent of this book is to present recent results on theoretical and numerical aspects of simulation and optimization for piecewise deterministic Markov processes (PDMPs). To date, few works have been devoted to the development of numerical methods dedicated to PDMPs in the literature. This book is the first monograph to address such problems. We attempt to provide a systematic framework for an understanding of the main concepts and tools associated with this problem, based on previous works of the authors. We tried to keep an equilibrium between the theoretical, numerical and practical sides related to PDMPs. This book also contains several motivational examples illustrating our approach and taken from the field of reliability.

Most of the material presented in this book was scattered throughout a variety of sources, which included journal articles and conference proceedings papers. This motivated the authors to write this text, systematically putting together these results with consistent and unified notation.

The book is targeted primarily for advanced students and practitioners of numerical simulation and/or optimization of stochastic processes. In particular, we hope that the book will be a valuable source for experts in the field of Markov processes. Moreover, we believe that the book should be suitable for certain advanced courses or seminars. As a background, we need to acquaint ourselves with the theory of probability and some knowledge of stochastic processes. No prerequisite knowledge on PDMPs is necessary.

The authors are indebted to many people and institutions who have contributed in many ways to the writing of this book. We gratefully acknowledge the University of Bordeaux, Bordeaux INP, the University of Montpellier and Inria Bordeaux Sud-Ouest, and in particular, *Institut*

Mathématiques de Bordeaux (IMB, Mathematics Institute of Bardeaux) and *Institut Montpelliéran Alexander Grothendieck*. This book owes much to the Inria Team CQFD and to our industrial partners Charles Elegbede (Airbus Defence & Space), Didier Bihannic and Michel Prenat (Thales Optronique SAS) to whom we are immensely grateful. Many thanks also go to the former PhD students of the Inria team CQFD Karen Gonzalez, Adrien Brandejsky and Camille Baysse in chronological order. We gratefully acknowledge the series editor Nikolaos Limnios for his encouragement, time and consideration. During the years of our study, this research was supported in part by the French National Research Agency (ANR), program ARPEGE, project FAUTOCOES, grant number ANR-09-SEGI-004 and program *Investments for the future*, IdEx Bordeaux – CPU, number ANR-10-IDEX-03-02.

Last, but not least, we are very grateful to our families for their continuing and unwavering support. To them we dedicate this book.

Benoîte DE SAPORTA
François DUFOUR
Huilong ZHANG
October 2015

Introduction

I.1. Preliminaries

This book focuses on the theoretical and numerical aspects of simulation and optimization for piecewise deterministic Markov processes (PDMPs). PDMPs were first introduced in the literature [DAV 84] as a general class of stochastic hybrid models. They form a family of Markov processes involving deterministic motion punctuated by random jumps. The motion of a PDMP includes both continuous and discrete variables. The continuous state variable represents the physical parameters of the system under consideration. The discrete mode characterizes the regimes of operation of the physical system and/or the environment. The process is defined through three local characteristics, namely: the flow describing the deterministic trajectory between two consecutive jumps, the intensity function giving the jump frequency and the Markov kernel specifying the post-jump location. A suitable choice of the state space and these local characteristics provides stochastic models covering a large number of problems such as engineering systems, operation research, economics, management science, biology, internet traffic, networks and reliability. The class of PDMPs is thus considered and recognized as a powerful modeling tool for complex systems as described in [DAV 93].

However, surprisingly few works are devoted to the development of numerical methods for PDMPs to solve problems of practical importance such as evaluation and optimization of functionals of the process. For the numerical aspects, stochastic control theory has, up to now, been mainly concerned with stochastic differential equations (SDEs in short). The numerical developments for this class of processes are extensive and largely complete. It capitalizes on the connection between SDEs and second order partial differential equations (PDEs in short) and the fact that the properties of the latter equations are very well understood. It is, however, hard to deny that the development of computational methods for the control of PDMPs has received little attention. One of the main reasons is that the role played by the familiar PDEs in the diffusion models is played here by certain systems of integro-differential equations for which there is not (and cannot be) a unified theory such as for PDEs as emphasized by Davis in his book [DAV 93]. The main objective of our book consists of presenting mathematical tools recently developed by the authors to address such problems. This book is not only focused on theoretical aspects such as proof of convergence of the approximation procedures but is also concerned with its applicability to practical problems. The approach we are proposing is general enough to be applied to several application domains. In particular, our results are illustrated by examples from the field of reliability.

Our approximation technique is based on the special structure of PDMPs. Embedded into the continuous time process defining a PDMP, there exists a discrete time Markov chain given by the post-jump locations and jump times of the PDMP. Our approach then consists of the following steps:

1) First, rewrite the quantity to be approximated in terms of the discrete-time chain. This is always possible as this chain contains all the randomness of the process, the motion of the PDMP being deterministic between jumps. We can often obtain a recursive formulation due to the Markov property.

2) Second, replace the post-jump locations and jump times by random variables taking only finitely many values. This can be done using optimal quantization techniques. We thus obtain a fully computable approximation. We can also estimate the sharpness of this approximation.

This strategy enables us to address a large class of numerical problems. In particular, in this book we focus on the computation of expectation of functionals of PDMPs with applications to the evaluation of service times, and we are also interested in solving optimal control problems with applications to maintenance optimization.

The objective here is twofold. On the one hand, we will rigorously construct the numerical approximation described above and give proof of its convergence under reasonably weak assumptions. On the other hand, we will demonstrate its high practical applicability through the numerous examples developed in this book.

I.2. Overview of the chapters

The book is organized in three main parts.

The first part is dedicated to the main mathematical concepts illustrated by examples. In Chapter 1, we introduce some notations and provide the definition of a PDMP as well as their main properties that will be used repeatedly in this book. Here, we present the main assumptions that will be used throughout the book. The important discrete-time Markov chain naturally embedded in a PDMP is presented and studied in detail since this chain is at the heart of our numerical approximations. Finally, some simple examples of PDMPs are also presented that will serve to illustrate our results throughout the book. Chapter 2 is devoted to the presentation of two examples of complex systems arising in dynamic reliability problems that can be modeled as PDMPs. The first example concerns a metallic structure subject to corrosion. This example was provided to the authors by Astrium Space Transportation. The second example is the so-called heated hold-up tank. It is a well known case test of dynamic reliability. These two models will also serve to illustrate our numerical procedures developed throughout the following chapters. The basic properties of quantization methods are introduced in Chapter 3. It is important to emphasize that this powerful numerical technique plays a pivotal role in the development of our numerical tools. In particular, quantization of a single random variable and the quantization of a Markov chain are presented with their pseudo-code. Finally, the simulation of PDMPs is discussed in the context of quantization.

The second part is devoted to the study and the simulation of expectations of functionals of PDMPs. Chapter 4 focuses on the analysis of practical numerical methods for the approximation of generic expectation of function of PDMPs. Such quantities may correspond to reliability indices and also appear as cost or reward functions in optimization problems. The objective of Chapter 5 is to propose numerical methods for the evaluation of the survival function, the moments and the distribution of exit times. From the applications point of view, such random times, such as the time to reach a given threshold, are crucial in the calculation of service time of complex structures. For all our numerical procedures, we also provide the corresponding pseudo-code. Finally, the different approximation schemes proposed in the two previous chapters are numerically tested on the corrosion example in Chapter 6.

The third part is dedicated to the development of numerical techniques for optimal control problems such as perfectly observed and partially observed optimal stopping problems and impulse control problems, in Chapters 7, 8 and 10 respectively. These chapters are essentially organized in the same manner. The dynamic programming equation associated with the optimization problem under consideration is first studied. Based on this formulation, an approximation procedure is proposed to obtain computable estimations of the value function and of a ϵ-optimal strategy. As in the previous parts, we also provide the corresponding pseudo-code. The last step consists of analyzing the convergence of our numerical approximation scheme as well as in providing rates of convergence. The objective of Chapter 9 is to demonstrate the high practical power of our approach described in Chapter 7 applied to the challenging problem of maintenance optimization.

PART 1

Piecewise Deterministic Markov Processes and Quantization

1

Piecewise Deterministic Markov Processes

1.1. Introduction

In 1980, Davis [DAV 93] introduced in probability theory the class of piecewise deterministic Markov processes (PDMPs) as a general class of models suitable for formulating optimization problems in queuing and inventory systems, maintenance-replacement models, investment scheduling and in many other areas of operation research. These models are described by two variables. To the usual Euclidean variable representative of the state of the process, we add a discrete variable, called regime or mode that takes values in a finite or countable set. In this context, the state variable represents the physical parameters of the system under consideration. For example, it can be the position or the orientation of a satellite or the pressure in a tank. The mode characterizes the regimes of operation of the physical system from nominal to failure regime.

From a mathematical point of view, the notion of a PDMP is very intuitive and simple to describe. Starting from a point of the state space, the process follows a deterministic trajectory, namely a flow indexed by the mode, until the first jump time, which occurs either spontaneously in a random manner or when the trajectory hits the boundary of the state space. Between two jumps, the mode is assumed to be constant. In both cases, a new point and a new regime are selected by a random operator and the process restarts from this new point under this new mode. There exist two types of jumps. The first type is deterministic. From the mathematical point of view, it is given by the trajectory hitting the boundary of the state space. From the physical point of view, it can be seen as a modification of the mode of operation when a physical parameter reaches a prescribed level, for example when the pressure of a tank reaches the critical value. The second type is stochastic. It models the random nature of failures or inputs that modify the mode of operation of the system.

This chapter is dedicated to the definition of PDMPs and statement of the main properties that we will use throughout the book. It is organized as follows. Section 1.2 presents general notations. In section 1.3, we give the formal definition of a PDMP. In section 1.4, we state and comment the main technical assumptions that will be required in this book. This section may be skipped by readers who are only interested in applications. In section 1.5, we define a time-augmented PDMP and establish that the properties of the original PDMP transfer to the time-augmented one. In section 1.6, we define and study a discrete-time Markov chain naturally embedded into a PDMP. This chain is at the heart of our numerical approximations. Section 1.7 provides some technical properties of the stopping times of a PDMP. This section may also be skipped at first reading. Finally, in section 1.8, we give some simple examples of PDMPs that will serve to illustrate our results throughout the book. More involved examples are detailed in Chapter 2. The readers interested in examples outside the reliability scope or further technical details on PDMPs are referred to [DAV 93].

1.2. Notation

The purpose of this section is to standardize some terminology and notation that will be used throughout the book. The set of non-negative integers is denoted by $\mathbb{N}$, that of positive integers by $\mathbb{N}^*$, $\mathbb{R}$ is the set of real numbers and we use $\mathbb{R}^+$ for the non-negative real numbers. The d-dimensional Euclidean space is denoted by $\mathbb{R}^d$. For $a, b \in \mathbb{R}$, $a \wedge b = \min(a, b)$ (respectively, $a \vee b = \max(a, b)$) is the minimum (respectively, maximum) of a and b, and $a^+ = a \vee 0$. By convention, set $\inf \emptyset = +\infty$.

Let $\mathcal{X}$ be a metric space with distance $d_{\mathcal{X}}$. For a subset A of $\mathcal{X}$, ∂A is the boundary of A, $\overline{A}$ is its closure and A^c is its complement. We denote by $\mathfrak{B}(\mathcal{X})$ the Borel σ-field of $\mathcal{X}$ and by $\mathbf{B}(\mathcal{X})$ the set of real-valued, bounded and measurable functions defined on $\mathcal{X}$. For any function $w \in \mathbf{B}(\mathcal{X})$, we write C_w for its upper bound, that is:

$$C_w = \sup_{x \in \mathcal{X}} |w(x)|.$$

Let $\mathbf{L}(\mathcal{X})$ be the subset of $\mathbf{B}(\mathcal{X})$ of Lipschitz-continuous functions, and for any function $w \in \mathbf{L}(\mathcal{X})$, denote by L_w its Lipschitz constant:

$$L_w = \sup_{x \neq x' \in \mathcal{X}} \frac{|w(x) - w(x')|}{d_{\mathcal{X}}(x, x')}.$$

for a Markov kernel P on $(\mathcal{X}, \mathfrak{B}(\mathcal{X}))$ and functions w and w' in $\mathbf{B}(\mathcal{X})$, set:

$$Pw(x) = \int_{\mathcal{X}} w(y)P(x, dy), \quad (wPw')(x) = w(x) \int_{\mathcal{X}} w'(y)P(x, dy),$$

for any $x \in \mathcal{X}$.

1.3. Definition of a PDMP

PDMPs are defined as follows. Let M be the finite set of the possible modes of the system. For all modes m in M, let E_m be an open subset in $\mathbb{R}^d$ endowed with the usual Euclidean norm $|\cdot|$. Set:

$$E = \{(m, \zeta), m \in M, \zeta \in E_m\}.$$

Define on E the following distance, for $x = (m, \zeta)$ and $x' = (m', \zeta') \in E$:

$$|x - x'| = |\zeta - \zeta'| \mathbb{1}_{\{m = m'\}} + \infty \mathbb{1}_{\{m \neq m'\}}.$$

A PDMP on the state space E is defined by three local characteristics (Φ, λ, Q) where:

– the flow $\Phi(x, t) = (m, \Phi_m(\zeta, t))$ for all $x = (m, \zeta)$ in E and $t \geq 0$, where $\Phi_m : \mathbb{R}^d \times \mathbb{R} \to \mathbb{R}^d$ is continuous and for all $s, t \geq 0$, we have $\Phi_m(\cdot, t + s) = \Phi_m(\Phi_m(\cdot, s), t)$. It describes the deterministic trajectory of the process between jumps. Set:

$$t^*(x) = t^*_m(\zeta) = \inf\{t > 0 : \Phi_m(\zeta, t) \in \partial E_m\},$$

the time the flow takes to reach the boundary of the domain starting from position ζ in mode m;

– the jump intensity $\lambda : \overline{E} \to \mathbb{R}^+$ is measurable and has the following integrability property: for any $x = (m, \zeta)$ in E, there exists $\epsilon > 0$ such that:

$$\int_0^\epsilon \lambda\big(m, \Phi_m(\zeta, t)\big)dt < +\infty.$$

It characterizes the frequency of jumps. For all $x = (m, \zeta)$ in E, and $t \leq t^*(x)$, set:

$$\Lambda(m, \zeta, t) = \int_0^t \lambda\big(m, \Phi_m(\zeta, s)\big)ds;$$

– the Markov kernel Q on $\big(\overline{E}, \mathfrak{B}(\overline{E})\big)$ represents the transition measure of the process and allows us to select the new location and mode after each jump. It satisfies:

$$Q(x, \{x\} \cup \partial E) = 0$$

for all x in $\overline{E}$, meaning that each jump is made in E and changes the location and/or the mode of the process.

From these characteristics, it can be shown that there exists a filtered probability space $(\Omega, \mathcal{F}, (\mathcal{F}_t)_{t\geq 0}, (\mathbb{P}_x)_{x\in E})$ on which a process $(X_t)_{t\geq 0}$ can be defined as a strong Markov process. The process $(X_t)_{t\geq 0}$ has two components $X_t = (m_t, \zeta_t)$ where the first component m_t is usually called the mode or the regime and the second component ζ_t is the so-called Euclidean variable. The motion of this process can then be defined iteratively. Starting at an initial point $X_0 = (m_0, \zeta_0)$ with $m_0 \in M$ and $\zeta_0 \in E_{m_0}$, the first jump time T_1 is determined by the following survival function:

$$\mathbb{P}_{(m_0,\zeta_0)}(T_1 > t) = \mathrm{e}^{-\Lambda(m_0,\zeta_0,t)}\mathbb{1}_{\{t<t^*(m_0,\zeta_0)\}}. \qquad [1.1]$$

On the interval $[0, T_1)$, the process $(X_t)_{t\geq 0}$ follows the deterministic trajectory given by $\Phi_{m_0}(\zeta_0, t)$ and the regime m_t is constant and equal to m_0. At the random time T_1, a jump occurs. A jump can produce either a discontinuity in the Euclidean variable ζ_t and/or a change of mode. The process restarts at a new mode and/or position $X_{T_1} = (m_{T_1}, \zeta_{T_1})$, according to the distribution $Q(m_0, \Phi_{m_0}(\zeta_0, T_1), \cdot)$. An inter-jump time $T_2 - T_1$ is then selected in a similar way to equation [1.1], and on the interval $[T_1, T_2)$ the process follows the path $m_t = m_{T_1}$ and $\zeta_t = \Phi_{m_{T_1}}(\zeta_{T_1}, t - T_1)$. The process $(X_t)_{t\geq 0}$ thus defined is a PDMP on the state space E (see section 1.6 for an explicit construction). An example of such a trajectory is given in Figure 1.1. The first jump is random, but the second jump is forced (or deterministic) since the process hits the boundary of the domain.

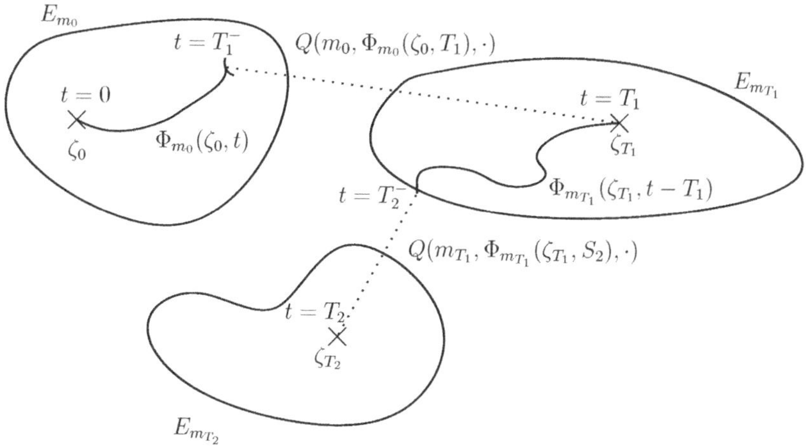

Figure 1.1. *A sample trajectory of a generic PDMP*

In order to avoid any technical problems due to the possible explosion of the process, we make the following standard assumption (see section 24 in [DAV 93]).

ASSUMPTION 1.1.– For all $(x, t) \in E \times \mathbb{R}^+$, $\mathbb{E}_x\left[\sum_{n=1}^{\infty} \mathbb{1}_{\{T_n < t\}}\right] < +\infty$.

We will also assume throughout the book that the deterministic time to reach the boundary of the state space is bounded.

ASSUMPTION 1.2.– The exit time t^* is in $\mathbf{B}(E)$ and bounded by C_{t^*}.

In most practical applications, the physical properties of the system ensure that either t^* is bounded, or the problem has a natural deterministic time horizon T. In the latter case, there is no loss of generality in considering that t^* is bounded by this deterministic time horizon. This leads to replacing C_{t^*} by T (see also section 2.2.2).

For notational convenience, we set $T_0 = 0$. The sequence $(Z_n)_{n\in\mathbb{N}}$ with $Z_n = X_{T_n}$ describes the post jump locations of the process $(X_t)_{t\geq 0}$ and

$(S_n)_{n\in\mathbb{N}}$ with $S_n = T_n - T_{n-1}$ for $n \geq 1$ and $S_0 = 0$ gives the times between two consecutive jumps.

1.4. Regularity assumptions

Here we introduce some technical regularity assumptions on the local characteristics of the PDMP that will be useful in the following chapters in order to prove the convergence of our approximation procedures; this section may be skipped at first reading. We start by introducing new sets of functions that have some regularity property along the flow.

1.4.1. *Lipschitz continuity along the flow*

Denote by $\mathbf{L}_\Phi(E)$ the set of functions $w \in \mathbf{B}(E)$ that are Lipschitz continuous along the flow, i.e. the real-valued, bounded, measurable functions defined on E and satisfying the following conditions:

– for all $x \in E$, the map $w(\Phi(x,\cdot))$: $[0, t^*(x)) \to \mathbb{R}$ is continuous, the limit $\lim_{t\to t^*(x)} w(\Phi(x,t))$ exists and is denoted by $w\big(\Phi(x, t^*(x))\big)$;

– there exists $[w]_1^E \in \mathbb{R}^+$ such that for $x, y \in E$ and $t \in [0, t^*(x) \wedge t^*(y)]$, we have

$$|w(\Phi(x,t)) - w(\Phi(y,t))| \leq [w]_1^E |x-y|;$$

– there exists $[w]_2^E \in \mathbb{R}^+$ such that for all $x \in E$ and $t, t' \in [0, t^*(x)]$, we have:

$$|w(\Phi(x,t)) - w(\Phi(x,t'))| \leq [w]_2^E |t-t'|;$$

– there exists $[w]_*^E \in \mathbb{R}^+$ such that for all $x, y \in E$, we have:

$$|w(\Phi(x,t^*(x))) - w(\Phi(y,t^*(y)))| \leq [w]_*^E |x-y|.$$

Denote by $\mathbf{L}_\Phi(\partial E)$ the subset of $\mathbf{B}(\partial E)$ of real-valued, bounded, measurable functions defined on ∂E for which there exists $[w]_*^{\partial E} \in \mathbb{R}^+$ such that for $x, y \in E$, we have:

$$|w(\Phi(x,t^*(x))) - w(\Phi(y,t^*(y)))| \leq [w]_*^{\partial E} |x-y|.$$

Whenever there is no ambiguity, we will use the simpler notation $[w]_i$ instead of $[w]_i^E$ for $i \in \{1, 2, *\}$ and $[w]_*$ instead of $[w]_*^{\partial E}$.

We also need a notion of local Lipschitz-continuity along the flow up to a given time horizon that may be smaller than the exit time t^*. For all $u \geq 0$, denote by $\mathbf{L}_\Phi^u(E)$ the set of functions $w \in \mathbf{B}(E)$ that are Lipschitz continuous along the flow until time u, i.e. the real-valued, bounded, measurable functions defined on E and satisfying the following conditions:

– for all $x \in E$, the map $w(\Phi(x, \cdot))$: $[0, t^*(x) \wedge u) \to \mathbb{R}$ is continuous and $\lim_{t \to t^*(x) \wedge u} w(\Phi(x, t))$ exists and is denoted by $w\big(\Phi(x, t^*(x) \wedge u)\big)$;

– there exists $[w]_1^{E,u} \in \mathbb{R}^+$ such that for all $x, y \in E$ and $t \in [0, t^*(x) \wedge t^*(y) \wedge u]$, we have:

$$|w(\Phi(x, t)) - w(\Phi(y, t))| \leq [w]_1^{E,u} |x - y|;$$

– there exists $[w]_2^{E,u} \in \mathbb{R}^+$ such that for $x \in E$ and $t, t' \in [0, t^*(x) \wedge u]$, we have:

$$|w(\Phi(x, t)) - w(\Phi(x, t'))| \leq [w]_2^{E,u} |t - t'|;$$

– there exists $[w]_*^{E,u} \in \mathbb{R}^+$ such that for all $x, y \in E$, if $t^*(x) \leq u$ and $t^*(y) \leq u$, we have:

$$|w(\Phi(x, t^*(x))) - w(\Phi(y, t^*(y)))| \leq [w]_*^{E,u} |x - y|.$$

We will also drop the superscript E in the Lipschitz constants whenever there is no ambiguity regarding the state space. Note that for all $u \leq u'$, we have $\mathbf{L}_\Phi^{u'}(E) \subset \mathbf{L}_\Phi^u(E)$ with $[w]_i^{E,u} \leq [w]_i^{E,u'}$, $i \in \{1, 2, *\}$. As the exit time t^* is bounded by C_{t^*}, then $\mathbf{L}_\Phi^u(E) = \mathbf{L}_\Phi(E)$ for all $u \geq C_{t^*}$.

Note that any function w in $\mathbf{L}_\Phi(E)$ is also in $\mathbf{L}(E)$ with $L_w \leq [w]_1$ as for all $x \in E$, $x = \Phi(x, 0)$. Conversely, if the flow Φ and the exit time t^* are bounded and Lipschitz continuous: $\Phi \in \mathbf{L}(E \times \mathbb{R}^+)$ and $t^* \in \mathbf{L}(E)$, then any function $w \in \mathbf{L}(E)$ is in $\mathbf{L}_\Phi(E)$ with $[w]_1 \leq L_w L_\Phi$, $[w]_2 \leq L_w L_\Phi$ and $[w]_* \leq L_w L_\Phi (L_{t^*} + 1)$.

1.4.2. *Regularity assumptions on the local characteristics*

We gather here the assumptions that are going to be needed in the following chapters. They may not be all required at the same time. We make no special assumption on the flow as we will rather assume that the functions of interest are Lipschitz continuous along the flow. However, we may need some regularity for the exit time.

ASSUMPTION 1.3.– The exit time is Lipschitz continuous: $t^* \in \mathbf{L}(E)$.

We require some boundedness and weaker form of Lipschitz continuity along the flow for the jump intensity.

ASSUMPTION 1.4.– The jump intensity λ is in $\mathbf{B}(E)$ and bounded by C_λ.

ASSUMPTION 1.5.– There exists $[\lambda]_1 \in \mathbb{R}^+$ such that for all $x, y \in E$ and $t \in [0, t^*(x) \wedge t^*(y)]$, we have:

$$|\lambda(\Phi(x,t)) - \lambda(\Phi(y,t))| \leq [\lambda]_1 |x-y|.$$

Finally, we need some global and local Lipschitz continuity properties for the Markov kernel Q.

ASSUMPTION 1.6.– The Markov kernel Q is Lipschitz in the following sense: there exists $[Q] \in \mathbb{R}^+$ such that for all functions $w \in \mathbf{L}_\Phi(E)$:

1) for all $x, y \in E$ and $t \in [0, t^*(x) \wedge t^*(y))$, we have

$$|Qw(\Phi(x,t)) - Qw(\Phi(y,t))| \leq [Q][w]_1 |x-y|;$$

2) for all $x, y \in E$, we have

$$|Qw(\Phi(x,t^*(x))) - Qw(\Phi(y,t^*(y)))| \leq [Q][w]_* |x-y|.$$

ASSUMPTION 1.7.– The Markov kernel Q is locally Lipschitz in the following sense: there exists $[Q] \in \mathbb{R}^+$ such that for all $u \geq 0$ and for all functions $w \in \mathbf{L}_\Phi^u(E)$:

1) for all $x, y \in E$ and $t \in [0, t^*(x) \wedge t^*(y) \wedge u)$, we have

$$|Qw(\Phi(x,t)) - Qw(\Phi(y,t))| \leq [Q][w]_1^u |x-y|;$$

2) for all $x, y \in E$ such that $t^*(x) \vee t^*(y) \leq u$, we have

$$|Qw(\Phi(x,t^*(x))) - Qw(\Phi(y,t^*(y)))| \leq [Q]\big([w]_*^u + [w]_1^u\big) |x-y|.$$

The necessity for the weaker notion of local Lipschitz continuity will be made clear in the next section. Our last assumption is often satisfied in applications. It states that the jumps cannot send the process too close to the boundary of E.

ASSUMPTION 1.8.– There exists $\epsilon > 0$ such that for all x in $E, Q(x, B_\epsilon) = 1$, where:

$$B_\epsilon = \{x \in E : t^*(x) \geq \epsilon\}.$$

1.5. Time-augmented process

In some applications, it may be useful to consider the time-augmented process $(X_t, t)_{t\geq 0}$, for instance to compute expectations of time-dependent functionals, see section 4.5. The formal definition of this process comes from [DAV 93] as well as the proof that it is still a PDMP. The new state space is:

$$\widetilde{E} = E \times \mathbb{R}^+,$$

equipped with the distance defined, for all $\xi = (x, t), \xi' = (x', t') \in \widetilde{E}$, by:

$$|\xi - \xi'| = |x - x'| + |t - t'|.$$

On this state space, we define the process:

$$\widetilde{X}_t = (X_t, t),$$

for $t \geq 0$. The local characteristics of $(\widetilde{X}_t)_{t\geq 0}$, denoted by $(\widetilde{\Phi}, \widetilde{\lambda}, \widetilde{Q})$, are given for all $\xi = (x, t) \in \widetilde{E}$, $s \leq t^*(x)$ and $A \in \mathfrak{B}(E)$ by:

$$\begin{aligned}
\widetilde{\Phi}(\xi, s) &= (\Phi(x, s), t + s), \\
\widetilde{\lambda}(\xi) &= \lambda(x), \\
\widetilde{Q}(\xi, A \times \{t\}) &= Q(x, A), \\
\widetilde{t}^*(\xi) &= \inf\{t > 0 : \widetilde{\Phi}(\xi, s) \in \partial\widetilde{E}\} = t^*(x).
\end{aligned}$$

As both processes $(X_t)_{t\geq 0}$ and $(\widetilde{X}_t)_{t\geq 0}$ have the same jump times, assumption 1.1 clearly holds for $(\widetilde{X}_t)_{t\geq 0}$. The interesting question is whether the Lipschitz properties on the local characteristics of the original process

$(X_t)_{t\geq 0}$ transfer to that of $(\widetilde{X}_t)_{t\geq 0}$ or not. As $\widetilde{t}^*(\xi) = t^*(x)$ and $\widetilde{\lambda}(\xi) = \lambda(x)$, the answer is yes for t^* and λ.

LEMMA 1.1.– If assumption 1.2 (1.3, 1.4 and 1.5, respectively) holds for $(X_t)_{t\geq 0}$, then it holds for $(\widetilde{X}_t)_{t\geq 0}$ and $C_{\widetilde{t}^*} = C_{t^*}$ ($L_{\widetilde{t}^*} = L_{t^*}$, $C_{\widetilde{\lambda}} = C_\lambda$ and $[\widetilde{\lambda}]_1 = [\lambda]_1$ respectively).

In order to investigate continuity properties for $\widetilde{Q}$, we need to be able to compare functions continuous along the flow Φ of the original PDMP with functions continuous along the flow $\widetilde{\Phi}$ of the time-augmented process.

LEMMA 1.2.– Set $u, t \geq 0$ and $w \in \mathbf{L}^u_{\widetilde{\Phi}}(\widetilde{E})$. Denote by w_t the function of $\mathbf{B}(E)$ defined by $w_t(x) = w(x,t)$ for all $x \in E$. Under assumption 1.3, w_t is in $\mathbf{L}^{u\wedge t}_\Phi(E)$ with:

$$[w_t]_1^{E,t\wedge u} \leq [w]_1^{\widetilde{E},u}, \quad [w_t]_2^{E,t\wedge u} \leq [w]_1^{\widetilde{E},u} + [w]_2^{\widetilde{E},u},$$
$$[w_t]_*^{E,t\wedge u} \leq (1 + L_{t^*})[w]_*^{\widetilde{E},u}.$$

PROOF.– For $x, x' \in E$ and $s \leq t^*(x) \wedge t^*(x') \wedge t \wedge u$, we have:

$$\begin{aligned}\big|w_t(\Phi(x,s)) - w_t(\Phi(x',s))\big| &= \big|w\big(\widetilde{\Phi}((x,t-s),s)\big) \\ &\quad - w\big(\widetilde{\Phi}((x',t-s),s)\big)\big|.\end{aligned}$$

We now use the fact that $w \in \mathbf{L}^u_{\widetilde{\Phi}}(\widetilde{E})$ which yields since $s \leq u$:

$$\begin{aligned}\big|w_t(\Phi(x,s)) - w_t(\Phi(x',s))\big| &\leq [w]_1^{\widetilde{E},u}\big|(x,t-s) - (x',t-s)\big| \\ &= [w]_1^{\widetilde{E},u}\big|x - x'\big|.\end{aligned}$$

Hence, $[w_t]_1^{E,t\wedge u} \leq [w]_1^{\widetilde{E},u}$. Similarly, for $s, s' \leq t^*(x) \wedge t^*(x') \wedge t \wedge u$, we have:

$$\begin{aligned}\big|w_t(\Phi(x,s)) - w_t(\Phi(x,s'))\big| &= \big|w\big(\widetilde{\Phi}((x,t-s),s)\big) - w\big(\widetilde{\Phi}((x,t-s'),s')\big)\big| \\ &\leq [w]_1^{\widetilde{E},u}|s - s'| + [w]_2^{\widetilde{E},u}|s - s'|.\end{aligned}$$

Finally, for $x, x' \in E$ such that $t^*(x) \vee t^*(x') \leq t \wedge u$, we have:

$$
\begin{aligned}
&|w_t(\Phi(x,t^*(x))) - w_t(\Phi(x',t^*(x')))| \\
&\quad =\big|w\big(\widetilde{\Phi}((x,t-t^*(x)),t^*(x))\big) - w\big(\widetilde{\Phi}((x',t-t^*(x')),t^*(x'))\big)\big| \\
&\quad =\big|w\big(\widetilde{\Phi}((x,t-t^*(x)),\widetilde{t}^*(x,t-t^*(x)))\big) \\
&\qquad - w\big(\widetilde{\Phi}((x',t-t^*(x')),\widetilde{t}^*(x',t-t^*(x')))\big)\big| \\
&\quad \leq [w]_*^{\widetilde{E},u}|(x,t-t^*(x)) - (x',t-t^*(x'))| \\
&\quad \leq [w]_*^{\widetilde{E},u}(1+L_{t^*})|x-x'|,
\end{aligned}
$$

as $w \in \mathbf{L}_{\widetilde{\Phi}}^u(\widetilde{E})$ and $\widetilde{t}^*(x,t-t^*(x)) \vee \widetilde{t}^*(x',t-t^*(x')) \leq u$ and using the Lipschitz continuity assumption 1.3 on t^*. □

Because of the dependence in t, the global Lipschitz continuity along the flow $\widetilde{\Phi}$ for w does not in general imply the global Lipschitz continuity of w_t along the flow Φ. As a result, in general, if the Markov kernel Q satisfies the global Lipschitz assumption 1.6, then $\widetilde{Q}$ may not satisfy it. This is why we introduced the weaker notion of local Lipschitz continuity. This one remains valid for the time-augmented process, provided that the Lipschitz continuity assumption on t^* also holds.

LEMMA 1.3.– If assumptions 1.7 and 1.3 hold for $(X_t)_{t\geq 0}$, then assumption 1.7 holds for $(\widetilde{X}_t)_{t\geq 0}$ and:

$$[\widetilde{Q}] \leq ([Q] \vee 1)(1 + L_{t_*}).$$

PROOF.– For $\xi = (x,t) \in \widetilde{E}$ and $w \in \mathbf{L}_{\widetilde{\Phi}}^u(\widetilde{E})$, we have, by definition of $\widetilde{Q}$:

$$\widetilde{Q}w(\xi) = \int_{\xi' \in \widetilde{E}} w(\xi')\widetilde{Q}\big((x,t),d\xi'\big) = \int_{z\in E} w(z,t)Q\big(x,dz\big) = Qw_t(x).$$

Let $\xi = (x,t)$ and $\xi' = (x',t') \in \widetilde{E}$. Let $s \in [0; \widetilde{t}^*(\xi) \wedge \widetilde{t}^*(\xi') \wedge u]$. We have:

$$
\begin{aligned}
\big|\widetilde{Q}w\big(\widetilde{\Phi}(\xi,s)\big) - \widetilde{Q}w\big(\widetilde{\Phi}(\xi',s)\big)\big| &= \big|\widetilde{Q}w\big(\Phi(x,s),t+s\big) \\
&\quad - \widetilde{Q}w\big(\Phi(x',s),t'+s\big)\big| \\
&= \big|Qw_{t+s}\big(\Phi(x,s)\big) - Qw_{t'+s}\big(\Phi(x',s)\big)\big|.
\end{aligned}
$$

We split it into the sum of two differences:

$$\begin{aligned}&\left|Qw_{t+s}\big(\Phi(x,s)\big) - Qw_{t'+s}\big(\Phi(x',s)\big)\right|\\ &\leq \left|Qw_{t+s}\big(\Phi(x,s)\big) - Qw_{t+s}\big(\Phi(x',s)\big)\right| + \left|Q(w_{t+s} - w_{t'+s})\big(\Phi(x',s)\big)\right|.\end{aligned}$$

Since $s \leq (t+s) \wedge u$, we may use lemma 1.2 and the Lipschitz continuity assumption 1.7 on Q to obtain:

$$\begin{aligned}\left|Qw_{t+s}\big(\Phi(x,s)\big) - Qw_{t+s}\big(\Phi(x',s)\big)\right| &\leq [Q][w_{t+s}]_1^{E,(t+s)\wedge u}|x-x'|\\ &\leq [Q][w]_1^{\widetilde{E},u}|x-x'|.\end{aligned}$$

For the second term, we have, from the definition of Q,:

$$\left|Q(w_{t+s} - w_{t'+s})\big(\Phi(x',s)\big)\right| \leq [w]_1^{\widetilde{E},u}|t-t'|,$$

hence, the first result:

$$\left|\widetilde{Q}w\big(\widetilde{\Phi}(\xi,s)\big) - \widetilde{Q}w\big(\widetilde{\Phi}(\xi',s)\big)\right| \leq ([Q]\vee 1)[w]_1^{\widetilde{E},u}|\xi-\xi'|.$$

We now reason similarly to bound $|\widetilde{Q}w(\widetilde{\Phi}(\xi,\widetilde{t}^*(\xi))) - \widetilde{Q}w(\widetilde{\Phi}(\xi',\widetilde{t}^*(\xi')))|$ where $\xi = (x,t)$ and $\xi' = (x',t') \in \widetilde{E}$ are such that $\widetilde{t}^*(\xi) \vee \widetilde{t}^*(\xi') \leq u$. Assuming, without loss of generality, that $t^*(x) \geq t^*(x')$, we have:

$$\begin{aligned}&\left|\widetilde{Q}w\ \ \big(\widetilde{\Phi}(\xi,\widetilde{t}^*(\xi))\big) - \widetilde{Q}w\big(\widetilde{\Phi}(\xi',\widetilde{t}^*(\xi'))\big)\right|\\ &\quad = \left|Qw_{t+t^*(x)}\big(\Phi(x,t^*(x))\big) - Qw_{t'+t^*(x')}\big(\Phi(x',t^*(x'))\big)\right|\\ &\quad \leq \left|Qw_{t+t^*(x)}\big(\Phi(x,t^*(x))\big) - Qw_{t+t^*(x)}\big(\Phi(x',t^*(x'))\big)\right|\\ &\quad +\left|Q(w_{t+t^*(x)} - Qw_{t'+t^*(x')})\big(\Phi(x',t^*(x'))\big)\right|\\ &\quad \leq [Q]\big((1+L_{t_*})[w]_*^{\widetilde{E},u} + [w]_1^{\widetilde{E},u}\big)|x-x'| + [w]_1^{\widetilde{E},u}(|t-t'|\\ &\quad +L_{t_*}|x-x'|),\end{aligned}$$

hence, the result. □

1.6. Embedded Markov chain

Associated with a PDMP $(X_t)_{t\geq 0}$, there exists a natural embedded discrete-time Markov chain $(\Theta_n)_{n\in\mathbb{N}}$ with $\Theta_n = (Z_n, S_n)$, the sequence of post-jump locations and inter-arrival times of the process. This Markov chain contains all the information about the random part of the process $(X_t)_{t\geq 0}$. Indeed, if we know the jump times and the positions after each jump, then we can reconstruct the deterministic part of the trajectory between jumps. As we will see in Chapter 3, $(\Theta_n)_{n\in\mathbb{N}}$ will be the cornerstone of our approach in order to develop numerical techniques for PDMPs.

This section gathers technical results about this Markov chain. The first result comes directly from the definition of the dynamics of the PDMP. It specifies the law of (Z_1, S_1) given $Z_0 = X_0 = x$ for all x in E.

LEMMA 1.4.– For all w in $\mathbf{B}(E)$, w' in $\mathbf{B}(\mathbb{R}^+)$ and x in E, we have:

$$\begin{aligned}
&\mathbb{E}_x[w(Z_1)w'(S_1)]\\
&= \int_0^{t^*(x)} \lambda Qw(\Phi(x,s))\mathrm{e}^{-\Lambda(x,s)}w'(s)ds\\
&\qquad + Qw(\Phi(x,t^*(x)))\mathrm{e}^{-\Lambda(x,t^*(x))}w'(t^*(x)).
\end{aligned}$$

In particular, note that the law of (Z_{n+1}, S_{n+1}) only depends on Z_n.

We now study the distribution of the jump times $(T_n)_{n\geq 1}$. For this, we need an explicit path-wise construction of the trajectories of the PDMP. Suppose, a sequence $(U_n)_{n\in\mathbb{N}}$ of independent random variables with uniform distribution on $[0;1]$ is defined on $(\Omega, \mathcal{F}, \mathbb{P})$. Let $x \in E$, $\omega \in \Omega$ and $t \geq 0$ and let us focus on the construction of the trajectory $(X_t(\omega))_{t\geq 0}$ of the process starting from point x. Let:

$$F(x,t) = \exp(-\Lambda(x,t))\mathbb{1}_{\{t<t^*(x)\}},$$

be the survival function of the first jump time T_1. Define its generalized inverse as:

$$F^{\leftarrow}(x,u) = \inf\{t \geq 0 : F(x,t) \leq u\}.$$

Let then $S_1(\omega) = T_1(\omega) = F^{\leftarrow}(x, U_1(\omega))$ and for all $t < T_1(\omega)$, set:

$$X_t(\omega) = \Phi(x,t).$$

If $T_1(\omega) < +\infty$, choose X_{T_1} with distribution $Q(\cdot, \Phi(x, T_1(\omega)))$. Assume that the trajectory is constructed until time $T_n(\omega)$. If $T_n(\omega) < +\infty$, set:

$$S_{n+1}(\omega) = F^{\leftarrow}(X_{T_n}, U_n(\omega)), \quad T_{n+1}(\omega) = T_n(\omega) + S_{n+1}(\omega).$$

If $T_{n+1}(\omega) < +\infty$, choose $X_{T_{n+1}}$ with distribution $Q(\cdot, \Phi(X_{T_n}, S_{n+1}))$. The trajectory is finally constructed by induction. With this construction, we can bound the probability that the jump times are smaller than a given threshold.

LEMMA 1.5.– Let H be a survival function such that for all $t \in \mathbb{R}^+$ and for all $x \in E$, $H(t) \leq F(x,t)$. Then, there exists a sequence of independent random variables $(\widetilde{S}_n)_{n\in\mathbb{N}}$ with distribution H and such that for all $T > 0$ and $N \in \mathbb{N}$:

$$\mathbb{P}(T_N < T) \leq \mathbb{P}(\widetilde{T}_N < T),$$

where $\widetilde{T}_N = \sum_{n=0}^{N} \widetilde{S}_n$.

PROOF.– Let H be such a survival function and let $\widetilde{F}^{\leftarrow}$ be its generalized inverse, i.e.:

$$\widetilde{F}^{\leftarrow}(u) = \inf\{t \geq 0 : H(t) \leq u\}.$$

The assumption made on H yields for all $x \in E$, $\widetilde{F}^{\leftarrow}(u) \leq F^{\leftarrow}(x,u)$. For all $n \in \mathbb{N}$ and for all $\omega \in \Omega$, set:

$$\widetilde{S}_n(\omega) = \widetilde{F}^{\leftarrow}(U_n(\omega)).$$

Notice that we are using the same U_n as in the definition of S_n, allowing us to write that $\widetilde{S}_n \leq S_n$ a.s. and therefore $\widetilde{T}_n \leq T_n$ a.s. The result follows. □

See the crack propagation model in section 1.8.4 for an example of a practical construction of H and an explicit bound for $\mathbb{P}(T_N < T)$.

1.7. Stopping times

For the sake of completeness, we present now some technical results related to the special structure of stopping times of PDMPs. This section can be skipped by readers mostly interested in applications.

For $n \in \mathbb{N}$, let $\mathcal{M}_n$ be the family of all $(\mathcal{F}_t)$-stopping times which are dominated by T_n and for $n < p$, let $\mathcal{M}_{n,p}$ be the family of all $(\mathcal{F}_t)$-stopping times ν satisfying $T_n \leq \nu \leq T_p$.

Let τ be an $(\mathcal{F}_t)_{t\in\mathbb{R}_+}$-stopping time. Let us recall the important result from [DAV 93].

THEOREM 1.1.– There exists a sequence $(R_n)_{n\in\mathbb{N}^*}$ of non-negative random variables such that R_n is $\mathcal{F}_{T_{n-1}}$-measurable and $\tau \wedge T_{n+1} = (T_n + R_{n+1}) \wedge T_{n+1}$ on $\{\tau \geq T_n\}$.

In this section, important sharp properties of stopping times for PDMPs are established and discussed. They will be used in sections 7.6 and 8.2.

LEMMA 1.6.– Define $\overline{R}_1 = R_1$, and $\overline{R}_k = R_k \mathbb{1}_{\{S_{k-1} \leq \overline{R}_{k-1}\}}$. Then, we have:

$$\tau = \sum_{n=1}^{\infty} \overline{R}_n \wedge S_n.$$

PROOF.– Clearly, on $\{T_k \leq \tau < T_{k+1}\}$, we have $R_j \geq S_j$ and $R_{k+1} < S_{k+1}$ for all $j \leq k$. Consequently, by definition $\overline{R}_j = R_j$ for all $j \leq k+1$, whence:

$$\begin{aligned}\sum_{n=1}^{\infty} \overline{R}_n \wedge S_n &= \sum_{n=1}^{k} \overline{R}_n \wedge S_n + \{\overline{R}_{k+1} \wedge S_{k+1}\} + \sum_{n=k+2}^{\infty} \overline{R}_n \wedge S_n \\ &= T_k + R_{k+1} + \sum_{n=k+2}^{\infty} \overline{R}_n \wedge S_n.\end{aligned}$$

Since $\overline{R}_{k+1} = R_{k+1} < S_{k+1}$, we have $\overline{R}_j = 0$ for all $j \geq k+2$. Therefore, $\sum_{n=1}^{\infty} \overline{R}_n \wedge S_n = T_k + R_{k+1} = \tau$, showing the result. □

There exists a sequence of measurable mappings $(r_k)_{k\in\mathbb{N}_*}$ defined on $E \times (\mathbb{R}_+ \times E)^{k-1}$ with value in $\mathbb{R}_+$ satisfying:

$$\begin{aligned}R_1 &= r_1(Z_0), \\ R_k &= r_k(Z_0, \Gamma_{k-1}),\end{aligned}$$

where $\Gamma_k = (S_1, Z_1, \ldots, S_k, Z_k)$.

DEFINITION 1.1.– *Consider $p \in \mathbb{N}_*$. Let $(\widehat{R}_{p,k})_{k\in\mathbb{N}_*}$ be a sequence of mappings defined on $E \times (\mathbb{R}_+ \times E)^p \times \Omega$ with value in $\mathbb{R}_+$ defined by:*

$$\widehat{R}_{p,1}(y, \gamma, \omega) = r_{p+1}(y, \gamma),$$

and for $k \geq 2$:

$$\widehat{R}_{p,k}(y,\gamma,\omega) = r_{p+k}(y,\gamma,\Gamma_{k-1}(\omega))\mathbf{1}_{\{S_{k-1} \leq \widehat{R}_{p,k-1}\}}(y,\gamma,\omega).$$

PROPOSITION 1.1.– Assume that $\tau \leq T_N$. Then, we have:

$$\tau = T_p + \widehat{\tau}(Z_0, \Gamma_p, \theta_{T_p}),$$

on $\{T_p \leq \tau\}$ where $\widehat{\tau}\colon E \times (\mathbb{R}_+ \times E)^p \times \Omega \to \mathbb{R}_+$ is defined by:

$$\widehat{\tau}(y,\gamma,\omega) = \sum_{n=1}^{N-p} \widehat{R}_{p,n}(y,\gamma,\omega) \wedge S_n(\omega). \quad [1.2]$$

PROOF.– First, let us prove by induction that for $k \in \mathbb{N}_*$, we have:

$$\widehat{R}_{p,k}(Z_0, \Gamma_p, \theta_{T_p}) = \overline{R}_{p+k}. \quad [1.3]$$

Indeed, $\widehat{R}_{p,1}(Z_0, \Gamma_p, \theta_{T_p}) = R_{p+1}$, and on the set $\{\tau \geq T_p\}$, $R_{p+1} = \overline{R}_{p+1}$. Consequently, $\widehat{R}_{p,1}(Z_0, \Gamma_p) = \overline{R}_{p+1}$. Now, assume that $\widehat{R}_{p,k}(Z_0, \Gamma_p, \theta_{T_p}) = \overline{R}_{p+k}$. Then, we have:

$$\widehat{R}_{p,k+1}(Z_0(\omega), \Gamma_p(\omega), \theta_{T_p}(\omega)) = r_{p+k+1}(Z_0(\omega), \Gamma_p(\omega), \Gamma_k(\theta_{T_p}(\omega))) \times \mathbf{1}_{\{S_k \leq \widehat{R}_k\}}(Z_0(\omega), \Gamma_p(\omega), \theta_{T_p}(\omega)).$$

By definition, we have $\Gamma_k(\theta_{T_p}(\omega)) = \big(S_{p+1}(\omega), Z_{p+1}(\omega), \ldots, S_{p+k}(\omega), Z_{p+k}(\omega)\big)$ and $\mathbb{1}_{\{S_k \leq \widehat{R}_k\}}(Z_0(\omega), \Gamma_p(\omega), \theta_{T_p}(\omega)) = \mathbb{1}_{\{S_{p+k} \leq \overline{R}_{p+k}\}}(\omega)$ by the induction hypothesis. Therefore, we get $\widehat{R}_{p,k+1}(Z_0, \Gamma_p, \theta_{T_p}) = \overline{R}_{p+k+1}$, showing [1.3]. Combining equations [1.2] and [1.3] yields:

$$\widehat{\tau}(Z_0, \Gamma_p, \theta_{T_p}) = \sum_{n=1}^{N-n} \overline{R}_{p+n} \wedge S_{p+n}. \quad [1.4]$$

However, we have already seen that on the set $\{T \geq T_p\}$, we have $R_k = \overline{R}_k \geq S_k$, for $k \leq p$. Consequently, using equation [1.4], we obtain:

$$T_p + \widehat{\tau}(Z_0, \Gamma_p, \theta_{T_p}) = \sum_{k=1}^{p} S_k + \sum_{k=p+1}^{N} \overline{R}_k \wedge S_k = \sum_{k=1}^{N} \overline{R}_k \wedge S_k.$$

Since $\tau \leq T_N$, we obtain from lemma 1.6 and its proof that $\tau = \sum_{n=1}^{N} \overline{R}_n \wedge S_n$, showing the result. □

PROPOSITION 1.2.– Let $(U_n)_{n\in\mathbb{N}^*}$ be a sequence of non-negative random variables such that U_n is $\mathcal{F}_{T_{n-1}}$-measurable and $U_{n+1} = 0$ on $\{S_n > U_n\}$, for all $n \in \mathbb{N}_*$. Set:

$$U = \sum_{n=1}^{\infty} U_n \wedge S_n.$$

Then, U is an $(\mathcal{F}_t)_{t\in\mathbb{R}_+}$-stopping time.

PROOF.– Assumption 1.1 yields:

$$\{U \leq t\} = \bigcup_{n=0}^{\infty} \Big[\Big(\{T_n \leq U < T_{n+1}\} \cap \{U \leq t\} \cap \{t < T_{n+1}\} \Big) \cup \Big(\{T_n \leq U < T_{n+1}\} \cap \{U \leq t\} \cap \{T_{n+1} \leq t\} \Big) \Big]. \quad [1.5]$$

From the definition of U_n, we have $\{U \geq T_n\} = \{U_n \geq S_n\}$, hence we have:

$$\{T_n \leq U < T_{n+1}\} \cap \{U \leq t\} \cap \{t < T_{n+1}\} = \{S_n \leq U_n\} \cap \{T_n + U_{n+1} \leq t\} \cap \{T_n \leq t\} \cap \{t < T_{n+1}\}.$$

On the one hand, $\{S_n \leq U_n\} \cap \{T_n + U_{n+1} \leq t\} \cap \{T_n \leq t\} \in \mathcal{F}_t$ by theorem 2.10 $ii)$ in [ELL 82], thus we have:

$$\{T_n \leq U < T_{n+1}\} \cap \{U \leq t\} \cap \{t < T_{n+1}\} \in \mathcal{F}_t. \quad [1.6]$$

On the other hand, we have:

$$\{T_n \leq U < T_{n+1}\} \cap \{U \leq t\} \cap \{T_{n+1} \leq t\} = \{S_n \leq U_n\} \cap \{U_{n+1} < S_{n+1}\} \cap \{T_{n+1} \leq t\}.$$

Hence, theorem 2.10 $ii)$ in [ELL 82] again yields:

$$\{T_n \leq U < T_{n+1}\} \cap \{U \leq t\} \cap \{T_{n+1} \leq t\} \in \mathcal{F}_t. \quad [1.7]$$

Combining equations [1.5], [1.6] and [1.7], we obtain the result. □

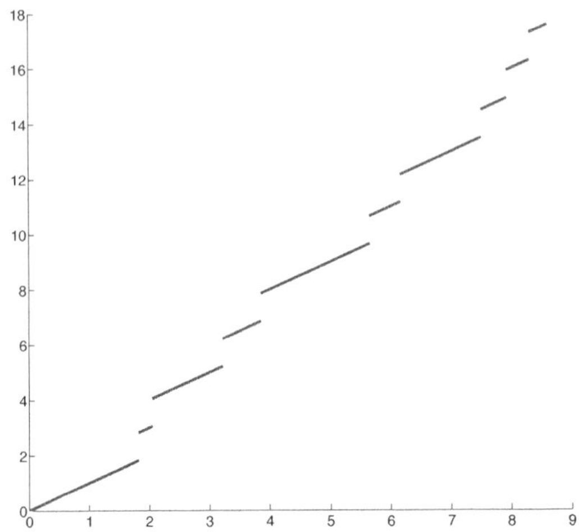

Figure 1.2. *A trajectory of the Poisson process with trend up to the* 10*-th jump time*

1.8. Examples of PDMPs

In this section, we give a few examples of simple PDMPs that will serve to illustrate our numerical results in the following chapters. For more involved examples in the field of reliability, the readers are referred to Chapter 2.

1.8.1. *Poisson process with trend*

Pure jump processes are among the simplest examples of PDMPs. Let $(N_t)_{t\geq 0}$ be a Poisson process with intensity $\lambda = 1$ and set $X_t = t + N_t$. The process $(X_t)_{t\geq 0}$ is a PDMP with state space $E = \mathbb{R}^+$ (with no boundary) and intensity $\lambda = 1$ as the inter-jump times are independent and identically distributed with exponential distribution of parameter λ. The flow is defined on $(\mathbb{R}^+)^2$ by $\Phi(x,t) = x + t$. The jump kernel satisfies $Q(x, x+1) = 1$. An example of trajectory of the process is represented in Figure 1.2. This example is further discussed in section 5.5.

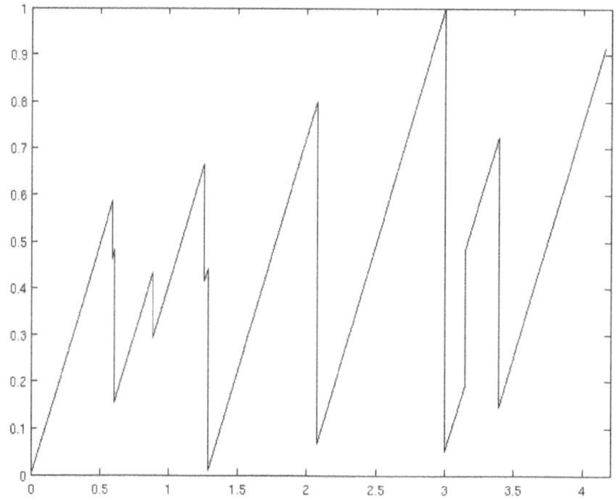

Figure 1.3. *One trajectory of the TCP process up to the* 10*-th jump*

1.8.2. *TCP*

The Transmission Control Protocol (TCP) process appears in the modeling of the famous transmission control protocol used for data transmission over the Internet. It has been designed to adapt to the various traffic conditions of the actual network. The state variable is the number of packets to be transmitted. It increases linearly at rate v until the congestion level is reached. Then, the next packet size to be transmitted is reduced at least by half.

We consider here that the state space $E = [0, 1)$ and the congestion level is $\partial E = \{1\}$. The flow is $\Phi(x, t) = x + vt$ for some positive v. We also allow random jumps with rate $\lambda(x) = \beta x^\alpha$ for some $\beta > 0$ and $\alpha \geq 1$, so that the probability to jump is higher as the boundary is closer. When a jump occurs, either randomly or at the boundary, the new position is selected according to $Q(x, \cdot)$ that is uniformly distributed on $[0, 1/2]$.

Figure 1.3 shows one sample trajectory of this process for $X_0 = 0$, $v = \alpha = 1$ and $\beta = 3$ and up to the 10-th jump. This example will be further discussed in sections 7.7 and 10.5 in relation with optimization problems.

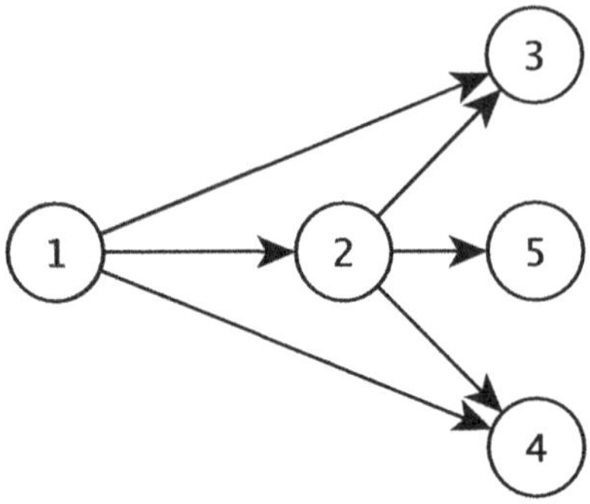

Figure 1.4. *Possible transitions for the air conditioning process*

1.8.3. *Air conditioning unit*

This example was provided by Thales Optronique SAS. We consider an air conditioning unit that can be in 5 different states:

1) stable;

2) degraded ball bearing;

3) failed electrovalve;

4) electronic failure;

5) ball bearing failure.

State 1 is the nominal state, state 2 is a functioning though degraded state and states 3 to 5 are failure states where the air conditioning unit no longer works. They are distinguished as they lead to different repair costs. The possible transitions between the states are given in Figure 1.4. The possible modes are naturally the 5 possible states of the system. The Euclidean variable is simply the running time in modes 1 and 2 and a cemetery state Δ in the other modes. The failure rates are constant for the ball bearing and electronic failures, but time-dependent according to a Weibull distribution for the electrovalve failure and the degradation of the ball bearing, leading to a time-dependent intensity. Note that this process has at most two jumps before reaching the cemetery state. The equipment has a lifetime of 10^5 h. Thus, the state state space is $E = \{1,2\} \times [0, 10^5) \cup \{1,2,3,4,5\} \times \{\Delta\}$. The flow in modes $m \in \{1,2\}$ is $\Phi_m(\zeta, t) = \zeta + t$, 0 otherwise. The jump intensity in modes 1 and 2 is:

$$\lambda(1,\zeta) = \lambda_2(\zeta) + \lambda_3(\zeta) + \lambda_4,$$
$$\lambda(2,\zeta) = \lambda_3(\zeta) + \lambda_4 + \lambda_5,$$

where λ_2 and λ_3 correspond to Weibull distributions. The values of the parameters are confidential. The Markov jump kernel can be deduced from the intensities and the graph in Figure 1.4. This example will be further discussed in Chapter 9 where we study a problem of maintenance for this equipment.

1.8.4. *Crack propagation model*

We give here a simple version of the process studied by Chiquet and Limnios in [CHI 08], which models a crack propagation. Let $(Y_t)_{t\geq 0}$ be a real-valued process representing the crack size satisfying $Y_0 > 0$ and:

$$\frac{dY_t}{dt} = A_t Y_t,$$

for all $t \geq 0$, where $(A_t)_{t\geq 0}$ is a pure jump Markov process with state space $\{\alpha, \beta\}$ where $0 < \alpha \leq \beta$ and infinitesimal generator $\mathcal{A} = (a_{i,j}, i, j \in \{\alpha, \beta\})$. Consider the PDMP $(X_t)_{t\geq 0}$ defined by $X_t = (A_t, Y_t)$, where A_t represents the mode at time t. The state space is $E = \{\alpha, \beta\} \times (0, +\infty)$ (with no boundary). The flow is:

$$\Phi_m(\zeta, t) = (\zeta e^{mt}),$$

for $\{\alpha, \beta\}$, $\zeta > 0$ and $t \geq 0$. The intensity in mode m is constant equal to $-a_{mm}$, thus assumption 1.4 holds. The jump kernel is:

$$Q((m,\zeta),(m',\zeta')) = \mathbb{1}_{\zeta=\zeta'}\mathbb{1}_{m\neq m'}.$$

Let us compute a bound for the probability that the jump times are smaller than a given threshold T. Set $H(t) = e^{-C_\lambda t}$, for $t \geq 0$. The survival function H represents, generally speaking, the worst distribution of the inter-jump times S_n in the sense that it is the one that implies the most frequent jumps. Lemma 1.5 provides a random variable $\widetilde{T}_N = \sum_{n=0}^{N} \widetilde{S}_n$ where $(\widetilde{S}_n)$ are independent and have survival function H such that:

$$\mathbb{P}(T_N < T) \leq \mathbb{P}(\widetilde{T}_N < T).$$

We now bound $\mathbb{P}(\widetilde{T}_N < T)$. Standard computations with exponential laws yield $\mathbb{E}[\widetilde{T}_N] = Nm$ and $\text{var}[\widetilde{T}_N] = N\sigma^2$ where:

$$m = \mathbb{E}[\widetilde{S}_1] = C_\lambda^{-1}, \quad \sigma^2 = \text{var}[\widetilde{S}_1] = C_\lambda^{-2}.$$

Assume now that N is chosen such that $Nm > T$ and note that:

$$\mathbb{P}(\widetilde{T}_N < T) \leq \mathbb{P}(|\widetilde{T}_N - \mathbb{E}[\widetilde{T}_N]| > \mathbb{E}[\widetilde{T}_N] - T).$$

Chebychev inequality then yields:

$$\mathbb{P}(\widetilde{T}_N < T) \leq N\sigma^2(Nm - T)^{-2}.$$

and that the right-hand side term goes to zero when N goes to infinity.

1.8.5. *Repair workshop model*

This repair workshop model is adapted from [DAV 93, section 2.1]. In a factory, a machine produces goods whose daily value is $r(x)$ where $x \in [0; 1]$ represents a parameter of evolution of the machine, a setting chosen by the operator. For instance, x may be some load or some pace imposed on the machine. This machine, initially working, may break down with age-dependent hazard rate $\lambda(t)$ and is then sent to the workshop for reparation. Moreover, the direction of the factory has decided that, whenever the machine has worked for a whole year without requiring reparation, it is sent to the workshop for maintenance. We assume that after a reparation or a maintenance, that both last a fixed time s, the machine is totally repaired and is not worn down.

We, therefore, consider three modes: the machine is working ($m = 1$), being repaired ($m = 2$) or undergoing maintenance ($m = 3$). The state of the process at time t will be denoted by $X_t = (m_t, \zeta_t)$ where ζ_t is the time since the last change of mode. This component is required since the hazard rate is age-dependent. The state space is:

$$E = (\{1\} \times [0; 365]) \cup (\{2\} \times [0; s]) \cup (\{3\} \times [0; s]).$$

In each mode m, the flow is $\Phi_m(\zeta, t) = \zeta + t$, so that we have:

$$t^*((1, \zeta)) = 365 - \zeta, \qquad t^*((2, \zeta)) = t^*((3, \zeta)) = s - \zeta.$$

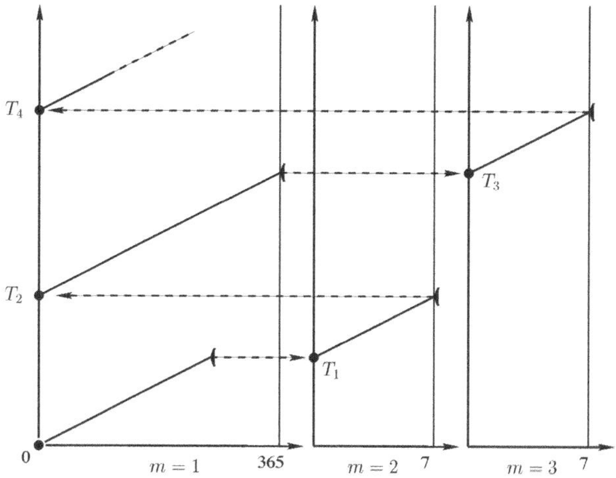

Figure 1.5. *Trajectory of the repair workshop model*

Therefore, the deterministic time to reach the boundary is bounded by $365 \vee s$. Concerning the transition kernel, from the point $(1, \zeta)$, the process can jump to the point $(2, 0)$ if $\zeta < 365$ and the jump is forced to $(3, 0)$ if $\zeta = 365$, thus we have for $\zeta < 365$:

$$Q((1,\zeta),(2,0)) = 1, \qquad Q((1,365),(3,0)) = 1,$$
$$Q((2,s),(1,0)) = 1, \qquad Q((3,s),(1,0)) = 1.$$

Figure 1.5 presents a sample trajectory of the process. The abscissa is the time since the last change of mode ζ_t and the coordinate is the total time elapsed since the beginning. This model will be further studied in section 4.7 where we solve an optimization problem to select parameter x so as to maximize some reward function.

We fully discuss more involved examples of PDMPs related to reliability problems in Chapter 2. Nowadays, another important field developing applications of PDMPs is biology. Although we will not focus on this field in this book, some of the ideas and techniques presented here may also be applied in that context. The interested readers may consult, for instance, the following works and references therein (not exhaustive) [AZA 14, DOU 15, FRI 14, PAK 10, RIE 12, RIE 13, ROB 14].

2

Examples in Reliability

2.1. Introduction

In dependability and safety theory, modeling is a key step to study the properties of the physical processes involved. Nowadays, it appears necessary to take into account in a realistic and explicit way the dynamic interactions existing between the physical parameters (for example, pressure, temperature, flow rate, liquid level, etc.) of the system and the functional and dysfunctional behavior of its components. Classically, the models described in dependability and safety literature do not take such interactions into account. A first set of methods used in reliability theory is the so-called combinatory approaches (fault trees, event trees, reliability diagrams and networks, etc.) which can be used to identify and evaluate the combinations of events leading to the occurrence of another event, desirable or not. These powerful methods suffer from the fact that such combinations do not take the order of occurrence into account, in that sense they eliminate any notion of dependency between events. A second set of methods is described by finite state Markov (or semi-Markov) models. In this context, the system is described by a fixed number of components which can be in different states. For any component, the set of its possible states is assumed to be finite (generally, it contains only two elements: an operational or a failure state). One of the main limitations encountered with such models is their difficulties to correctly model physical processes involving deterministic behavior. To overcome such difficulties, dynamic reliability was introduced in 1980 as a powerful mathematical framework capable of explicitly handling interactions between components and process variables. Nowadays in the literature, the multi-model approach appears as a natural framework to formulate dynamic reliability problems. The behavior of the physical model is then described by different modes of operation from nominal to failure states with intermediate dysfunctional regimes. For a large class of industrial processes, the layout of operational or accident sequences generally comes from the occurrence of two types of events:

1) The first type is directly linked to a deterministic evolution of the physical parameters of the process.

2) The second type is purely stochastic. It usually corresponds to random demands or failures of system components.

In both cases, these events will induce jumps in the behavior of the system leading to unstable or stable trajectories for the process.

Piecewise Deterministic Markov Processes (PDMPs) is a powerful modeling tool especially suitable for dynamic reliability problems. The key asset of this mathematical model is that it naturally takes into account the two kinds of events described above. Their dynamic character, allowing explicit time dependencies, in opposition to piecewise constant jump Markov processes, is a very attractive property for modeling real phenomenons of dynamic reliability, see, for example, [DEV 97] and [DUF 02].

We present in this chapter two examples of complex systems that can be modeled as PDMPs. The first example, described in section 2.2, concerns a metallic structure subject to corrosion. This example was provided to the authors by Astrium Space Transportation and first published in [SAP 12c]. The second example presented in section 2.3 is the so-called heated hold-up tank. It is a well-known case test of dynamic reliability (see [DEV 97, MAR 95, MAR 96, ZHA 08]). Both these examples are interesting in terms of reliability and also present non-trivial features as PDMPs. We present the two models in detail in this chapter. These models will also serve to illustrate our numerical procedures in Chapters 3, 6 and 9.

2.2. Structure subject to corrosion

Here we describe the evolution of the thickness of an aluminum metallic structure subject to corrosion, specifically a small homogeneous structure within a strategic ballistic missile. The missile is stored first in a workshop, then in a nuclear submarine missile launcher in operation or in the submarine in dry-dock. These various environments are more or less corrosive and the structure is inspected at each change of environment. It is made to withstand potentially long storage durations. The requirement for security is very strong. The mechanical stress exerted on the structure depends in part on its thickness. A loss of thickness will cause an overconstraint and therefore an increase in risk of rupture. It is thus crucial to monitor the evolution of the thickness of the structure over time, and to intervene before possible failure.

Here we describe the usage profile of the missile in more detail. It is stored in three different environments: the workshop, the submarine in operation and the submarine in dry-dock. This is because the structure must be equipped and used in a given order. Then, it goes back to the workshop and so on. The missile stays in each environment during a random duration with exponential distribution. Its parameter depends on the environment. At the beginning of its service time, the structure is treated against corrosion. The period of effectiveness of this protection is also random, with a Weibull distribution. The thickness loss only begins when this initial protection is gone. The degradation law for the thickness loss then depends on the environment through two parameters, a deterministic transition period and a random corrosion rate uniformly distributed within a given range. Typically, the workshop and dry-dock are the more corrosive environments. The randomness of the corrosion rate accounts for small variations and uncertainties in the corrosiveness of each environment.

2.2.1. *PDMP model*

The dynamics of the thickness loss for the structure can be described by a PDMP. The finite set of modes is

$$M = \{1, 2, 3\} \times \{0, 1\},$$

where the first coordinate corresponds to the environment: 1 for the workshop, 2 for the submarine in operation and 3 for the dry-dock, and the second coordinate states whether the corrosion has started (1) or not (0). The state space in modes $m \in \{1, 2, 3\} \times \{0\}$ is

$$E_m = (0, +\infty) \times \mathbb{R}^+,$$

with boundary $\partial E_m = \{0\} \times \mathbb{R}^+$, and the Euclidean variables are (γ_t, a_t) the remaining time of effectivity of the protection and the time since the last change of mode. The state space in modes $m = (i, 1) \in \{1, 2, 3\} \times \{1\}$ is

$$E_m = \mathbb{R}^+ \times [\rho_i^-, \rho_i^+] \times \mathbb{R}^+,$$

with empty boundary and the Euclidean variables are (d_t, ρ_t, a_t) the thickness loss, the corrosion rate of the given environment and time since the last change of mode.

Originally at time 0, the mode is $m_0 = (1, 0)$ and $\zeta_0 = (\gamma_0, 0)$, where γ_0 is drawn according to a Weibull distribution with parameters $\alpha = 2.5$ and $\beta = 11800$ hours^{-1}. In other words, the process starts in the workshop environment with a random protection that will be effective during γ_0 hours. The flow in modes $m \in \{1, 2, 3\} \times \{0\}$ is

$$\Phi(m, (\gamma, a), s) = (m, \gamma - s, a + s),$$

with $t^*(m, \gamma, a) = \gamma$, meaning that the boundary is reached when the the protection is gone. In modes $m = (i, 1) \in \{1, 2, 3\} \times \{1\}$, the flow is

$$\Phi(m, (d, \rho, a), s) = (m, d + d_i(\rho, a + s) - d_i(\rho, a), \rho, a + s),$$

where

$$d_i(\rho, s) = \rho\big(s + \eta_i(\mathrm{e}^{-s/\eta_i} - 1)\big),$$

is the dynamics of the thickness loss in environment i. The parameters are given in Table 2.1. All the parameters come from expert opinion. The boundary is empty in this case, so that $t^*(m, d, \rho, a) = +\infty$. We will discuss in the next section how to change the value of t^* so that our boundedness assumption 1.2 may hold true.

The jump intensity in environment i is a constant λ_i, whose value is given in Table 2.1. The jump kernel is as follows

– if $m = (i, 0), \zeta = (\gamma, a)$ with $\gamma > 0$,

$$Q((m, \zeta), \{m'\} \times \{(\gamma, 0)\}) = \mathbb{1}_{\{m'=(i+1[3],0)\}},$$

with $i + 1[3] = i + 1$ if $i \leq 2$ and $3 + 1[3] = 1$;

– if $m = (i, 0), \zeta = (0, a)$, and A is in $\mathfrak{B}(\mathbb{R}^+)$

$$Q((m, \zeta), \{m'\} \times \{0\} \times A \times \{0\}) = \mathbb{1}_{\{m'=(i,1)\}} \frac{1}{\rho_i^+ - \rho_i^-} \int_{\rho_i^-}^{\rho_i^+} \mathbb{1}_A(s) ds,$$

where the parameters $\rho_i^{\pm}$ are given in Table 2.1;

– if $m = (i, 1), \zeta = (d, \rho, a)$, and A is in $\mathfrak{B}(\mathbb{R}^+)$

$$Q((m, \zeta), \{(j, 1)\} \times \{d\} \times A \times \{0\}) = \mathbb{1}_{\{j=i+1[3]\}} \frac{1}{\rho_j^+ - \rho_j^-} \int_{\rho_j^-}^{\rho_j^+} \mathbb{1}_A(s) ds.$$

	environment 1	environment 2	environment 3
λ_i (h^{-1})	$(17520)^{-1}$	$(131400)^{-1}$	$(8760)^{-1}$
η_i (h)	30,000	200,000	40,000
ρ_i^- ($mm{\cdot}h^{-1}$)	10^{-6}	10^{-7}	10^{-6}
ρ_i^+ ($mm{\cdot}h^{-1}$)	10^{-5}	10^{-6}	10^{-5}

Table 2.1. *Numerical values of the parameters of the corrosion model*

Figure 2.1 shows some simulated trajectories of the thickness loss along time. The slope changes correspond to changes of environment. The observed dispersion is characteristic of the random nature of the phenomenon, and illustrates the poorness of an average rule to monitor such a process. We need to make decisions adapted to each specific trajectory. The structure is considered unusable if the loss of thickness reaches 0.2 mm.

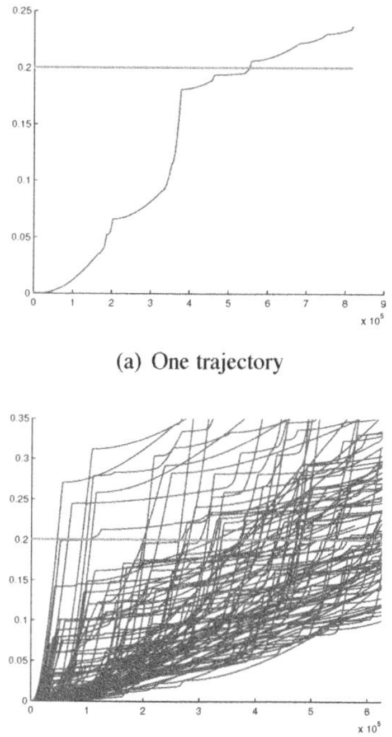

(a) One trajectory

(b) 100 trajectories

Figure 2.1. *Simulated trajectories of the thickness loss along time for the corrosion process*

2.2.2. *Deterministic time to reach the boundary*

Because the deterministic exit time is unbounded in modes $m \in \{1,2,3\} \times \{1\}$, the PDMP model described above does not fit our framework. There are two main ways to overcome this difficulty.

The first way consists of killing the process when the thickness lost by corrosion reaches the prescribed level of 0.2 mm. The state space in modes $m = (i,1) \in \{1,2,3\} \times \{1\}$ thus becomes

$$E_m = [0, 0.2) \times [\rho_i^-, \rho_i^+] \times \mathbb{R}^+,$$

with boundary $\partial E_m = \{0.2\} \times [\rho_i^-, \rho_i^+] \times \mathbb{R}^+$, and deterministic time to boundary is

$$t^*((i,1), d, \rho, a) = \inf\{t > 0 : d + d_i(\rho, a+t) - d_i(\rho, a) = 0.2\}.$$

Thus, t^* is uniformly bounded on E_m for $m \in \{1,2,3\} \times \{1\}$. Rigorously, this does not hold true for all points in E_m with $m \in \{1,2,3\} \times \{0\}$ as the support of the Weibull law is unbounded. However, the initial protection is drawn only once and for a given initial protection time γ, the quantities $t^*(X_t)$ are bounded uniformly for all t, which is actually all that is required in our proofs.

The second way to bound the deterministic time to the boundary is by considering the time-augmented process $(\widetilde{X}_t)_{t \geq 0} = (X_t, t)_{t \geq 0}$ with a finite fixed horizon T for the trajectories. Thus, the state space becomes

$$E_m = (0, +\infty) \times \mathbb{R}^+ \times [0, T),$$

in modes $m \in \{1,2,3\} \times \{0\}$, with boundary

$$\partial E_m = \{0\} \times \mathbb{R}^+ \times [0,T) \cup (0,+\infty) \times \mathbb{R}^+ \times \{T\} \cup \{0\} \times \mathbb{R}^+ \times \{T\},$$

and deterministic time to reach the boundary

$$t^*(m, \gamma, a, t) = \gamma \wedge (T - t).$$

The state space in modes $m = (i,1) \in \{1,2,3\} \times \{1\}$ becomes

$$E_m = \mathbb{R}^+ \times [\rho_i^-, \rho_i^+] \times [0,T) \times [0,T),$$

with boundary

$$\partial E_m = \mathbb{R}^+ \times [\rho_i^-, \rho_i^+] \times [0, T) \times \{T\},$$

and deterministic time to reach the boundary

$$t^*(m, d, \rho, a, t) = (T - t),$$

as $a \leq t$ by construction, the boundary may not be reached by a. Thus, t^* is bounded by T uniformly in all the modes and state variables.

Depending on the applications we have in mind, we will select either one of these options. Refer to section 6.1 for the computation of the mean thickness loss in environment 2 for a horizon of $T = 18$ years. In this case, we naturally use the time-augmented process with finite horizon. In section 6.2, we compute the service time of the structure up to the threshold 0.2 mm of thickness loss. In this case, it is more natural to kill the process when it reaches the threshold.

This corrosion example will be further studied in Chapters 3, 6 and 9.

2.3. The heated hold-up tank

The heated hold-up tank is a well-known test case for dynamic reliability (see, for example, [DEV 97, MAR 95, MAR 96, ZHA 08]). The system consists of a tank containing a fluid whose level is controlled by three components: two inlet pumps and one outlet valve. A thermal power source heats up the fluid. The failure rate of the components depends on the temperature, the position of the three components monitors the liquid level in the tank, and in turn, the liquid level determines the temperature. The main characteristic of this system is that it can be modeled by a PDMP, where the discrete and continuous parts interact in a closed loop. As a result, simulating this process and computing related reliability indices has been a challenge for the dynamic reliability community.

The tank model was first introduced by Aldemir [ALD 87] where only one continuous variable (liquid level) is taken into account, and then in [MAR 94] and [MAR 95] where the second variable (temperature) is introduced. They have tested various Monte Carlo approaches to simulate the process to compute reliability and safety indices. In [TOM 96], the authors have used the same system to present continuous cell-to-cell mapping Markovian approach (CCCMT) still to simulate the process. The simulation of the hold-up tank example has been and is still widely studied in the literature (see (not exhaustive) [SIU 94, COJ 96, DUT 97, SCH 06, LI 11]). We have kept the values of the parameters defined in those papers.

2.3.1. *Tank dynamics*

The system is represented in Figure 2.2. It consists of a tank containing a fluid whose level is controlled by three components: two inlet pumps (units 1 and 2) and one outlet valve (unit 3). A thermal power source heats up the fluid. The variables of interest are the liquid level h, the liquid temperature θ, the state of the three components and the controller. Each component has four states: ON, OFF, Stuck ON or Stuck OFF. Once a unit is stuck (either on or off), it cannot change state. The possible transitions between these four states are given in Figure 2.3(a). Thus, by a random transition, a working unit can only become stuck (either on or off). The initial state of the components is ON for units 1 and 3 and OFF for unit 2. The intensity of jumps λ^i for unit i depends on the temperature through the equation $\lambda^i = a(\theta)l^i$ with $a(\theta)$ given in equation [2.1] (see [MAR 94, MAR 95]).

$$a(\theta) = \frac{b_1 \exp\big(b_c(\theta - 20)\big) + b_2 \exp\big(-b_d(\theta - 20)\big)}{b_1 + b_2}. \qquad [2.1]$$

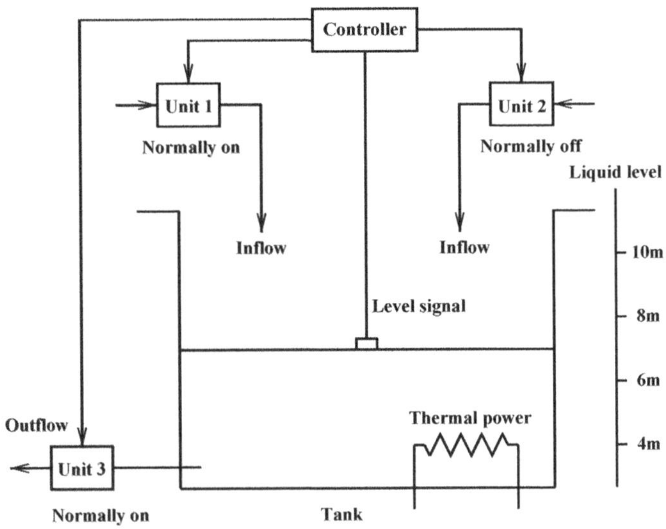

Figure 2.2. *The heated hold-up tank*

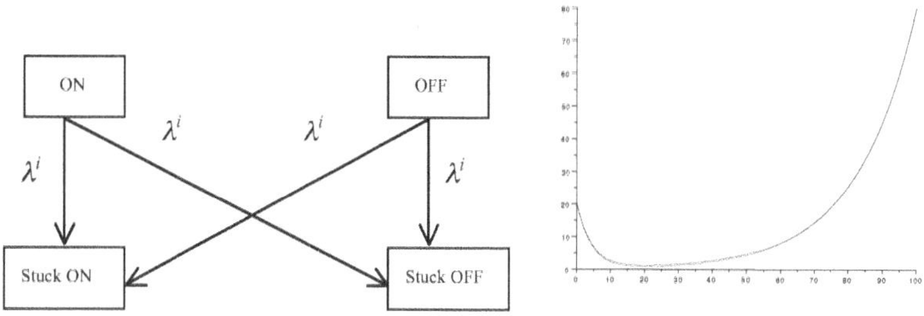

(a) Transitions for unit i (b) $a(\theta)$ as a function of θ

Figure 2.3. *Transitions diagram and failure rate for the three components of the tank*

Function $a(\theta)$ is represented in Figure 2.3(b), where the different parameters come from [MAR 94, MAR 95] and are given in Table 2.2. The special form of the failure rate λ^i as a product of a constant depending on i and a function of the temperature allows for all three units to have failure rates with the same dependence on the temperature, but different scaling parameters. Indeed, at the reference temperature of 20°C, the mean time to failure of unit 1 is 438 h, for unit 2 it is 350 h and for unit 3 it is 640 h. Furthermore, the shape for function $a(\theta)$ was chosen in the original benchmark so that there is a very high failure rate when the temperature is high. More specifically, the parameters are chosen such that $a(\theta)$ is the lowest (equal to 1) when the temperature is equal to a reference temperature of 20°C, it equals 20 when the temperature is 0°C and it is the highest (equal to 80) when the temperature equals the critical value of 100°C. The exponential functions are chosen in order to enable this very high dependence with the temperature. Generally speaking, the units fail 80 times more often when the temperate is 100°C than when it is 20°C.

Parameter	Value	Parameter	Value
b_1	3.0295	l^1	$2.2831 \cdot 10^{-3}\ h^{-1}$
b_2	0.7578	l^2	$2.8571 \cdot 10^{-3}\ h^{-1}$
b_c	0.05756	l^3	$1.5625 \cdot 10^{-3}\ h^{-1}$
b_d	0.2301	G	$1.5\ mh^{-1}$
θ_{in}	15°C	K	$23.88915\ m°Ch^{-1}$

Table 2.2. *Parameters for the tank dynamics*

Control laws are used to modify the state of the components to keep the liquid within two acceptable limits: 6 and 8 m. If the liquid level drops under 6 m, the components 1, 2, 3 are put, respectively, in state ON, ON and OFF (provided that they are not stuck). If the liquid level rises above 8 m, the components 1, 2, 3 are put, respectively, in the state OFF, OFF and ON (provided that they are not stuck). Unlike the classical model presented in [DEV 97, MAR 95, MAR 96, ZHA 08], we also allow the control unit to fail. At each solicitation, the control may succeed with probability $p = 0.8$ independently from previous successes. Once it has failed, it will never succeed again. Therefore, the control unit has two possible states: working 1 or failed 0.

The evolution of the liquid level h depends on the position of the three components through the differential equation [2.2]:

$$\frac{\partial h}{\partial t} = (\nu_1 + \nu_2 - \nu_3)G, \qquad [2.2]$$

where $\nu_i = 1$ if component i is ON or Stuck ON, $\nu_i = 0$ otherwise, and G is the common flow of the three components and is given in Table 2.2. The initial level is $h_0 = 7$ m. Equation [2.2] simply means that each pump in state ON contributes to rise the liquid level, whereas if the outlet valve is on, it contributes to the decrease in the liquid level. The temperature θ depends on the liquid level through the differential equation [2.3]:

$$\frac{\partial \theta}{\partial t} = \big((\nu_1 + \nu_2)G(\theta_{in} - \theta) + K\big)h^{-1}, \qquad [2.3]$$

where θ_{in} is the temperature of the incoming fluid and K is a constant of the tank, the values of these parameters are given in Table 2.2. As the tank is heated and the incoming liquid has a constant temperature θ_{in}, equation [2.3] reflects that the temperature converges to an equilibrium state as long as there is some incoming fluid. The temperature can increase to the threshold 100°C if there is no incoming fluid. The initial temperature is $\theta_0 = 30.9261$°C, so that the system is at an equilibrium state, and nothing happens until the failure of one of the components. The system stops as soon as one of the top events is reached: dry out ($h < 4$ m), overflow ($h > 10$ m) or hot temperature ($\theta > 100$°C).

2.3.2. *PDMP model*

As in [ZHA 08], we model the tank by a PDMP. The modes correspond to the possible positions of the inlet pumps, outlet valve and control unit. The components can be ON, OFF, Stuck ON or Stuck OFF, the control unit can be in position 0 or 1. Therefore, there are 128 possible modes for our system (but only 74 can actually be reached from the equilibrium starting point, see Figure 3.6).

$$M = \{\text{ON}, \text{OFF}, \text{Stuck ON}, \text{Stuck OFF}\}^3 \times \{0, 1\}.$$

For all modes m in M, we have

$$E_m \subset (4, 10) \times [15, 100) \times [0, +\infty),$$

depending on m, the three variables representing, respectively, the liquid level, liquid temperature and running time. The precise geometry of each state space and their boundaries are highly mode-dependent and will not be detailed here. Simply note that the boundary is one of the thresholds 4 m, 6 m, 8 m, 10 m, 100°C or 1000 h for the running time, so that the deterministic time to reach the boundary always satisfies

$$t^*(h, \theta, t) \leq 1000 - t \leq 1000,$$

Assumption 1.2 holds. The flow is given by equations [2.2] and [2.3] for the liquid level and temperature, respectively, and the running time is increasing with speed 1. The jump intensity given a mode m is the sum of the intensities λ_i for the remaining possible jumps of the three units. The Markov kernel Q acts only on the mode components and leaves the liquid level h, temperature θ and running time unchanged. It selects one of the remaining possible failures of the three components, or corresponds to an attempted control law. See [GON 10] for a detailed description of the state space and local characteristics in each mode (in the case where the control until cannot fail).

The optimization problem for this model is detailed in Chapter 9.

3

Quantization Technique

3.1. Introduction

Since the middle of the 20th Century, optimal quantization has been an important field of information theory. Its original aim is to convert a continuous signal into a discrete signal taking only finitely many values. Quantization methods have been developed recently in numerical probability or optimal stochastic control with applications in finance. For examples, see [BAL 03, PAG 04b, BAL 05], the survey [PAG 15] and the Website http://www.quantize.maths-fi.com.

The quantization of a random variable X consists of finding a finite grid such that the projection $\widehat{X}$ of X on this grid minimizes some L_p norm of the difference $X - \widehat{X}$. Generally speaking, such a grid will have more points in the areas of high density of X and fewer points in the areas of low density of X. Replacing X by $\widehat{X}$ turns integrals of the form $\mathbb{E}[f(X)]$ into finite sums and makes numerical computations possible. This chapter is dedicated to a quick exposition of the optimal quantization algorithm and the theoretical properties of the optimal quantizers that will be used in this book. There exists extensive literature on quantization methods for random variables and processes. Interested readers may, for instance, consult [GRA 98, PAG 04b, PAG 15] and the references therein.

As explained, for instance, in [PAG 04b, section 3], under some Lipschitz-continuity conditions, bounds for the rate of convergence of functionals of the quantized process toward the original process are available, which makes this approach appealing both on numerical and theoretical points of view. To build quantization grids, we must be able to simulate the underlying process, therefore, we also give details regarding the simulation of Piecewise Deterministic Markov Processes (PDMPs).

This chapter is organized as follows. In section 3.2, we introduce the quantization of a single random variable and the quantization of a Markov chain. In section 3.3, we explain how to simulate the PDMPs presented in this book. In the last section 3.4, we present how to quantize PDMPs and the associated practical problems encountered when we tried to quantize the PDMPs associated with the examples from the previous chapter.

3.2. Optimal quantization

We start with the definition of optimal quantization for a single random variable before turning to the extension of Markov chains.

3.2.1. *Optimal quantization of a random variable*

Let X be an $\mathbb{R}^d$-valued random variable defined on a probability space $(\mathbf{\Omega}, \mathcal{A}, \mathbf{P})$ with finite L_p-norm $\|X\|_p = \mathbf{E}[|X|^p]^{1/p} < \infty$. Let K be a fixed integer. The optimal L_p-quantization of the random variable X consists of finding the best possible L_p-approximation of X by a random vector $\widehat{X}$ taking at most K values. It can be shown that the determination of the optimal L_p-quantization of the random variable X is equivalent to finding a grid Γ of size K which achieves the infimum of

$$\|X - proj_\Gamma(X)\|_p,$$

where $proj_\Gamma$ denotes the projection on Γ following the closest-neighbor rule (also called the Voronoi quantizer). More precisely, the projection on Γ is defined by

$$proj_\Gamma(\cdot) = \sum_{i=1}^{K} x_i \mathbb{1}_{C_i(\Gamma)}(\cdot),$$

where $(C_i(\Gamma))_{1 \leq i \leq K}$ is the Voronoi tessellation of $\Gamma = \{x_1, \ldots, x_K\}$, i.e. a Borel partition of $\mathbb{R}^d$ satisfying

$$C_i(\Gamma) \subset \{y \in \mathbb{R}^d : |y - x_i| = \min_{1 \leq j \leq K} |y - x_j|\}.$$

The asymptotic properties of the optimal L_p-quantization are given by Zador's theorem (see, for example, [GRA 00, PAG 04b]). There also exists a non-asymptotic version of this result, known as Pierce Lemma (see [GRA 00]).

THEOREM 3.1.– If $\mathbb{E}[|X|^{p+\eta}] < +\infty$ for some $\eta > 0$, then we have

$$\lim_{K\to\infty} K^{p/d} \min_{|\Gamma|\leq K} \|X - \widehat{X}^{\Gamma}\|_p^p = J_{p,d} \int |h|^{d/(d+p)}(u)du,$$

where the law of X can be decomposed with respect to the Lebesgue measure λ_d on $\mathbb{R}^d$ as $\mathcal{L}_X(du) = h(u)\lambda_d(du) + \nu(du)$, h being the density of its absolutely continuous part, ν its singular part and $J_{p,d}$ is a universal constant.

Note that X needs to have finite moments up to the order $p + \eta$ to ensure the above convergence. Another interesting property of quantization is related to convexity. We recall proposition 11 from [PAG 98].

PROPOSITION 3.1.– If the support of the distribution of X is a convex set C, then $\widehat{X}$ also lies in C.

For the numerical results presented in this book, we used the competitive learning vector quantization algorithm (CLVQ) described, for example, in [BAL 03, Section 3]. This algorithm provides a (sub)optimal grid $\Gamma^* = \{x_1^*, \ldots, x_K^*\}$ for the L_2-quantization of a random variable. The CLVQ algorithm is described in algorithm 3.1. The requirements on the sequence $(\gamma_n)_{n\in\mathbb{N}}$ are that the γ_n must be positive, $\sum_{n\in\mathbb{N}} \gamma_n = +\infty$ and $\sum_{n\in\mathbb{N}} \gamma_n^2 < +\infty$. Note that the weights for each point of the optimal grid Γ^* can be obtained by Monte Carlo simulations of X and projection on Γ^*, see algorithm 3.2. See also Table 3.1 for our choice of sequence $(\gamma_n)_{n\in\mathbb{N}}$ in the numerical applications presented in this book.

Figure 3.1 shows the density (a) and the corresponding quantization grid (b) with the Voronoi tesselation obtained with the CLVQ algorithm for the two-dimensional Gaussian distribution with zero mean and covariance matrix

$$\Sigma = \begin{pmatrix} 1 & 0.4 \\ 0.4 & 1 \end{pmatrix}.$$

Other examples of quantization grids are given in section 3.4.

Algorithm 3.1: CLVQ algorithm to compute an optimal quantization grid

input : Number of points K, Number of runs NR, Initial grid Γ^0 with K points,
Simulator of target law ν, Sequence (γ_n)
output: Optimized grid Γ^*

begin
  for $m \leftarrow 0$ **to** $NR - 1$ **do**
    simulate x according to law ν
    competitive phase select y as the closest neighbor of x in Γ^m
    learning phase set $y' = y - \gamma_{m+1}(y - x)$
    $\Gamma^{m+1} \leftarrow \Gamma^m \cup \{y'\} \backslash \{y\}$
  end
  return: Γ^{NR}
end

Algorithm 3.2: Algorithm to compute the weights of the quantization grid

input : Grid Γ with K points, Number of Monte Carlo runs NMC,
Simulator of target law ν
output: Weight vector

begin
  count $\leftarrow (0, 0, \ldots, 0)$ (size K)
  for $n \leftarrow 1$ **to** NMC **do**
    simulate x according to law ν
    select i in $\{1, \ldots, K\}$ such that x_i is the closest neighbor of x in Γ
    count(i) $\leftarrow$ count(i)+1
  end
  count$\leftarrow$count/NMC
  return: count
end

3.2.2. *Optimal quantization of a Markov chain*

In this section, we are interested in the optimal quantization of a Markov chain and an associated numerical method, the so-called extended CLVQ algorithm. Let $(X_n)_{1 \leq n \leq N}$ be an $\mathbb{R}^d$-valued $(\mathcal{A}_n)_{0 \leq n \leq N}$-Markov chain with initial distribution μ defined on a filtered probability space $(\boldsymbol{\Omega}, \mathcal{A}, (\mathcal{A}_n)_{0 \leq n \leq N}, \mathbf{P})$ where N denotes the time horizon. The optimal quantization of the Markov chain $(X_n)_{0 \leq n \leq N}$ consists of finding an approximation of the distribution of $(X_n)_{0 \leq n \leq N}$ by the distribution of a process $(\widehat{X}_n)_{0 \leq n \leq N}$ taking values on finite grids. According to the approach developed in the previous section, a possible method would be to use the optimal quantization of the vector $(X_n)_{0 \leq n \leq N}$ at each step $0 \leq n \leq N$ on a grid Γ_n of size K. This would lead to an approximation error of order $K^{-1/Nd}$. This direct method has the drawback to depend drastically on the *length* N of the time horizon. In [PAG 04a], two different approaches are proposed with a better rate of convergence of order $N^{1+1/d}K^{-1/d}$. The first method is based upon the quantization of the random variable X_n at each step n. It is called the *marginal quantization* in the literature. The second approach is called the *Markovian quantization*. One of its important features is that it keeps the Markovian property of the process.

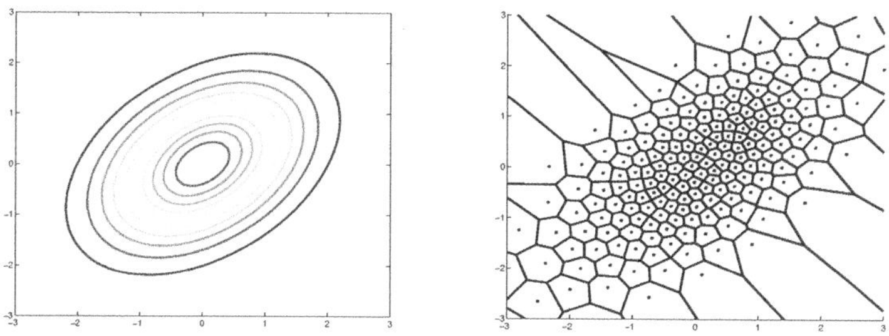

(a) Density (b) Quantization grid (200 points)

Figure 3.1. *Density and optimal quantization grid for a centered Gaussian law with covariance matrix* Σ

In this book, we use the marginal quantization. Let us describe this approximation. Let $\widehat{X}_n$ be the optimal L_p-quantization of the random variable X_n for $0 \leq n \leq N$ by a random variable taking K points. Let us denote by $\Gamma_n = \{x_1^n, \dots, x_K^n\}$ the grid at step n and the associated Voronoi tessellation $C_1(\Gamma_n), \dots, C_K(\Gamma_n)$. Clearly, the process $(\widehat{X}_n)_{1 \leq n \leq N}$ is not a Markov chain. However, at each step $1 \leq n \leq N$, we can compute

$$\begin{aligned}
\widehat{p}_n(i,j) &= \mathbf{P}\big(\widehat{X}_n = x_j^n \big| \widehat{X}_{n-1} = x_i^{n-1}\big) \\
&= \frac{\mathbf{P}\big(\widehat{X}_{n-1} = x_i^{n-1}, \widehat{X}_n = x_j^n\big)}{\mathbf{P}\big(\widehat{X}_{n-1} = x_i^{n-1}\big)} \\
&= \frac{\mathbf{P}\big(X_{n-1} \in C_i(\Gamma_{n-1}), X_n \in C_j(\Gamma_n)\big)}{\mathbf{P}\big(\widehat{X}_{n-1} \in C_i(\Gamma_{n-1})\big)},
\end{aligned}$$

for $1 \leq i, j \leq N$. The marginal quantization approximation of the Markov chain $(X_n)_{0\leq n\leq N}$ is then defined by the Markov chain $(\widetilde{X}_n)_{0\leq n\leq N}$ whose transition matrix at step n is given by $\left(\widehat{p}_n(i,j)\right)_{1\leq i,j\leq N}$, and the initial distribution of $\widetilde{X}_0$ is given by that of $\widehat{X}_0$. It can be shown (for all the details see [PAG 04a, Theorem 3.1]) that the joint distribution of $(X_n)_{0\leq n\leq N}$ can be approximated by that of $(\widetilde{X}_n)_{0\leq n\leq N}$ with a rate of convergence of order $N^{1+1/d}K^{-1/d}$.

Algorithm 3.3 describes how to adapt the CLVQ technique to the context of an optimal quantization of a Markov chain. We refer the readers to [BAL 03, section 3.3] for further details. The computation of the weights and transition matrices by Monte Carlo simulations is similar to algorithm 3.2.

Algorithm 3.3: Extended CLVQ algorithm to quantize a Markov chain

input : Number of points K, Number of runs NR, Horizon N,
Initial grids (Γ_n^0) $0 \leq n \leq N$ with K points each, Sequence (γ_n)
Simulator of trajectories of target Markov chain

output: Optimized grids $(\Gamma_n^*), 0 \leq n \leq N$

begin
  for $m \leftarrow 0$ **to** $NR - 1$ **do**
    simulate trajectory $(x_0, x_1, \ldots, x_N)$ according to law of Markov chain
    for $n \leftarrow 0$ **to** N **do**
      competitive phase select y closest neighbor of x_n in Γ_n^m
      learning phase set $y' = y - \gamma_{n+1}(y - x_n)$
      $\Gamma_n^{m+1} \leftarrow \Gamma_n^m \cup \{y'\} \backslash \{y\}$
    end
  end
  return: $(\Gamma_n^{NR}), 0 \leq n \leq N$
end

3.3. Simulation of PDMPs

In this book, we only present examples of PDMPs where the flow Φ has an explicit analytic expression. We briefly discuss at the end of this section how to deal with flows only defined as solution of a system of ordinary differential equations.

The main difficulty in simulating a PDMP with explicit flow thus lies in the simulation of the jump mechanism and more specifically the jump times when the intensity is state dependent. In this section, we will denote $\lambda(t) = \lambda(\Phi(x, t))$, where λ is the intensity of the PDMP, to emphasize the fact that simulating a state-dependent intensity is equivalent to simulating a time-dependent intensity.

3.3.1. *Simulation of time-dependent intensity*

The easiest case to simulate the jump times is when the cumulative distribution function

$$t \longmapsto 1 - \mathrm{e}^{-\Lambda(x,t)}$$

is analytically invertible. Direct exact simulation is then available. This is the case for standard exponential or Weibull distributions as in the Poisson process with trend and the air conditioning unit from section 1.8 or the corrosion process from Chapter 2. The cumulative distribution function of the transmission control protocol (TCP) process (see section 1.8) is

$$t \longmapsto 1 - \exp\left(\beta \frac{(x+tv)^{\alpha+1} - x^{\alpha+1}}{v(\alpha+1)}\right),$$

that can also be inverted analytically.

Otherwise, we use a reject algorithm based on the Poisson distribution as described in [COC 97, Proposition 1.21], detailed in algorithm 3.4 and illustrated in Figure 3.2. Basically, we draw a (Poisson) random number of points according to the uniform distribution on a rectangle locally covering the intensity curve, and select the points that fall below the curve. We only need to have local upper bounds for the intensity.

When the jumps mechanism is a competition between several time-dependent intensities, for instance, several components may fail as in the tank example (see Chapter 2, where the actual jump time is the smallest failure time). When the number of components is reasonably small, as throughout this book, we can just simulate all the failure times and select the smallest one. When the number of competitive failures is very high, we cannot compute all the failure times, and should instead simulate the global intensity as above or use a Gillespie-type algorithm (see, for example, [GIL 77, GIL 01, RIE 13, FRI 14, VEL 15]).

3.3.2. *Simulation of trajectories*

Once we can simulate the jump times, the simulation of trajectories follows the recursive definition of PDMPs.

1) Choose the starting point x.

2) Simulate the next jump time T_1 according to algorithm 3.4.

3) Compute the flow on $[0, T_1]$.

4) Draw the next starting point Z_1 from distribution $Q(\Phi(x, T_1), \cdot)$.

5) Iterate steps 2–5 until the desired number of jumps or the time horizon is reached.

Algorithm 3.4: Reject algorithm to simulate a time-dependent intensity

input : Intensity function $\lambda(t)$, positive real number A
output: One realization of the random variable with intensity $\lambda(t)$

```
begin
    a ← 0
    S ← ∅
    while S ≠ ∅ do
        select L ≥ sup_{a≤t≤a+A} λ(t)
        draw n from a Poisson distribution of parameter L × A
        if n ≠ 0 then
            for 1 ≤ k ≤ n do
                draw u_k from a uniform distribution on [a; a + A]
                draw v_k from a uniform distribution on [0; L]
                if v_k ≤ λ(u_k) then
                    S = S ∪ {u_k}
                end
            end
        end
        a ← a + A
    end
    return: min S
end
```

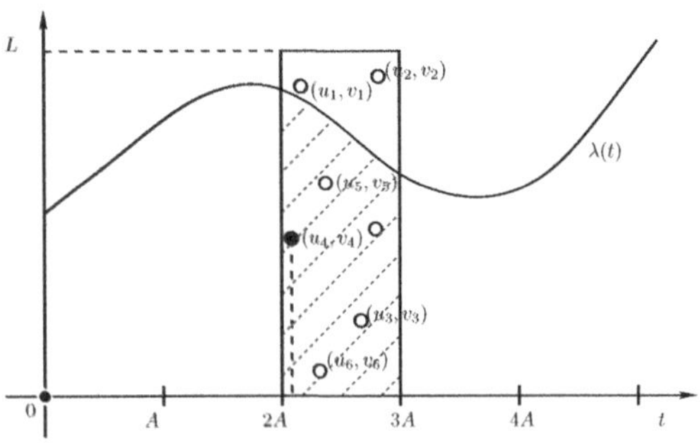

Figure 3.2. *Illustration of the reject algorithm to simulate a time-dependent intensity*

We used this procedure to simulate all the examples of PDMPs presented in this book. When the flow is not explicit, very few theoretical results concerning simulation are available. One possible policy is to compute and approximation of the flow using any efficient numerical scheme and plug the approximate flow into the procedure described above. This should work reasonably well, but very few results concerning the error thus made are available in the literature, see [RIE 13] for a special case of PDMP. Some fully deterministic schemes are also available, but for special cases of PDMPs, see [EYM 08].

3.4. Quantization of PDMPs

To discretize PDMPs, we can use their special recursive structure. As explained in section 1.6, a nice property, especially suitable for the applications targeted in this book, is that all the randomness of the process is contained in the embedded discrete-time Markov chains (Z_n, S_n), or (Z_n, T_n). Therefore, to obtain discretized versions of PDMPs, we chose to quantize the Markov chain $\Theta_n = (Z_n, S_n)$ (or (Z_n, T_n), depending on applications).

We denote by $\Gamma_n^\Theta \subset E \times \mathbb{R}^+$ the optimal grid for $\Theta_n = (Z_n, S_n)$, for some $n \geq 0$, Γ_n^Z is its projection on the state space E and Γ_n^S is its projection on $\mathbb{R}^+$. Set $\widehat{\Theta}_n = (\widehat{Z}_n, \widehat{S}_n) = proj_{\Gamma_n^\Theta}(\Theta_n)$. Note that $\widehat{Z}_n$ and $\widehat{S}_n$ depend on both Z_n and S_n through the projection operator on the joint grid Γ_n^Θ.

We first simulate the PDMP as described above, and save only the values of (Z_n, S_n) or (Z_n, T_n). Then, we run the quantization algorithm 3.3 with the parameters value given in Table 3.1. Figure 3.3 shows an example of resulting quantization grid for (Z_2, S_2) for the TCP process described in section 1.8.2 with 500 points in the grid.

Parameter	Value
Number of points in the grids	K
Number of runs	$NR = K \times 1000$
Sequence (γ_n)	$\gamma_n = a(a + b \times n)^{-1}$, $a = 4\sqrt{K}$, $b = \pi^2 K^{-1}$

Table 3.1. *Parameters for the CLVQ quantization algorithm for K points in the grids*

When we ran the extended CLVQ quantization algorithm on the PDMPs described in Chapter 2, we encountered some extra difficulties due to their specific nature. We detail them and how we solved them from the practical point of view in the following sections.

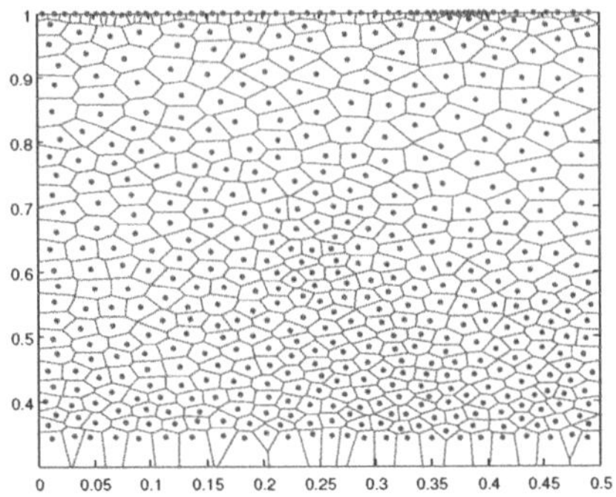

Figure 3.3. *Quantization grids (500 points) with Voronoi tesselation for* (Z_2, S_2) *for the TCP process*

3.4.1. *Scale of coordinates of the state variable*

For the corrosion example of section 2.2, the main difficulty lies in the fact that the variables of interest have very different scales: from about 10^{-6} for the corrosion rate ρ to 10^5 for the average time spent in environment 2. This poses a problem in the classical quantization algorithm as searching for the nearest neighbor and gradient calculations are done in Euclidean norm, regardless of the magnitudes of the components. Figure 3.4 illustrates this problem by presenting two examples of quantization grids for a uniform distribution on $[0, 1] \times [0, 5000]$. Figure 3.4(a) shows the result obtained by the conventional algorithm, while Figure 3.4(b) is obtained by weighting the Euclidean norm to renormalize each variable on the same scale.

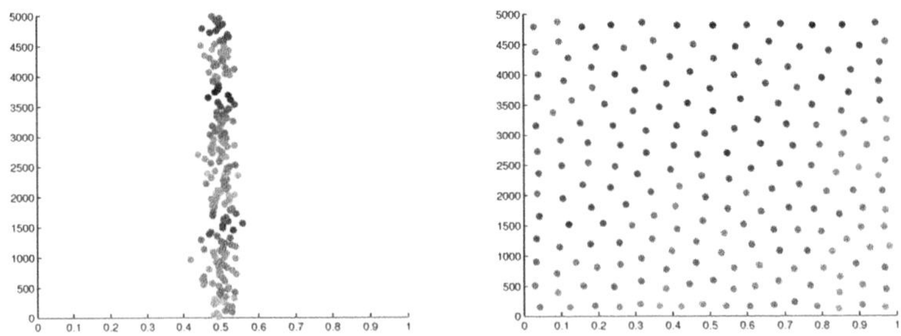

(a) Standard CLVQ algorithm (b) Weighted-norm CLVQ algorithm

Figure 3.4. *Quantization grids (2,000 points) for the distribution on* $[0, 1] \times [0, 5000]$

It is clear from this example that the conventional method is not satisfactory, because the grid obtained is far from uniform. This defect is corrected by a renormalization of the variables. Therefore, a weighted Euclidean norm is used to quantify the Markov chain associated with the corrosion process to rescale the variables. Projections of the quantization grids obtained for 2000 points on the variables interarrival times S_n (abscissa) and thickness loss at the jump times d_{T_n} (ordinate) are given in Figure 3.5. Notice how different the grids are from one jump time to the other. Note also that the Markov chain does not seem to follow any well-known probability distribution, and that if we had used a Cartesian grid on the rectangle, a huge number of points would have been useless. Thus, the quantization approach is especially suitable as it allows for dynamically changing grids along time, instead of a single fixed grid.

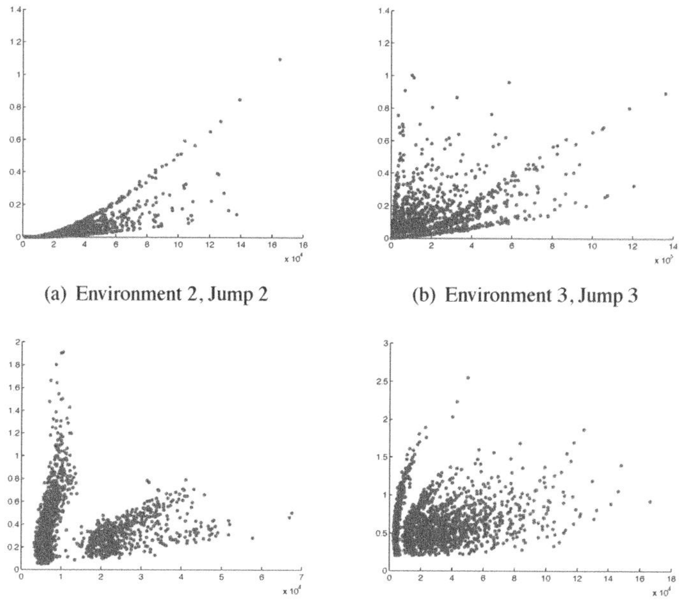

(a) Environment 2, Jump 2

(b) Environment 3, Jump 3

(c) Environment 1, Jump 16

(d) Environment 2, Jump 26

Figure 3.5. *Quantization grids (2,000 points) for the corrosion process: interarrival times (abscissa) and thickness loss at the jump times (ordinate)*

3.4.2. *Cardinality of the mode variable*

Regarding the heated hold-up tank example described in section 2.3, we also used a weighted norm for the CLVQ algorithm, but encountered a new difficulty of a different nature, due to the high cardinality of the mode set and possibly low probability of reaching some of the modes. Our PDMP model for the dynamics of the tank is naturally hybrid: there is a discrete mode variable (the positions of the components and state of the control unit) and a continuous variable (liquid level, temperature and running time). Of course, we need not discretize the mode variable as it can already take only finitely many values. Our procedure requires one discretization grid at each jump time of the process. However, at a given jump time, several modes can appear. Figure 3.6 illustrates the complexity of the modes set, and the possible transitions between modes.

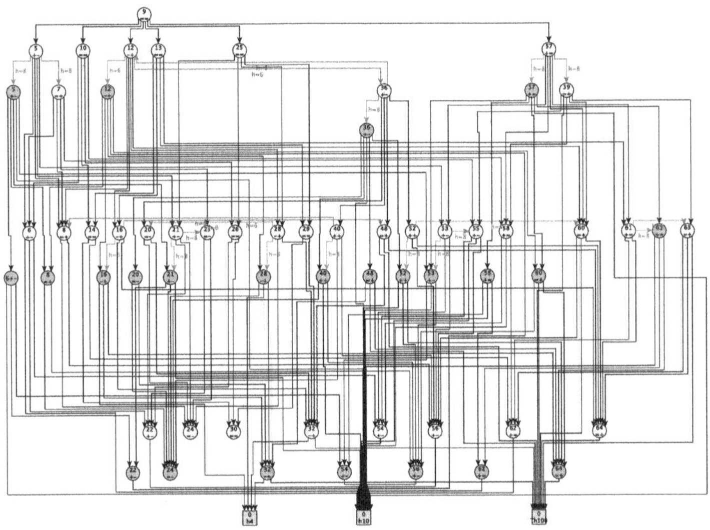

Figure 3.6. *Reachable modes and their transitions for the tank process*

For instance, at time 0, the starting mode is $(ON, OFF, ON, 1)$. After the first jump time, one of the components has failed, so there are now six possible modes:

– $(Stuck\ ON, OFF, ON, 1)$;

– $(Stuck\ OFF, OFF, ON, 1)$;

– $(ON, Stuck\ ON, ON, 1)$;

– $(ON, Stuck\ OFF, ON, 1)$;

– $(ON, OFF, Stuck\ ON, 1)$;

– $(ON, OFF, Stuck\ OFF, 1)$.

Table 3.2 gives the theoretical number of possible modes at each time step as well as the observed one for $3 \cdot 10^9$ simulated trajectories. After five jumps, the theoretical number of modes is constantly equal to 18, but all the 18 modes can actually be observed only as long as the controller unit does not fail. As the probability for the controller to remain in its operational state decreases with the number of trials of the control laws, the 18 modes become increasingly rare and by the 25th jump time are not observed anymore, which means that the system has reached one of the top events and stopped.

Jump	Theory	Simulations	Jump	Theory	Simulation
n=0	1	1	n=20	18	17
n=1	6	6	n=21	18	16
n=2	18	18	n=22	18	14
n=3	30	30	n=23	18	7
n=4	25	25	n=24	18	1
n=5 to n=19	18	18	n=25	18	0
			n=26	18	0

Table 3.2. *Theoretical and observed number of modes at each jump time step for $3 \cdot 10^9$ Monte Carlo simulations for the tank process*

The comparatively rare events are problematic for the implementation of the quantization algorithm. Indeed, we first choose the number K of discretization points and then we usually initialize the algorithm by throwing K trajectories of the process at random. Thus, some rare modes may not be reached by the initial simulations, and the algorithm will not perform well when new trajectories are thrown reaching these modes. Indeed, the algorithm is based on a nearest neighbor search, within the nodes having the same mode as the original point. When no such mode is present, the algorithm provides highly unsatisfactory results. Therefore, we had to find a way to ensure that the initializing grids have at least one point in each observed mode for each time step. To do so, at each time step, we allocated the K points to the possible modes proportionally to their frequency (computed with $3 \cdot 10^9$ Monte Carlo simulations) and forcing 1 point for the modes with frequency less than $1/K$.

The quantization grids for the corrosion and tank examples will be used in Chapters 6 and 9, respectively. Note that the grids only depend on the dynamics of the process and not a specific functional or optimization problem we may want to solve or compute. Although it may take a lot of time to compute them, this computation can be done offline and the same grids may be used to solve different problems related to the same process.

PART 2

Simulation of Functionals

4

Expectation of Functionals

4.1. Introduction

The aim of this chapter is to propose a practical numerical method to approximate some expectations related to Piecewise Deterministic Markov Processes (PDMPs) using the quantization of the discrete-time Markov chain naturally embedded within the continuous-time process as described in section 1.6. More precisely, we are interested in the approximation of expectations of functions the form

$$J_N(l,c)(x) = \mathbb{E}_x\Big[\int_0^{T_N} l(X_t)dt + \sum_{n=1}^{N} c(X_{T_n^-})\mathbb{1}_{\{X_{T_n^-}\in\partial E\}}\Big], \qquad [4.1]$$

where $(X_t)_{\{t\geq 0\}}$ is a PDMP and l and c are some non-negative functions in $\mathbf{B}(E)$ and $\mathbf{B}(\partial E)$, respectively. Such expectations are discussed in [DAV 93, Chapter 3]. They often appear as *cost* or *reward* functions in optimization problems. The first term is referred to as the *running cost*, while the second term is the *boundary jump cost*. Moreover, they are quite general since M.H.A. Davis shows how a "*wide variety of apparently different functionals*" can be obtained from the above specific form. For example, this wide variety includes quantities such as a mean exit time and even, for any fixed $t \geq 0$, the distribution of X_t, i.e. $\mathbb{E}_x[\mathbb{1}_A(X_t)]$ for any measurable subset A of E.

There are surprisingly few works in the literature devoted to the actual computation of such expectations, using other means than direct Monte Carlo simulations. M.H.A Davis showed that these expectations satisfy integro-differential equations. However, the set of partial differential equations that is obtained is unusual. Generally speaking, these differential equations are basically transport equations with a non-constant velocity and they are coupled by the boundary conditions and by some integral terms involving kernels that are derived from the properties of the underlying stochastic process. The main difficulty comes from the fact that the domains on which the equations have to be solved vary from one equation to another making their numerical resolution highly problem specific. See [GON 10, BOY 10, BOY 12] for a detailed exposition of this approach on the tank example presented in section 2.3. Another similar approach has been investigated in [COC 06, EYM 08]. It is based on a discretization of the Chapman–Kolmogorov equations satisfied by the distribution of the process $(X_t)_{\{t\geq 0\}}$. The authors propose an approximation of such expectations based on finite volume methods. Unfortunately, their method is only valid if there are no jumps at the boundary.

Our approach is completely different and does not rely on differential equations, but on the fact that such expectations can be computed by iterating an integral operator G. This operator only involves the embedded Markov chain $(Z_n, S_n)_{n\in\mathbb{N}}$ and conditional expectations. It is, therefore, natural to propose a computational method based on the quantization of this Markov chain. In order to derive some error bounds for this approximation, some Lipschitz-continuity conditions are required. To achieve this aim, we replace the indicator function $\mathbb{1}_{\{X_{T_n^-}\in\partial E\}}$ contained within the functional by some Lipschitz continuous approximation.

Our results can also be extended to the more general framework of time-dependent cost functions of the form

$$\mathbb{E}_x\Big[\int_0^{T_N} l(X_t,t)dt + \sum_{n=1}^{N} c(X_{T_n^-},T_n)\mathbb{1}_{\{X_{T_n^-}\in\partial E\}}\Big].$$

This important generalization has huge applicative consequences. For instance, it allows for discounted cost/reward functions such as $l(x,t) = e^{-\delta t}l(x)$ and $c(x,t) = e^{-\delta t}c(x)$ for some discount rate δ. To compute such expectations, our strategy consists of considering the time augmented process $\widetilde{X}_t = (X_t, t)$ introduced in section 1.5.

The second important generalization is to consider the deterministic time horizon problem. Indeed, it seems crucial, regarding applications, to be able to approximate

$$\mathbb{E}_x\Big[\int_0^{T} l(X_t,t)dt + \sum_{T_n\leq T} c(X_{T_n^-},T_n)\mathbb{1}_{\{X_{T_n^-}\in\partial E\}}\Big],$$

for some fixed $T > 0$ regardless of how many jumps occurred before this deterministic time. Our strategy consists of bounding T by some jump time T_N with high enough probability and multiply by the indicator functions $\mathbb{1}_{\{t \leq T\}}$ and $\mathbb{1}_{\{T_n \leq T\}}$, respectively, the cost functions l and c. In both these extensions, the crucial point is to check the Lipschitz-continuity properties of the time augmented process and new cost functions.

An important advantage of our method is that it is flexible. Indeed, as pointed out in [BAL 03], a quantization-based method is "*obstacle free*" which means, in our case, that it produces, once and for all, a discretization of the process independently of the functions l and c since the quantization grids merely depend on the dynamics of the process. They are only computed once, stored offline and may therefore serve many purposes. Once they have been obtained, we are able to approximate very easily and quickly any of the expectations described earlier. This flexibility is definitely an important advantage of our scheme over standard methods such as Monte Carlo simulations since, with such methods, we would have to run the whole algorithm for each expectation we want to compute.

The main material of this chapter was originally published as [BRA 12b]. This chapter is organized as follows. In section 4.2, we explain how the expectation can be computed by iterating an operator G, and give the approximation scheme. In section 4.3, we establish the stability of $\mathbf{L}_\Phi(E)$ by G. In section 4.4, we prove the convergence of our numerical scheme, with error bounds. In section 4.5, we explain how our method can be adapted to encompass time-dependent functionals. Section 4.6 deals with the case of functionals with deterministic horizon. Finally, we give some numerical results on the repair-workshop example in section 4.7.

4.2. Recursive formulation

The first step to obtain a computable approximation of the expectation [4.1] is to rewrite it as the solution of a recursion involving only the embedded Markov chain (Z_n, S_n). Since the boundary jumps occur exactly at the deterministic exit times from E, we have,

$$\begin{aligned} J_N(l,c)(x) = \mathbb{E}_x\Bigg[&\int_0^{T_N} l(X_t)dt \\ &+ \sum_{n=1}^{N} c(\Phi(Z_{n-1}, t^*(Z_{n-1}))\mathbb{1}_{\{S_n = t^*(Z_{n-1})\}} \Bigg]. \end{aligned}$$

4.2.1. *Lipschitz continuity*

The Lipschitz-continuity property is a crucial point when it comes to proving the convergence of our approximation scheme. We thus make the assumption that the cost functions l and c have suitable Lipschitz-continuity properties.

ASSUMPTION 4.1.– The cost function l is in $\mathbf{L}_\Phi(E)$ and c is in $\mathbf{L}_\Phi(\partial E)$.

For the same purpose, we replace the indicator function in $J_N(l,c)$ by a Lipschitz continuous approximation. Let δ^a be the triangular approximation of the indicator function defined on $E \times \mathbb{R}$ by

$$\delta^a(x,t) = \begin{cases} at - (at^*(x) - 1) & \text{for } t \in [t^*(x) - 1/a; t^*(x)], \\ -at + (at^*(x) + 1) & \text{for } t \in [t^*(x); t^*(x) + 1/a], \\ 0 & \text{otherwise.} \end{cases}$$

Figure 4.1 gives the graphical representation of function $\delta^a(x,\cdot)$. For all $x \in E$ and $t \geq 0$, $\delta^a(x,t)$ goes to $\mathbb{1}_{\{t=t^*(x)\}}$ when a goes to infinity. This approximation was chosen such that $\delta^a(x,\cdot)$ is Lipschitz continuous with Lipschitz constant a. Moreover, it has the following straightforward properties.

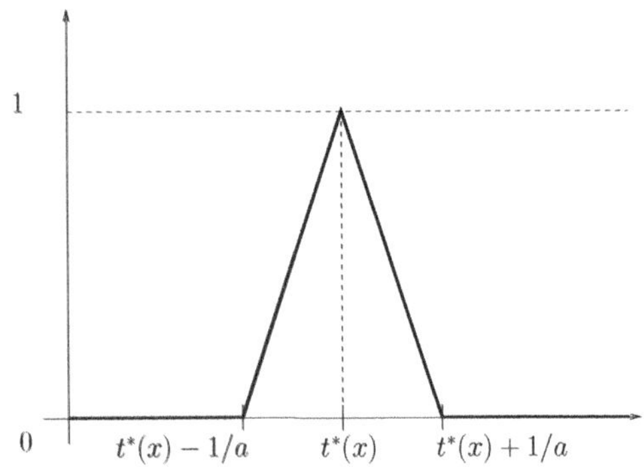

Figure 4.1. *Graphical representation of the triangular function* $\delta^a(x,\cdot)$

LEMMA 4.1.– Under assumption 1.3, the function δ^a is in $\mathbf{L}(E \times \mathbb{R}^+)$ with $L_{\delta^a} \leq aL_{t^*}$. More precisely, for all $x, y \in E$ and $t, u \in \mathbb{R}^+$, we have

$$|\delta^a(x,t) - \delta^a(y,u)| \leq aL_{t^*}|x-y| + a|t-u|.$$

In addition, for all $x \in E$ and $t, s \in \mathbb{R}^+$ such that $t + s \leq t^*(x)$, we have

$$\delta^a(\Phi(x, s), t) = \delta^a(x, s + t).$$

PROOF.– The first result is straightforward from the definition of δ^a. The semi-group property of the flow Φ yields $t^*(\Phi(x, s)) = t^*(x) - s$. Hence, the second result. □

We can now define a first approximation of the expectation $J_N(l, c)(x)$. For all $x \in E$ and $a > 0$, set

$$J_N^a(l, c)(x) = \mathbb{E}_x\Big[\int_0^{T_N} l(X_t)dt + \sum_{n=1}^{N} c(\Phi(Z_{n-1}, t^*(Z_{n-1}))\delta^a(Z_{n-1}, S_n)\Big].$$

The following proposition proves the convergence of $J_N^a(l, c)(x)$ toward $J_N(l, c)(x)$ with an error bound.

PROPOSITION 4.1.– Under assumptions 1.4 and 4.1, for all $x \in E$, $a > 0$ and $N \in \mathbb{N}^*$, we have

$$|J_N^a(l, c)(x) - J_N(l, c)(x)| \leq a^{-1} N C_c C_\lambda.$$

PROOF.– For all $x \in E$, we have

$$\begin{aligned}
|J_N^a & (l, c)(x) - J_N(l, c)(x)| \\
&= \Big|\mathbb{E}_x\Big[\sum_{n=1}^{N} c(\Phi(Z_{n-1}, t^*(Z_{n-1})) \times \big(\delta^a(Z_{j-1}, S_n) - \mathbb{1}_{\{S_n = t^*(Z_{n-1})\}}\big)\Big]\Big| \\
&\leq C_c \sum_{n=1}^{N} \mathbb{E}_x\left[\left|\delta^a(Z_{n-1}, S_n) - \mathbb{1}_{\{S_n = t^*(Z_{n-1})\}}\right|\right] \\
&\leq C_c \sum_{n=1}^{N} \mathbb{E}_x\big[\mathbb{E}[|\delta^a(Z_{n-1}, S_n) - \mathbb{1}_{\{S_n = t^*(Z_{n-1})\}}||Z_{n-1}]\big],
\end{aligned}$$

using the boundedness of c given by assumption 4.1. Recall from section 1.6 that the conditional law of S_n with respect to Z_{n-1} has density

$$s \longmapsto \lambda\big(\Phi(Z_{n-1}, s)\big)e^{-\Lambda(Z_{n-1}, s)}$$

on $[0; t^*(Z_{n-1}))$ and puts the weight $e^{-\Lambda(Z_{n-1}, t^*(Z_{n-1}))}$ on the point $t^*(Z_{n-1})$. As λ is bounded due to assumption 1.4, we have

$$
\begin{aligned}
&|J_N^a(l,c)(x) - J_N(l,c)(x)| \\
&\leq C_c \sum_{n=1}^{N} \mathbb{E}_x \left[\int_{t^*(Z_{n-1})-1/a}^{t^*(Z_{n-1})} \delta^a(Z_{n-1}, s)\lambda(\Phi(Z_{n-1}, s)) e^{-\Lambda(Z_{j-1}, s)} ds \right] \\
&\leq a^{-1} N C_c C_\lambda.
\end{aligned}
$$

Hence, the result. □

Consequently, we consider, from now on, the approximation of $J_N^a(l,c)(x)$ for some fixed a, large enough to ensure that the previous error is as small as required. The suitable choice of a is discussed at the end of section 4.4.

4.2.2. *Iterated operator*

As shown in [DAV 93, Section 32], the expectation $J_N^a(l,c)(x)$ that we are interested in approximating is obtained by iterating an integral operator G defined below.

DEFINITION 4.1.– *Define functions L and C for all $x \in E$ and $t \in [0; t^*(x)]$ by*

$$
L(x,t) = \int_0^t l(\Phi(x,s)) ds, \quad C(x,t) = c(\Phi(x, t^*(x)))\delta^a(x,t),
$$

and define the integral operator $G : \mathbf{B}(E) \to \mathbf{B}(E)$ by

$$
Gw(x) = \mathbb{E}_x \left[L(x, S_1) + C(x, S_1) + w(Z_1) \right].
$$

Next, we define the iterates of operator G starting from the null function. For convenience of notation that will be explained below, we choose a backward numeration.

DEFINITION 4.2.– *Define the sequence of functions $(v_n)_{0 \leq n \leq N}$ in $\mathbf{B}(E)$ by a backward recursion as follows. For all $x \in E$ and $0 \leq n \leq N-1$, set*

$$
v_N(x) = 0, \qquad v_n(x) = Gv_{n+1}(x).
$$

The strong Markov property together with an induction argument yield

$$v_{N-n}(x) = \mathbb{E}_x\Big[\int_0^{T_n} l(X_t)dt + \sum_{k=1}^{n} c\big(\Phi(Z_{k-1},t^*(Z_{k-1}))\big)\delta^a(Z_{k-1},S_k)\Big] = J_n^a(l,c)(x),$$

for all $0 \le n \le N-1$. Note that, due to the Markov property of the chain $(Z_n, S_n)_{n\in\mathbb{N}}$, we have

$$Gw(x) = \mathbb{E}\left[L(Z_n, S_{n+1}) + C(Z_n, S_{n+1}) + w(Z_{n+1})\middle|Z_n = x\right]. \qquad [4.2]$$

for all $0 \le n \le N-1$. Let $V_n = v_n(Z_n)$ for all $0 \le n \le N$. Hence, we have

$$V_N = 0, \qquad V_n = \mathbb{E}\left[L(Z_n, S_{n+1}) + C(Z_n, S_{n+1}) + V_{n+1}\middle|Z_n\right],$$

for all $0 \le n \le N-1$. We chose the backward numeration to match the indices of v_n and Z_n. To obtain an approximation of $J_N^a(l,c)(x) = V_0$, we need an approximation of the sequence of random variables $(V_n)_{0\le n\le N}$. This sequence satisfies a recursion that only depends on the chain $(Z_n, S_n)_{0\le n\le N}$, therefore, it appears natural to propose an approximation scheme based on the discretization by quantization of this chain.

4.2.3. *Approximation scheme*

We now use the optimal quantization $(\widehat{\Theta}_n)_{n\le N} = (\widehat{Z}_n, \widehat{S}_n)_{n\le N}$ of the process $(\Theta_n)_{n\le N} = (Z_n, S_n)_{n\le N}$ up to horizon N, with the same notation as in section 3.4. In order to approximate the random variables $(V_n)_{n\le N}$, we introduce quantized versions of operator G. Notice that the quantized process is no longer a homogeneous Markov chain so that we have different operators for each time step n. Their definitions naturally arise from equation [4.2].

DEFINITION 4.3.– *For all $1 \le n \le N$, $w \in \mathbf{B}(\Gamma_n^Z)$ and $z \in \Gamma_{n-1}^Z$, let*

$$\widehat{G}_n w(z) = \mathbb{E}\left[L(z, \widehat{S}_n) + C(z, \widehat{S}_n) + w(\widehat{Z}_n)\middle|\widehat{Z}_{n-1} = z\right].$$

Introduce the functions $(\widehat{v}_n)_{0\le k\le N}$ as $\widehat{v}_N(z) = 0$ for all $z \in \Gamma_N^Z$ and

$$\widehat{v}_n(z) = \widehat{G}_{n+1}\widehat{v}_{n+1}(z),$$

for all $0 \leq n \leq N-1$ and $z \in \Gamma_n^Z$ .For all $0 \leq n \leq N$, let

$$\widehat{V}_n = \widehat{v}_n(\widehat{Z}_n).$$

The conditional expectation in $\widehat{G}_n w(z)$ is a finite sum. Thus, the numerical computation of the sequence $(\widehat{V}_n)_{n \leq N}$ is easily performed as soon as the quantized process $(\widehat{\Theta}_n)_{n \leq N}$ has been obtained. The procedure is detailed in algorithm 4.1.

Algorithm 4.1: Approximation of $J_N^a(l,c)(x)$

input : Quantization grids $(\Gamma_n)_{0 \leq n \leq N}$, Transition matrices $(\widehat{P}_n)_{1 \leq n \leq N}$
output: $\widehat{V}_0$

begin
  for $z \in \Gamma_N^Z$ **do**
    $\widehat{v}_N(z) \leftarrow 0$
  end
  for $n \leftarrow N-1$ **to** 0 **do**
    for $z \in \Gamma_n^Z$ **do**
      $\widehat{v}_n(z) \leftarrow \sum_{(z',s') \in \Gamma_{n+1}} (L(z,s') + C(z,s') + v_{n+1}(z'))\widehat{P}_{n+1}(z,(z',s'))$
    end
  end
  return: $\widehat{v}_0$
end

We have assumed that $Z_0 = x$ a.s. Thus, the quantization algorithm provides $\widehat{Z}_0 = x$ a.s. too. Consequently, the random variable $\widehat{V}_0 = \widehat{v}_0(\widehat{Z}_0)$ is, in fact, deterministic.

4.3. Lipschitz regularity

In order to prove the convergence of this scheme, we need to investigate the Lipschitz regularity of the operators involved in the recursion. We start with establishing the Lipschitz regularity of functions L and C.

LEMMA 4.2.– Under assumption 4.1, the function L is in $\mathbf{L}(E \times \mathbb{R}^+)$. More precisely, for all $x, y \in E$ and $t, u \in \mathbb{R}^+$, we have $C_L \leq C_{t^*} C_l$ and

$$|L(x,t) - L(y,u)| \leq [l]_1 C_{t^*} |x-y| + C_l |t-u|.$$

PROOF.– The first assertion is obvious from the boundedness of l and t^*. For the second assertion, the Lipschitz continuity of l along the flow and again the boundedness of t^*, yield

$$\begin{aligned}|L(x,t) - L(y,u)| &\leq \int_0^{t\wedge u} |l(\Phi(x,s)) - l(\Phi(y,s))|ds + \int_{t\wedge u}^{t\vee u} C_l ds \\ &\leq [l]_1 C_{t^*}|x-y| + C_l|t-u|,\end{aligned}$$

hence, the result. □

LEMMA 4.3.– Under assumptions 1.3 and 4.1, the function C is in $\mathbf{L}(E\times\mathbb{R}^+)$. More precisely, for all $x, y \in E$ and $t, u \in \mathbb{R}^+$, we have $C_C \leq C_c$ and

$$|C(x,t) - C(y,u)| \leq ([c]_1 + aC_c L_{t^*})|x-y| + aC_c|t-u|.$$

PROOF.– The first assertion is obvious from the boundedness of c. We use the Lipschitz continuity of c along the flow together with lemma 4.1 to obtain the second one

$$\begin{aligned}|C(x,t) - C(y,u)| &\leq |c(\Phi(x,t^*(x))) - c(\Phi(y,t^*(y)))| \\ &\quad + |\delta^a(x,t) - \delta^a(y,u)| \\ &\leq ([c]_1 + aC_c L_{t^*})|x-y| + aC_c|t-u|,\end{aligned}$$

hence, the result. □

In order to establish that operator G maps the set $\mathbf{L}_\Phi(E)$ onto itself, we first need the auxiliary general results below. Both results are straightforward consequence of the Lipschitz-continuity properties of w, t^* and λ and are thus stated without proof.

LEMMA 4.4.– Under assumptions 1.3 and 1.5, for all functions w in $\mathbf{L}_\Phi(E)$, $x, y \in E$ and $t \leq t^*(x) \wedge t^*(y)$, we have

$$\begin{aligned}&\left|\int_t^{t^*(x)} w(\Phi(x,s))\mathrm{e}^{-\Lambda(x,s)}ds - \int_t^{t^*(y)} w(\Phi(y,s))\mathrm{e}^{-\Lambda(y,s)}ds\right| \\ &\qquad\qquad \leq (C_{t^*}[w]_1 + (C_{t^*}^2[\lambda]_1 + L_{t^*})C_w)|x-y|.\end{aligned}$$

LEMMA 4.5.– Under assumptions 1.3, 1.4 and 1.5, for all functions w in $\mathbf{L}_\Phi(\partial E)$ and $x, y \in E$, we have

$$\left|e^{-\Lambda(x,t^*(x))} w\big(\Phi(x,t^*(x))\big) - e^{-\Lambda(y,t^*(y))} w\big(\Phi(y,t^*(y))\big)\right| \\ \leq \big([w]_* + C_w(C_{t^*}[\lambda]_1 + L_{t^*}C_\lambda)\big)|x-y|.$$

We now turn to the operator G. We define a new family of operators (G_t) by

$$G_t w(x) = \mathbb{E}_x[(L(x,S_1) + C(x,S_1) + w(Z_1))\mathbb{1}_{\{S_1 \geq t\}}],$$

for w in $\mathbf{L}_\Phi(E)$, $x \in E$ and $t \geq 0$, so that $G_0 = G$ and $G_t w(x) = 0$ for all $t > t^*(x)$.

LEMMA 4.6.– For all functions w in $\mathbf{L}_\Phi(E)$, $x \in E$ and $t \leq t^*(x)$, we have the following integral decomposition

$$G_t w(x) = \sum_{k=1}^{6} \Upsilon_k(x,t),$$

with

$$\Upsilon_1(x,t) = e^{-\Lambda(x,t)} \int_0^t l(\Phi(x,s))ds,$$

$$\Upsilon_2(x,t) = \int_t^{t^*(x)} l(\Phi(x,s))e^{-\Lambda(x,s)}ds$$

$$= \int_t^{t^*(x)} l(\Phi(x,s))e^{-\Lambda(x,s)}ds,$$

$$\Upsilon_3(x,t) = c(\Phi(x,t^*(x))) \int_t^{t^*(x)} \delta^a(x,s)\lambda(\Phi(x,s))e^{-\Lambda(x,s)}ds,$$

$$\Upsilon_4(x,t) = e^{-\Lambda(x,t^*(x))}c(\Phi(x,t^*(x))),$$

$$\Upsilon_5(x,t) = \int_t^{t^*(x)} \lambda Qw(\Phi(x,s))e^{-\Lambda(x,s)}ds$$

$$= \int_t^{t^*(x)} \lambda Qw(\Phi(x,s))e^{-\Lambda(x,s)}ds,$$

$$\Upsilon_6(x,t) = e^{-\Lambda(x,t^*(x))}Qw(\Phi(x,t^*(x))).$$

PROOF.– As a result of lemma 1.4 giving the law of (Z_1, S_1) conditionally to $Z_0 = x$, we obtain

$$G_t w(x) = \int_0^{t^*(x)} \lambda Q w(\Phi(x,s)) e^{-\Lambda(x,s)} \mathbb{1}_{\{s \geq t\}} ds$$

$$+ \int_0^{t^*(x)} \lambda(\Phi(x,s)) e^{-\Lambda(x,s)} \mathbb{1}_{\{s \geq t\}} \big(L(x,s) + C(x,s)\big) ds$$

$$+ e^{-\Lambda(x,t^*(x))} \big(Qw(\Phi(x,t^*(x))) + L(x,t^*(x)) + C(x,t^*(x)\big).$$

The terms in C and w straightforwardly turn into Υ_3 to Υ_6. Let us now deal with the terms in L. From definition 4.1 of L and Fubini's lemma , we get

$$\int_0^{t^*(x)} \lambda(\Phi(x,s)) e^{-\Lambda(x,s)} \mathbb{1}_{\{s \geq t\}} L(x,s) ds$$

$$= \int_0^{t^*(x)} l(\Phi(x,u)) \big(e^{-\Lambda(x,t\vee u)} - e^{-\Lambda(x,t^*(x))}\big) du$$

$$= e^{-\Lambda(x,t)} \int_0^t l(\Phi(x,u)) du + \int_t^{t^*(x)} l(\Phi(x,u)) e^{-\Lambda(x,u)} du$$

$$- e^{-\Lambda(x,t^*(x))} \int_0^{t^*(x)} l(\Phi(x,u)) du,$$

$$= \Upsilon_1(x,t) + \Upsilon_2(x,t) - e^{-\Lambda(x,t^*(x))} L(x,t^*(x)),$$

Hence, the result. □

Due to this integral decomposition, we can establish that operator G_t maps the set $\mathbf{L}_\Phi(E)$ onto $\mathbf{L}(E)$.

LEMMA 4.7.– Under assumptions 1.3, 1.4, 1.5, 1.7 and 4.1, for all functions w in $\mathbf{L}_\Phi(E)$ and $t \geq 0$, $G_t w$ is in $\mathbf{L}(E)$, and for all $x, y \in E$ and $t \leq t^*(x) \wedge t^*(y)$, we have

$$|G_t w(x)| \leq C_l C_{t^*} + C_c + C_w, \qquad |G_t w(x) - G_t w(y)| \leq \gamma(a,w)|x-y|,$$

with $\gamma(a,w) = E_1 + E_2 a + E_3 C_w + E_4 [w]_1 + [Q][w]_*$ and

$$E_1 = 2[l]_1 C_{t^*} + C_l\big(L_{t^*} + 2C_{t^*}^2[\lambda]_1\big) + [c]_*\big(1 + C_{t^*} C_\lambda\big)$$
$$+ C_c\big(2[\lambda]_1 C_{t^*} + C_\lambda C_{t^*}^2[\lambda]_1 + 2L_{t^*} C_\lambda\big),$$
$$E_2 = C_c C_{t^*} C_\lambda L_{t^*}, E_3 = 2C_\lambda L_{t^*} + C_{t^*}[\lambda]_1\big(2 + C_{t^*} C_\lambda\big),$$
$$E_4 = \big(1 + C_{t^*} C_\lambda\big)[Q].$$

PROOF.– The bound for $G_t w(x)$ comes from the definition of G_t and boundedness of w, L and C given by lemmas 4.2 and 4.3. From lemma 4.6, $|G_t w(x) - G_t w(y)|$ is less than a sum of six differences. The first one $|\Upsilon_1(x,t) - \Upsilon_1(y,t)|$ is bounded by

$$\begin{aligned}|\Upsilon_1(x,t) - \Upsilon_1(y,t)| &\leq C_{t^*} C_l |\mathrm{e}^{-\Lambda(x,t)} - \mathrm{e}^{-\Lambda(y,t)}| \\ &\quad + \int_0^t \big| l(\Phi(x,s)) - l(\Phi(y,s)) \big| ds \\ &\leq \big(C_{t^*}^2 C_l [\lambda]_1 + C_{t^*}[l]_1\big)|x-y|.\end{aligned}$$

The differences $|\Upsilon_2(x,t) - \Upsilon_2(y,t)|$ and $|\Upsilon_5(x,t) - \Upsilon_5(y,t)|$ are bounded due to lemma 4.4 for functions l and λQw. Notice that $C_{\lambda Qw} \leq C_\lambda C_w$ and $[\lambda Qw]_1 \leq C_\lambda [Q][w]_1 + C_w[\lambda]_1$. Similarly, the differences $|\Upsilon_4(x,t) - \Upsilon_4(y,t)|$ and $|\Upsilon_6(x,t) - \Upsilon_6(y,t)|$ are bounded by lemma 4.5 for functions c and Qw, respectively, taking into account that $C_{Qw} \leq C_w$ and $[Qw]_* \leq [Q]([w]_* + [w]_1)$ under the local Lipschitz-continuity assumption 1.7 on the kernel Q. Finally, to bound $|\Upsilon_3(x,t) - \Upsilon_3(y,t)|$, we assume without loss of generality that $t^*(x) \leq t^*(y)$ and we have

$$\begin{aligned}&|\Upsilon_3(x,t) - \Upsilon_3(y,t)| \\ &\leq C_c \int_t^{t^*(x)} \Big|\delta^a(x,s)\lambda(\Phi(x,s))\mathrm{e}^{-\Lambda(x,s)} - \delta^a(y,s)\lambda(\Phi(y,s))\mathrm{e}^{-\Lambda(y,s)}\Big| ds \\ &\quad + C_c \int_{t^*(x)}^{t^*(y)} \Big|\delta^a(y,s)\lambda(\Phi(y,s))\mathrm{e}^{-\Lambda(y,s)}\Big| ds + [c]_* C_{t^*} C_\lambda |x-y| \\ &\leq C_c \int_t^{t^*(x)} \Big(C_\lambda \big|\delta^a(x,s) - \delta^a(y,s)\big| + [\lambda]_1|x-y| + C_\lambda \big|\mathrm{e}^{-\Lambda(x,s)} \\ &\quad - \mathrm{e}^{-\Lambda(y,s)}\big|\Big) ds + C_c L_{t^*} C_\lambda |x-y| + [c]_* C_{t^*} C_\lambda |x-y| \\ &\leq \big(C_c C_{t^*}(a C_\lambda L_{t^*} + [\lambda]_1 + C_\lambda C_{t^*}[\lambda]_1) + C_c L_{t^*} C_\lambda + [c]_* C_{t^*} C_\lambda\big)|x-y|.\end{aligned}$$

The result follows. □

LEMMA 4.8.– Under assumptions 1.3, 1.4, 1.5, 1.7 and 4.1, for all functions w in $\mathbf{L}_\Phi(E)$, $x \in E$ and $s, t \leq t^*(x)$, we have

$$|G_t w(x) - G_s w(x)| \leq \big(2C_l + C_\lambda(C_l C_{t^*} + C_c + C_w)\big)|t-s|.$$

PROOF.– From Lemma 4.6, $|G_t w(x) - G_s w(x)|$ is less than a sum of six differences. Note that Υ_4 and Υ_6 do not depend on t, so only four differences need to be investigated. The first one $|\Upsilon_1(x,t) - \Upsilon_1(x,s)|$ is bounded by

$$\begin{aligned}|\Upsilon_1(x,t) - \Upsilon_1(x,s)| &\leq C_{t^*} C_l |\mathrm{e}^{-\Lambda(x,t)} - \mathrm{e}^{-\Lambda(x,s)}| + C_l |t-s| \\ &\leq (C_{t^*} C_l C_\lambda + C_l)|t-s|.\end{aligned}$$

Similarly, we obtain

$$\begin{aligned}&|\Upsilon_2(x,t) - \Upsilon_2(x,s)| \leq C_l |t-s|, \\ &|\Upsilon_3(x,t) - \Upsilon_3(x,s)| \leq C_c C_\lambda |t-s|, \\ &|\Upsilon_5(x,t) - \Upsilon_5(x,s)| \leq C_\lambda C_w |t-s|,\end{aligned}$$

by noticing that $C_{\lambda Q w} \leq C_\lambda C_w$. □

We now have all the tools to establish that operator G maps the set $\mathbf{L}_\Phi(E)$ onto itself. First, we investigate the expression of Gw along the flow, which will make clear the relationship between G and G_t.

LEMMA 4.9.– For all functions w in $\mathbf{L}_\Phi(E)$, $x \in E$ and $0 \leq t \leq t^*(x)$, we have

$$Gw\big(\Phi(x,t)\big) = \mathrm{e}^{\Lambda(x,t)} G_t w(x) - \int_0^t l\big(\Phi(x,s)\big) ds.$$

PROOF.– We want to use the integral decomposition of Gw given by lemma 4.6. First, note that the semi-group property of the flow Φ yields the following equalities

$$\begin{aligned}\Phi(\Phi(x,t),s) &= \Phi(x,t+s), \\ t^*(\Phi(x,t)) &= t^*(x) - t, \\ \Phi\big(\Phi(x,t), t^*(\Phi(x,t))\big) &= \Phi(x, t^*(x)), \\ \Lambda(\Phi(x,t),s) &= \Lambda(x,t+s) - \Lambda(x,t), \\ \Lambda\big(\Phi(x,t), t^*(\Phi(x,t))\big) &= \Lambda(x,t^*(x)) - \Lambda(x,t).\end{aligned}$$

Now, we apply lemma 4.6 to obtain

$$\begin{aligned}&Gw\big(\Phi(x,t)\big) \\ &= \mathrm{e}^{\Lambda(x,t)} \int_t^{t^*(x)-t} l(\Phi(x,t+s))\mathrm{e}^{-\Lambda(x,t+s)} ds\end{aligned}$$

$$+ c(\Phi(x,t^*(x)))\mathrm{e}^{\Lambda(x,t)}\int_t^{t^*(x)-t}\delta^a(x,t+s)\lambda(\Phi(x,t+s))\mathrm{e}^{-\Lambda(x,t+s)}ds,$$

$$+ \mathrm{e}^{\Lambda(x,t)}\mathrm{e}^{-\Lambda(x,t^*(x))}\big(c(\Phi(x,t^*(x))) + Qw(\Phi(x,t^*(x)))\big),$$

$$+ \mathrm{e}^{\Lambda(x,t)}\int_t^{t^*(x)-t}\lambda Qw(\Phi(x,t+s))\mathrm{e}^{-\Lambda(x,t+s)}ds,$$

and the result is obtained by the change of variable $u = t + s$ in the integrals and addition and subtraction of the missing term Υ_1. □

LEMMA 4.10.– Under assumptions 1.3, 1.4, 1.5, 1.7 and 4.1, for all functions w in $\mathbf{L}_\Phi(E)$, Gw is in $\mathbf{L}_\Phi(E)$ with

$$[Gw]_1 = \mathrm{e}^{C_{t^*}C_\lambda}\big(C_{t^*}C_l + C_c + C_w + \gamma(a,w)\big) + C_{t^*}[l]_1,$$

$$[Gw]_2 = 2\mathrm{e}^{C_{t^*}C_\lambda}\big(C_\lambda(C_{t^*}C_l + C_c + C_w) + C_l\big) + C_l,$$

$$[Gw]_* = [c]_* + [Q]([w]_* + [w]_1).$$

PROOF.– First, lemma 4.9 implies that the limit of $Gw\big(\Phi(x,t)\big)$ as t goes to $t^*(x)$ exists. For all x, y in E and $t \le t^*(x) \wedge t^*(y)$, we have

$$|Gw(\Phi(x,t)) - Gw(\Phi(y,t))|$$

$$\le |\mathrm{e}^{\Lambda(x,t)}G_t w(x) - \mathrm{e}^{\Lambda(y,t)}G_t w(y)| + \int_0^t \big|l(\Phi(x,s)) - l(\Phi(y,s))\big|ds.$$

Note that $(x,t) \mapsto \mathrm{e}^{\Lambda(x,t)}$ is in $\mathbf{L}(E)$ with $\mathrm{e}^{\Lambda(x,t)} \le \mathrm{e}^{C_{t^*}C_\lambda}$ and $|\mathrm{e}^{\Lambda(x,t)} - \mathrm{e}^{\Lambda(y,s)}| \le \mathrm{e}^{C_{t^*}C_\lambda}\big(C_{t^*}[\lambda]_1|x-y| + C_\lambda|t-s|\big)$, and we use lemma 4.7 to obtain the bound for $[Gw]_1$. Similarly, for $s \le t \le t^*(x)$, we have

$$|Gw(\Phi(x,t)) - Gw(\Phi(x,s))|$$

$$\le |\mathrm{e}^{\Lambda(x,t)}G_t w(x) - \mathrm{e}^{\Lambda(y,s)}G_s w(y)| + \int_s^t |l(\Phi(x,u))|du$$

$$\le \big(\mathrm{e}^{C_{t^*}C_\lambda}C_\lambda(C_{t^*}C_l + C_c + C_w)$$

$$+ \quad \mathrm{e}^{C_{t^*}C_\lambda}\big(2C_l + C_\lambda(C_lC_{t^*} + C_c + C_w)\big) + C_l\big)|t-s|,$$

by lemma 4.8 and the previous inequalities. Finally, for x, y in E, lemmas 4.6 for $t = t^*(x)$ and 4.9 yield

$$|Gw(\Phi(x,t^*(x))) - Gw(\Phi(y,t^*(y)))|$$

$$= |c(\Phi(x,t^*(x)) - c(\Phi(y,t^*(y))$$

$$+ Qw(\Phi(x,t^*(x)) - Qw(\Phi(y,t^*(y))|$$

hence, the result. □

4.4. Rate of convergence

We now establish the convergence of our approximation procedure described in section 4.2.3 and provide a bound for the convergence rate. First, the Lipschitz-continuity property of the sequence of functions (v_n) obtained by iterating operator G is a direct consequence of the previous lemma 4.10.

LEMMA 4.11.– Under assumptions 1.3, 1.4, 1.5, 1.7 and 4.1, for all $0 \leq n \leq N-1$, v_n is in $\mathbf{L}_\Phi(E)$ and $\mathbf{L}(E)$ with

$$C_{v_n} \leq (N-n)(C_{t^*}C_l + C_c), \quad L_{v_n} \leq \gamma(a, v_{n+1}),$$

$$[v_n]_1 \leq \mathrm{e}^{C_{t^*}C_\lambda}\left(\gamma(a, v_{n+1}) + (N-n-1)C_{t^*}[\lambda]_1(C_{t^*}C_l + C_c)\right) + C_{t^*}[l]_1,$$

$$[v_n]_2 \leq e^{C_{t^*}C_\lambda}\left(C_{t^*}C_lC_\lambda + 2C_l + C_\lambda C_c + 2(N-n-1)C_\lambda(C_{t^*}C_l + C_c)\right) + C_l,$$

$$[v_n]_* \leq [c]_* + [Q]([v_{n+1}]_* + [v_{n+1}]_1).$$

PROOF.– Recall that $v_N = 0$ is in $\mathbf{L}_\Phi(E)$. Thus, by induction, v_n is also in $\mathbf{L}_\Phi(E)$ for all n due to lemma 4.10. The exact values of the bounds are obtained in the same way, as all Lipschitz constants of v_N equal 0. □

THEOREM 4.1.– Under assumptions 1.3, 1.4, 1.5, 1.7 and 4.1, the approximation error satisfies

$$|J_N^a(l,c)(x) - \widehat{V}_0| \leq \varepsilon_N(l,c,a),$$

where

$$\varepsilon_N(l,c,A) = \sum_{n=0}^{N-1}\Big(2L_{v_{n+1}}\|Z_{n+1} - \widehat{Z}_{n+1}\|_p + (2L_{v_n} + F_1)\|Z_n - \widehat{Z}_n\|_p + F_2\|S_{n+1} - \widehat{S}_{n+1}\|_p\Big) + a^{-1}NC_cC_\lambda,$$

with

$$F_1 = C_{t^*}[l]_1 + [c]_* + aL_{t^*}C_c, \qquad F_2 = C_l + aC_c,$$

and the Lipschitz constants L_{v_n} from lemma 4.11.

PROOF.– The main idea of the proof is to split the difference into a sum of suitably chosen tractable terms. First, notice that

$$|J_N(l,c)(x) - \widehat{V}_0| \leq |J_N(l,c)(x) - V_0| + |V_0 - \widehat{V}_0|.$$

Since $V_0 = J_N^a(l, c)(x)$, the first term is bounded by proposition 4.1. To study the second term, we proceed by induction. Recall that $V_N = \widehat{V}_N = 0$, so that $\|V_N - \widehat{V}_N\|_p = 0$. Set $n < N$. We have

$$|V_n - \widehat{V}_n| \leq |v_n(Z_n) - v_n(\widehat{Z}_n)| + |v_n(\widehat{Z}_n) - \widehat{v}_n(\widehat{Z}_n)|. \quad [4.3]$$

The first term is easily bounded using the Lipschitz property of v_n from lemma 4.11. We split the second term of equation [4.3] into two terms

$$\begin{aligned} |v_n(\widehat{Z}_n) - \widehat{v}_n(\widehat{Z}_n)| =& |Gv_{n+1}(\widehat{Z}_n) - \widehat{G}_{n+1}\widehat{v}_{n+1}(\widehat{Z}_n)| \\ \leq& |Gv_{n+1}(\widehat{Z}_n) - \widehat{G}_{n+1}v_{n+1}(\widehat{Z}_n)| \\ &+ |\widehat{G}_{n+1}v_{n+1}(\widehat{Z}_n) - \widehat{G}_{n+1}\widehat{v}_{n+1}(\widehat{Z}_n)|. \end{aligned} \quad [4.4]$$

From definition 4.3 of $\widehat{G}_{n+1}$, the second term in equation [4.4] is bounded in L_p norm by

$$\begin{aligned} &\|\widehat{G}_{n+1}v_{n+1}(\widehat{Z}_n) - \widehat{G}_{n+1}\widehat{v}_{n+1}(\widehat{Z}_n)\|_p \\ &\quad \leq \|v_{n+1}(\widehat{Z}_{n+1}) - \widehat{v}_{n+1}(\widehat{Z}_{n+1})\|_p \\ &\quad \leq L_{v_{n+1}}\|Z_{n+1} - \widehat{Z}_{n+1}\|_p + \|V_{n+1} - \widehat{V}_{n+1}\|_p. \end{aligned}$$

Regarding the first term in equation [4.4], this is where we use the specific properties of quantization. We have

$$\begin{aligned} &|Gv_{n+1}(\widehat{Z}_n) - \widehat{G}_{n+1}v_{n+1}(\widehat{Z}_n)| \\ &\quad \leq |\mathbb{E}[L(Z_n, S_{n+1}) + C(Z_n, S_{n+1}) + v_{n+1}(Z_{n+1}) \mid Z_n = \widehat{Z}_n] \\ &\quad - \mathbb{E}[L(Z_n, S_{n+1}) + C(Z_n, S_{n+1}) + v_{n+1}(Z_{n+1}) \mid \widehat{Z}_n]| \quad [4.5] \\ &\quad + |\mathbb{E}[L(Z_n, S_{n+1}) + C(Z_n, S_{n+1}) + v_{n+1}(Z_{n+1}) \mid \widehat{Z}_n] \\ &\quad - \mathbb{E}[L(\widehat{Z}_n, \widehat{S}_{n+1}) + C(\widehat{Z}_n, \widehat{S}_{n+1}) + v_{n+1}(\widehat{Z}_{n+1}) \mid \widehat{Z}_n]|. \quad [4.6] \end{aligned}$$

On the one hand, the second difference [4.6] is bounded in L_p norm using the Lipschitz-continuity properties of L, C and v_{n+1} given in lemmas 4.2, 4.3 and 4.11, respectively. On the other hand, recall that by construction of the quantized process, we have the following property: $\sigma\{\widehat{Z}_n\} \subset \sigma\{Z_n, S_n\}$. By using the special structure of the PDMP $(X(t))_{t\geq 0}$, we have $\sigma\{Z_n, S_n\} \subset \mathcal{F}_{T_n}$. Now, by using the Markov property of the process $\{X(t)\}$, it follows

$$\begin{aligned}
&\mathbb{E}[L(Z_n, S_{n+1}) + C(Z_n, S_{n+1}) + v_{n+1}(Z_{n+1}) \mid \widehat{Z}_n] \\
&\qquad = \mathbb{E}\big[\mathbb{E}[L(Z_n, S_{n+1}) + C(Z_n, S_{n+1}) + v_{n+1}(Z_{n+1}) \mid \mathcal{F}_{T_n}] \mid \widehat{Z}_n\big] \\
&\qquad = \mathbb{E}\big[\mathbb{E}[L(Z_n, S_{n+1}) + C(Z_n, S_{n+1}) + v_{n+1}(Z_{n+1}) \mid Z_n] \mid \widehat{Z}_n\big].
\end{aligned}$$

Hence, the difference [4.5] can be rewritten as $|\mathbb{E}[Gv_{n+1}(\widehat{Z}_n) - Gv_{n+1}(Z_n) \mid \widehat{Z}_n]|$, which is bounded by lemma 4.7. Gathering all the bounds of equations [4.3]–[4.6], we obtain

$$\begin{aligned}
\|v_n(\widehat{Z}_n)-\widehat{v}_n(\widehat{Z}_n)\|_p \leq (2L_{v_n}+F_1)\|Z_n-\widehat{Z}_n\|_p+F_2\|S_{n+1}-\widehat{S}_{n+1}\|_p \\
+ 2L_{v_{n+1}}\|Z_{n+1} - \widehat{Z}_{n+1}\|_p + \|V_{n+1} - \widehat{V}_{n-1}\|_p,
\end{aligned}$$

hence, the result by backward induction on n. □

We now briefly discuss the optimal choice for the regularization constant a. Proposition 4.1 suggests that a should be as large as possible. However, the constants F_1, F_2 and L_{v_n} that appear in the bound of the approximation error proposed by the above theorem 4.1 grow linearly with a. Thus, in order to control this error, it is necessary that the order of magnitude of the quantization error $\|\Theta_n - \widehat{\Theta}_n\|_p$ be at most a^{-1}.

4.5. Time-dependent functionals

We now turn to the main contribution of this chapter and present two generalizations of the previous problem. On the one hand, we will consider time-depending functionals of the form

$$\mathbb{E}_x\left[\int_0^{T_N} l(X_t, t)dt + \sum_{j=1}^{N} c(X_{T_j^-}, T_j)\mathbb{1}_{\{X_{T_j^-}\in\partial E\}}\right], \qquad [4.7]$$

where l and c are the Lipschitz continuous functions on $E \times \mathbb{R}^+$ and $\partial E \times \mathbb{R}^+$, respectively. On the other hand, we wish to replace the random time horizon T_N by a deterministic one denoted by T, i.e.

$$\mathbb{E}_x\left[\int_0^{T} l(X_t, t)dt + \sum_{T_j\leq T} c(X_{T_j^-}, T_j)\mathbb{1}_{\{X_{T_j^-}\in\partial E\}}\right].$$

We will reason as follows. As it is suggested by Davis in [DAV 93], we will consider the time-augmented process $(\widetilde{X}_t) = (X_t, t)$ defined in section 1.5. Indeed, both the expectation of the time-depending functional and the one with deterministic time horizon are none other than expectations of time invariant functionals for the time augmented process $(\widetilde{X}_t)_{t\geq 0}$. As we have seen in section 1.5, the Lipschitz-continuity assumptions on the characteristics of $(X_t)_{t\geq 0}$ required in the previous sections are also valid for $(\widetilde{X}_t)_{t\geq 0}$. Consequently, we may apply our numerical method to the time augmented process $(\widetilde{X}_t)_{t\geq 0}$. In other words, for $l \in \mathbf{L}_{\widetilde{\Phi}}(\widetilde{E})$, $c \in \mathbf{L}_{\widetilde{\Phi}}(\partial\widetilde{E})$ and $\xi \in \widetilde{E}$, our approximation scheme may be used to compute

$$\widetilde{J}_N(l,c)(\xi) = \mathbb{E}_\xi\left[\int_0^{T_N} l(\widetilde{X}_t)dt + \sum_{j=1}^N c(\widetilde{X}_{T_j^-})\mathbb{1}_{\{\widetilde{X}_{T_j^-}\in\partial\widetilde{E}\}}\right]. \qquad [4.8]$$

We will now see that the time-depending functional and the deterministic time horizon problems boil down to computing such quantities $\widetilde{J}_N(l,c)(\xi)$ for suitably chosen functions l and c.

We first consider the time-depending functional problem with Lipschitz continuous cost functions. Thus, let then $l \in \mathbf{L}_{\widetilde{\Phi}}(\widetilde{E})$, $c \in \mathbf{L}_{\widetilde{\Phi}}(\partial\widetilde{E})$ and $x \in E$, we wish to compute the expectation [4.7]. It is straightforward to notice that this quantity may be expressed using the time augmented process starting from the point $\xi_0 = (x, 0)$. Indeed, we have

$$\widetilde{J}_N(l,c)(\xi_0) = \mathbb{E}_{\xi_0}\left[\int_0^{T_N} l(X_t,t)dt + \sum_{j=1}^N c(X_{T_j^-},T_j)\mathbb{1}_{\{X_{T_j^-}\in\partial E\}}\right].$$

where $\widetilde{J}_N(l,c)(\xi_0)$ is given by equation [4.8]. Although they are time depending, the cost functions l and c are seen, on the left-hand side term, as time invariant functions of the time augmented process. The expectation of the time-depending functional is, therefore, obtained by computing the expectation of a time invariant functional for the transformed PDMP due to the approximation scheme described in section 4.2.3. This yields to the following theorem, whose proof arises from the previous discussion.

THEOREM 4.2.– Suppose assumptions 1.3, 1.4, 1.5 and 1.7 hold and let $l \in \mathbf{L}_{\widetilde{\Phi}}(\widetilde{E})$ and $c \in \mathbf{L}_{\widetilde{\Phi}}(\partial\widetilde{E})$. Applying the approximation scheme described in section 4.2.3 to the time augmented process $(\widetilde{X}_t)_{t\geq 0}$, we have

$$\left|\mathbb{E}_x\left[\int_0^{T_N} l(X_t,t)dt + \sum_{j=1}^N c(X_{T_j^-},T_j)\mathbb{1}_{\{X_{T_j^-}\in\partial E\}}\right] - \widehat{V}_0\right| \leq \widetilde{\varepsilon}_N(l,c,a).$$

where we denoted $\widetilde{\varepsilon}_N(l,c,a)$ the bound of the approximation error provided by theorem 4.1 when our approximation scheme is applied with cost functions l and c to the time augmented process $(\widetilde{X}_t)_{t\geq 0}$.

The quantity $\widetilde{\varepsilon}_N(l,c,a)$ is computed with respect to the process $(\widetilde{X}_t)_{t\geq 0}$ instead of $(X_t)_{t\geq 0}$ so that we actually have

$$\widetilde{\varepsilon}_N(l,c,a) = \sum_{n=0}^{N-1} \big(2L^{\widetilde{E}}_{v_{n+1}} \|\widetilde{Z}_{n+1} - \widehat{\widetilde{Z}}_{n+1}\|_p + (2L^{\widetilde{E}}_{v_n} + \widetilde{F}_1)\|\widetilde{Z}_n - \widehat{\widetilde{Z}}_n\|_p$$

$$+ \widetilde{F}_2 \|\widetilde{S}_{n+1} - \widehat{\widetilde{S}}_{n+1}\|_p\big) + a^{-1} N C_c C_\lambda,$$

where $(\widetilde{Z}_n, \widetilde{S}_n)_{n\in\mathbb{N}}$ denotes the sequence of the post-jump locations and the inter-jump times of the time augmented process $(\widetilde{X}_t)_{t\geq 0}$, $(\widehat{\widetilde{Z}}_n, \widehat{\widetilde{S}}_n)_{0\leq n\leq N}$ is its quantized approximation,

$$\widetilde{F}_1 = C_{t^*}[l]^{\widetilde{E}}_1 + [c]^{\widetilde{E}}_* + aL_{t^*}C_c\widetilde{E}, \qquad \widetilde{F}_2 = C^{\widetilde{E}}_l + aC^{\widetilde{E}}_c,$$

the Lipschitz constants of functions (v_n) are

$$C^{\widetilde{E}}_{v_n} \leq (N-n)(C_{t^*}C^{\widetilde{E}}_l + C^{\widetilde{E}}_c), \quad L^{\widetilde{E}}_{v_n} \leq \widetilde{\gamma}(a, v_{n+1}),$$

$$[v_n]^{\widetilde{E}}_1 \leq e^{C_{t^*}C_\lambda}\big(\widetilde{\gamma}(a,v_{n+1}) + C_{t^*}[\lambda]_1 C^{\widetilde{E}}_{v_{n+1}}\big) + C_{t^*}[l]^{\widetilde{E}}_1,$$

$$[v_n]^{\widetilde{E}}_2 \leq e^{C_{t^*}C_\lambda}\big(C_{t^*}C^{\widetilde{E}}_l C_\lambda + 2C^{\widetilde{E}}_l + C_\lambda C^{\widetilde{E}}_c + 2C_\lambda C^{\widetilde{E}}_{v_{n+1}}\big) + C^{\widetilde{E}}_l,$$

$$[v_n]^{\widetilde{E}}_* \leq [c]^{\widetilde{E}}_* + ([Q]\vee 1)(1+C_{t^*})([v_{n+1}]^{\widetilde{E}}_* + [v_{n+1}]^{\widetilde{E}}_1),$$

and for all w in $\mathbf{L}_{\widetilde{\Phi}}(\widetilde{E})$,

$$\widetilde{\gamma}(a,w) = \widetilde{E}_1 + \widetilde{E}_2 a + \widetilde{E}_3 C^{\widetilde{E}}_w + \widetilde{E}_4[w]^{\widetilde{E}}_1 + ([Q]\vee 1)(1+C_{t^*})[w]^{\widetilde{E}}_*,$$

with

$$\widetilde{E}_1 = 2[l]^{\widetilde{E}}_1 C_{t^*} + C_l\big(L_{t^*} + 2C^2_{t^*}[\lambda]_1\big) + [c]^{\widetilde{E}}_*\big(1 + C_{t^*}C_\lambda\big)$$

$$+ C^{\widetilde{E}}_c\big(2[\lambda]_1 C_{t^*} + C_\lambda C^2_{t^*}[\lambda]_1 + 2L_{t^*}C_\lambda\big),$$

$$\widetilde{E}_2 = E_2 = C_c C_{t^*} C_\lambda L_{t^*}, \quad \widetilde{E}_3 = E_3 = 2C_\lambda L_{t^*} + C_{t^*}[\lambda]_1\big(2 + C_{t^*}C_\lambda\big),$$

$$\widetilde{E}_4 = \big(1 + C_{t^*}C_\lambda\big)([Q]\vee 1)(1+C_{t^*}).$$

4.6. Deterministic time horizon

In the context of applications, it seems relevant to consider a deterministic time horizon T. For instance, we may want to estimate a mean cost over a given period no matter how many jumps occur during this period. In fact, we will choose a time horizon of the form $T \wedge T_N$ with N large enough to ensure the N-th jump will occur after time T with high enough probability, i.e. that $\mathbb{P}_x(T_N < T)$ be close to zero. For a theoretical bound of the above probability, see lemma 1.5. Simply notice that in practice, this probability may be estimated through Monte Carlo simulations. We thus intend to approximate the following quantity for $l \in \mathbf{L}_{\widetilde{\Phi}}(\widetilde{E})$, $c \in \mathbf{L}_{\widetilde{\Phi}}(\partial\widetilde{E})$ and $x \in E$

$$\mathbb{E}_x\Big[\int_0^{T_N \wedge T} l(X_t,t)dt + \sum_{T_j \leq T} c(X_{T_j^-},T_j)\mathbb{1}_{\{X_{T_j^-} \in \partial E\}}\Big]$$

$$= \mathbb{E}_x\Big[\int_0^{T_N} l(X_t,t)\mathbb{1}_{\{t \leq T\}}dt + \sum_{j=1}^{N} c(X_{T_j^-},T_j)\mathbb{1}_{\{X_{T_j^-} \in \partial E\}}\mathbb{1}_{\{T_j \leq T\}}\Big].$$

The natural approach would consist of killing the process at time T as Davis suggests in [DAV 93, section 31], and applying our method to the new process. However, the killed process will not necessarily fulfill our Lipschitz-continuity assumptions because of the discontinuity introduced at time T.

A second idea is then to use the previous results, to consider the time augmented process, and to define $\widetilde{l}(x,t) = l(x,t)\mathbb{1}_{\{t \leq T\}}$ and $\widetilde{c}(x,t) = c(x,t)\mathbb{1}_{\{t \leq T\}}$. However, a similar problem appears. Indeed, such functions $\widetilde{l}$ and $\widetilde{c}$ are not Lipschitz continuous and our numerical method requires this assumption. In the remainder of this section, we will see how to overcome this drawback. On the one hand, we prove that the Lipschitz-continuity condition on l may be relaxed so that our numerical method may be used directly to approximate $\widetilde{J}_N(\widetilde{l},c)$ for any $c \in \mathbf{L}_{\widetilde{\Phi}}(\partial\widetilde{E})$. On the other hand, in the general case, we will deal with the non-Lipschitz continuity of $\widetilde{c}$ by bounding it between two Lipschitz continuous functions.

4.6.1. *Direct estimation of the running cost term*

Let us explain how the Lipschitz-continuity condition on the running cost function may be relaxed so that theorem 4.1, stating the convergence of our approximation scheme, remains true when the running cost function is $l(x,t)\mathbb{1}_{\{t \leq T\}}$ with $l \in \mathbf{L}_{\widetilde{\Phi}}(\widetilde{E})$ and the boundary jump cost function is

$c \in \mathbf{L}_{\widetilde{\Phi}}(\partial\widetilde{E})$, albeit with slightly different constants in the bound of the convergence rate. Indeed, the running cost function $\widetilde{l}$ appears inside an integral that has a regularizing effect allowing us to derive the required Lipschitz property of the functional in spite of the discontinuity of $\widetilde{l}$.

Define $\widetilde{l} \in \mathbf{B}(\widetilde{E})$ by

$$\widetilde{l}(\xi) = l(x,t)\mathbb{1}_{\{t \leq T\}}, \qquad [4.9]$$

for all $\xi = (x,t) \in \widetilde{E}$. We intend to prove that the convergence of our approximation scheme, stated by theorem 4.1, remains true if we choose $\widetilde{l}$ as the running cost function even though it does not fulfill the required Lipschitz conditions, i.e. $\widetilde{l} \notin \mathbf{L}_{\widetilde{\Phi}}(\widetilde{E})$. The Lipschitz continuity of l is used four times in the proof of the theorem, once in lemma 4.2, twice in lemma 4.7, when bounding the difference of the Υ_1 and Υ_2 terms, and once in lemma 4.10 when bounding $[Gw]_1$. In each case, the Lipschitz continuity of the running cost function l is used to bound a term of the form

$$\int_s^{s'} \left|l(\widetilde{\Phi}(\xi,u)) - l(\widetilde{\Phi}(\xi',u))\right| du, \qquad [4.10]$$

for $\xi, \xi' \in \widetilde{E}$ and $s, s' \in [0; \widetilde{t}^*(\xi) \wedge \widetilde{t}^*(\xi')]$, or of the form

$$\int_s^{\widetilde{t}^*(\xi)\wedge\widetilde{t}^*(\xi')} \left|l(\widetilde{\Phi}(\xi,u))\mathrm{e}^{-\widetilde{\Lambda}(\xi,u)} - l(\widetilde{\Phi}(\xi',u))\mathrm{e}^{-\widetilde{\Lambda}(\xi',u)}\right| du,$$

for $\xi, \xi' \in \widetilde{E}$ and $s \in [0; \widetilde{t}^*(\xi) \wedge \widetilde{t}^*(\xi')]$ and where we naturally denoted

$$\widetilde{\Lambda}(\xi,u) = \int_0^u \widetilde{\lambda}(\widetilde{\Phi}(\xi,v))dv.$$

Concerning this second form, notice that

$$\int_s^{\widetilde{t}^*(\xi)\wedge\widetilde{t}^*(\xi')} |l(\widetilde{\Phi}(\xi,u))\mathrm{e}^{-\widetilde{\Lambda}(\xi,u)} - l(\widetilde{\Phi}(\xi',u))\mathrm{e}^{-\widetilde{\Lambda}(\xi',u)}| du$$

$$\leq \int_s^{\widetilde{t}^*(\xi)\wedge\widetilde{t}^*(\xi')} |l(\widetilde{\Phi}(\xi,u)) - l(\widetilde{\Phi}(\xi',u))| du$$

$$+C_l \int_s^{\widetilde{t}^*(\xi)\wedge\widetilde{t}^*(\xi')} |\mathrm{e}^{-\widetilde{\Lambda}(\xi,u)} - \mathrm{e}^{-\widetilde{\Lambda}(\xi',u)}|du$$

$$\leq \int_s^{s'} \big|l(\widetilde{\Phi}(\xi,u)) - l(\widetilde{\Phi}(\xi',u))\big|du, +C_l C_{t^*}^2[\lambda]_1,$$

so that, to ensure that theorem 4.1 remains true with $\widetilde{l}$ as the running cost function, it is sufficient to be able to bound terms of the form [4.10].

LEMMA 4.12.– For all l in $\mathbf{L}_{\widetilde{\Phi}}(\widetilde{E})$ and $\xi = (x,t), \xi' = (x',t') \in \widetilde{E}$ and $s \in [0; \widetilde{t}^*(\xi) \wedge \widetilde{t}^*(\xi')]$, we have

$$\int_0^s \big|\widetilde{l}(\widetilde{\Phi}(\xi,u)) - \widetilde{l}(\widetilde{\Phi}(\xi',u))\big|du \leq (C_{t^*}[l]_1 + C_l)|\xi - \xi'|.$$

PROOF.– We have

$$\int_0^s \big|\widetilde{l}(\widetilde{\Phi}(\xi,u)) - \widetilde{l}(\widetilde{\Phi}(\xi',u))\big|du$$

$$\leq \int_0^s \big|l(\widetilde{\Phi}(\xi,u))\mathbb{1}_{\{t+u\leq T\}} - l(\widetilde{\Phi}(\xi',u))\mathbb{1}_{\{t'+u\leq T\}}\big|du$$

$$\leq \int_s^{s'} \big|l(\widetilde{\Phi}(\xi,u)) - l(\widetilde{\Phi}(\xi',u))\big|du + C_l \int_0^s |\mathbb{1}_{\{t+u\leq T\}} - \mathbb{1}_{\{t'+u\leq T\}}|du.$$

The first term is bounded by $C_{t^*}[l]_1|\xi - \xi'|$ since $l \in \mathbf{L}_{\widetilde{\Phi}}(\widetilde{E})$. For the second term, assume without loss of generality that $t \leq t'$, we have

$$\begin{aligned}|\mathbb{1}_{\{t+u\leq T\}} - \mathbb{1}_{\{t'+u\leq T\}}| &= |\mathbb{1}_{\{t-T\leq u\}} - \mathbb{1}_{\{t'-T\leq u\}}| \\ &= \mathbb{1}_{\{t-T\leq u< t'-T\}},\end{aligned}$$

so that the right-hand side term is bounded by $C_l|t - t'|$. The result follows. □

Thus, theorem 4.1 remains true if we choose $\widetilde{l}$ as the running cost function. We only need to slightly modify the Lipschitz constants: the terms $C_{t^*}[l]_1$ have to be replaced by $C_{t^*}[l]_1 + C_l$. As a result, our approximation scheme may be used to compute, for any $\xi \in \widetilde{E}$,

$$\widetilde{J}_N(\widetilde{l},c)(x,0) = \mathbb{E}_x\Big[\int_0^{T_N\wedge T} l(X_t,t)dt + \sum_{j=1}^N c(X_{T_j^-},T_j)\mathbb{1}_{\{X_{T_j^-}\in\partial E\}}\Big].$$

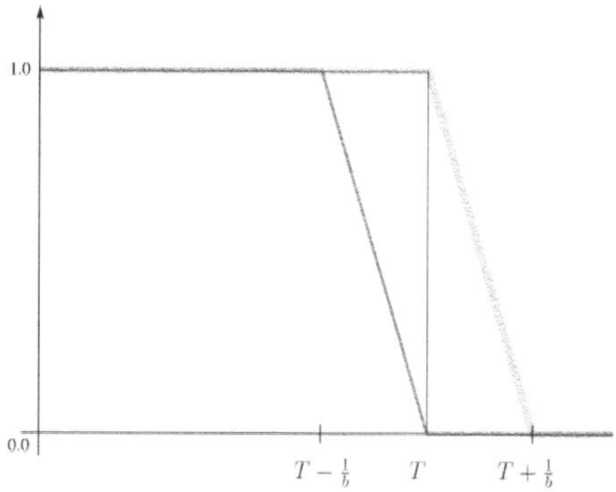

Figure 4.2. *Graphical representation of the functions $\underline{u}_b$ (thin) and $\overline{u}_b$ (thick)*

4.6.2. *Bounds of the boundary jump cost term*

We explained in the previous section how the Lipschitz-continuity condition on l may be relaxed. However, when it comes to c, this condition cannot be avoided and our numerical method cannot be used directly with $\widetilde{c}(x,t) = c(x,t)\mathbb{1}_{\{t\leq T\}}$. We overcome this drawback by using Lipschitz continuous approximations of the indicator function as in section 4.2.1. For $b > 0$, we introduce the real-valued functions $\underline{u}_b$ and $\overline{u}_b$ defined on $\mathbb{R}^+$ by

$$\underline{u}_b(t) = \mathbb{1}_{\{t<T-1/b\}} - b(t-T)\mathbb{1}_{\{T-1/b\leq t<T\}},$$
$$\overline{u}_b(t) = \mathbb{1}_{\{t<T\}} + (1-b(t-T))\mathbb{1}_{\{T\leq t<T+1/b\}}.$$

An illustration of these functions is given in Figure 4.2. They have the following properties.

LEMMA 4.13.– For all $t \geq 0$, $\lim_{b\to+\infty} \underline{u}_b(t) = \mathbb{1}_{[0;T[}(t)$ and $\lim_{b\to+\infty} \overline{u}_b(t) = \mathbb{1}_{[0;T]}(t)$. Furthermore, for all $b > 0$, $\underline{u}_b$ and $\overline{u}_b$ are in $\mathbf{L}(\mathbb{R}^+)$ with bound 1 and Lipschitz constant b. Moreover, we have $|\underline{u}_b - \mathbb{1}_{[0;T]}| \leq 1$, $|\overline{u}_b - \mathbb{1}_{[0;T]}| \leq 1$ and

$$\underline{u}_b \leq \mathbb{1}_{[0;T]} \leq \overline{u}_b.$$

Define for $c \in \mathbf{L}_{\widetilde{\Phi}}(\partial\widetilde{E})$

$$\underline{c}_b(x,t) = c(x,t)\underline{u}_b(t), \qquad \overline{c}_b(x,t) = c(x,t)\overline{u}_b(t). \qquad [4.11]$$

We now check that these functions satisfy our Lipschitz-continuity conditions.

LEMMA 4.14.– Under assumption 1.3, the functions $\underline{c}_b$ and $\overline{c}_b$ belong to $\mathbf{L}_{\widetilde{\Phi}}(\partial\widetilde{E})$ with $[\underline{c}_b]_*$ and $[\overline{c}_b]_*$ less than $[c]_* + bC_c(1 \vee L_{t^*})$.

PROOF.– We prove the result for $\underline{c}_b$, the other case being similar. For all $\xi = (x, t), \xi' = (x', t') \in \widetilde{E}$, we have

$$\begin{aligned}
&\left|\underline{c}_b\big(\widetilde{\Phi}(\xi, t^*(\xi))\big) - \underline{c}_b\big(\widetilde{\Phi}(\xi', t^*(\xi'))\big)\right| \\
&\quad = \left|c\big(\widetilde{\Phi}(\xi, \widetilde{t^*}(\xi))\big)\underline{u}_b(t + \widetilde{t^*}(\xi)) - c\big(\widetilde{\Phi}(\xi', \widetilde{t^*}(\xi'))\big)\underline{u}_b(t' + \widetilde{t^*}(\xi'))\right| \\
&\quad \leq [c]_*|\xi - \xi'| + C_c\left|\underline{u}_b(t + \widetilde{t^*}(\xi)) - \underline{u}_b(t' + \widetilde{t^*}(\xi'))\right| \\
&\quad \leq [c]_*|\xi - \xi'| + C_c b\big(|t - t'| + [\widetilde{t^*}]|x - x'|\big) \\
&\quad \leq \big([c]_* + C_c b(1 \vee L_{t^*})\big)|\xi - \xi'|.
\end{aligned}$$

Hence, the result. □

Therefore, the functions $\underline{c}_b$ and $\overline{c}_b$ are acceptable boundary jump cost functions and we may bound the deterministic horizon expectation by

$$\begin{aligned}
&\widetilde{J}_N(\widetilde{l}, \underline{c}_b)(x, 0) \\
&\quad \leq \mathbb{E}_x\left[\int_0^{T_N} l(X_t, t)\mathbb{1}_{\{t \leq T\}}dt + \sum_{j=1}^{N} c(X_{T_j^-}, T_j)\mathbb{1}_{\{X_{T_j^-} \in \partial E\}}\mathbb{1}_{\{T_j \leq T\}}\right] \\
&\quad \leq \widetilde{J}_N(\widetilde{l}, \overline{c}_b)(x, 0).
\end{aligned}$$

The following proposition provides the convergence of the bounds.

PROPOSITION 4.2.– Suppose assumptions 1.3, 1.4, 1.5 and 1.7 hold and let $l \in \mathbf{L}_{\widetilde{\Phi}}(\widetilde{E})$ and $c \in \mathbf{L}_{\widetilde{\Phi}}(\partial\widetilde{E})$. For all $x \in E$, we have

$$\begin{aligned}
&\lim_{b \to +\infty} \widetilde{J}_N(\widetilde{l}, \underline{c}_b)(x, 0) \\
&\quad = \lim_{b \to +\infty} \widetilde{J}_N(\widetilde{l}, \overline{c}_b)(x, 0) \\
&\quad = \mathbb{E}_x\left[\int_0^{T_N \wedge T} l(X_t, t)dt + \sum_{j=1}^{N} c(X_{T_j^-}, T_j)\mathbb{1}_{\{X_{T_j^-} \in \partial E\}}\mathbb{1}_{\{T_j \leq T\}}\right].
\end{aligned}$$

The convergence holds for every $T > 0$ in the case of $\widetilde{J}_N(\widetilde{l}, \overline{c}_b)(x, 0)$ but only for almost every $T > 0$ w.r.t. the Lebesgue measure on $\mathbb{R}$ in the case of $\widetilde{J}_N(\widetilde{l}, \underline{c}_b)(x, 0)$.

PROOF.– Let $x \in E$. We first consider $\widetilde{J}_N(\widetilde{l}, \overline{c}_b)(x, 0)$.

$$\left|\mathbb{E}_x\left[\sum_{j=1}^{N} c(X_{T_j^-}, T_j)\mathbb{1}_{\{X_{T_j^-} \in \partial E\}}\mathbb{1}_{\{T_j \leq T\}} - \sum_{j=1}^{N} \overline{c}_b(X_{T_j^-}, T_j)\mathbb{1}_{\{X_{T_j^-} \in \partial E\}}\right]\right|$$

$$\leq \mathbb{E}_x\left[\sum_{j=1}^{N} \left|c(X_{T_j^-}, T_j)\right| \left|\mathbb{1}_{\{T_j \leq T\}} - \overline{u}_b(T_j)\right|\right]$$

$$\leq C_c \mathbb{E}_x\left[\sum_{j=1}^{N} \mathbb{1}_{\{T < T_j \leq T + \frac{1}{b}\}}\right]$$

$$\leq C_c \sum_{j=1}^{N} (\varphi_j(T + b^{-1}) - \varphi_j(T)),$$

where φ_j is the distribution function of T_j. For all $j \leq N$, $\lim_{b \to +\infty} \big(\varphi_j(T + 1/b) - \varphi_j(T)\big) = 0$ since φ_j is right-continuous which shows the required convergence. We now turn to the case of $\widetilde{J}_N(\widetilde{l}, \underline{c}_b)(x, 0)$. Similar computations yield

$$\left|\mathbb{E}_x\left[\sum_{j=1}^{N} c(X_{T_j^-}, T_j)\mathbb{1}_{\{X_{T_j^-} \in \partial E\}}\mathbb{1}_{\{T_j \leq T\}} - \sum_{j=1}^{N} \underline{c}_b(X_{T_j^-}, T_j)\mathbb{1}_{\{X_{T_j^-} \in \partial E\}}\right]\right|$$

$$\leq C_c \sum_{j=1}^{N} \big(\varphi_j(T) - \varphi_j(T - b^{-1})\big).$$

We cannot conclude as in the previous case since φ_j is not necessary left-continuous. We, therefore, assume that T is not an atom of any of the laws of the random variables T_j. Then, for all $j \leq N$, $\lim_{b \to +\infty}(\varphi_j(T) - \varphi_j(T - 1/b)) = 0$ and the result follows. Indeed, the set of the atoms of T_j is at most countable so that the convergence holds for almost every T w.r.t. the Lebesgue measure on $\mathbb{R}$. □

4.6.3. *Bounds in the general case*

The previous results show that deterministic horizon expectations may be bounded by applying our numerical method with $\widetilde{l}$ and successively $\underline{c}_b$ and $\overline{c}_b$. In other words, we have shown the following theorem.

THEOREM 4.3.– Suppose assumptions 1.3, 1.4, 1.5 and 1.7 hold and let $l \in \mathbf{L}_{\widetilde{\Phi}}(\widetilde{E})$ and $c \in \mathbf{L}_{\widetilde{\Phi}}(\partial \widetilde{E})$. Let $(\underline{V}_{n,b})_{0 \leq n \leq N}$ (respectively, $(\overline{V}_{n,b})_{0 \leq n \leq N}$) be the sequence of random variables $(V_n)_{0\leq n\leq N}$ described in section 4.2.3 when applying our approximation scheme to the time augmented process $(\widetilde{X}_t)_{t\geq 0}$ with cost functions $\widetilde{l}$ and $\underline{c}_b$ (respectively, $\overline{c}_b$) defined by equations [4.9] and [4.11]. The bounds of the approximation error provided by theorem 4.1 are, respectively, denoted by $\widetilde{\varepsilon}_N(l, \underline{c}_b, a, b)$ and $\widetilde{\varepsilon}_N(l, \overline{c}_b, a, b)$. We have then

$$\begin{aligned}\underline{V}_{0,b}-&\widetilde{\varepsilon}_N(l, \underline{c}_b, a, b)\\ &\leq \mathbb{E}_x\Big[\int_0^{T_N\wedge T} l(X_t,t)dt + \sum_{j=1}^{N} c(X_{T_j^-}, T_j)\mathbb{1}_{\{X_{T_j^-}\in\partial E\}}\mathbb{1}_{\{T_j\leq T\}}\Big]\\ &\leq \overline{V}_{0,b} + \widetilde{\varepsilon}_N(l, \overline{c}_b, a, b).\end{aligned}$$

In the previous theorem, the quantities $\widetilde{\varepsilon}_N(l, \underline{c}_b, a, b)$ and $\widetilde{\varepsilon}_N(l, \overline{c}_b, a, b)$ are computed with respect to the time-augmented process $(\widetilde{X}_t)_{t\geq 0}$, so that we have

$$\begin{aligned}\widetilde{\varepsilon}_N(l, \underline{c}_b, a, b) = \sum_{n=0}^{N-1}\Big(&2L_{v_{n+1,b}}^{\widetilde{E}}\|\widetilde{Z}_{n+1}-\widehat{\widetilde{Z}}_{n+1}\|_p+(2L_{v_{n,b}}^{\widetilde{E}}+\widetilde{F}_1^b)\|\widetilde{Z}_n-\widehat{\widetilde{Z}}_n\|_p\\ &+\widetilde{F}_2^b\|\widetilde{S}_{n+1}-\widehat{\widetilde{S}}_{n+1}\|_p\Big) + a^{-1}NC_cC_\lambda,\end{aligned}$$

where $(\widetilde{Z}_n, \widetilde{S}_n)_{n\in\mathbb{N}}$ denotes the sequence of the post-jump locations and the inter-jump times of the time augmented process $(\widetilde{X}_t)_{t\geq 0}$, $(\widehat{\widetilde{Z}}_n, \widehat{\widetilde{S}}_n)_{0\leq n\leq N}$ is its quantized approximation,

$$\widetilde{F}_1^b = \widetilde{F}_1 + C_c(1\vee L_{t^*})b, \qquad \widetilde{F}_2^b = \widetilde{F}_2,$$

the Lipschitz constants of functions $(v_{n,b})$ are unchanged from theorem 4.2, except that $\widetilde{\gamma}(a, v_{n+1,b})$ is replaced by $\widetilde{\gamma}(a, b, v_{n+1,b})$ and for all w in $\mathbf{L}_{\widetilde{\Phi}}(\widetilde{E})$,

$$\widetilde{\gamma}(a,b,w) = \widetilde{E}_1 + \widetilde{E}_2 a + \widetilde{E}_5 b + \widetilde{E}_3 C_w^{\widetilde{E}} + \widetilde{E}_4[w]_1^{\widetilde{E}} + ([Q]\vee 1)(1+C_{t^*})[w]_*^{\widetilde{E}},$$

with $\widetilde{E}_5 = C_c(1\vee L_{t^*})(1+C_{t^*}C_\lambda)$.

It is important to stress the fact that applying twice our numerical method does not increase significantly the computational time. Indeed, the computation of the quantization grids is, by far, the most costly step. These grids, which only depend on the dynamics of the process, may be stored offline and used for the approximation of both bounds.

We now discuss the choice of the parameter b, the discussion is quite similar to the one concerning the choice of a at the end of section 4.4. Proposition 4.2 suggests that b should be chosen as large as possible. However, choosing a large value for b will lead to large Lipschitz constants that will decrease the sharpness of the bounds $\widetilde{\varepsilon}_N(l, \underline{c}_b, a, b)$ and $\widetilde{\varepsilon}_N(l, \overline{c}_b, a, b)$ for the approximation error provided by theorem 4.3. Indeed, it is easy to check that $L_{v_{n,b}}$ grows linearly with b (see the precise expressions of the Lipschitz constant $\widetilde{\gamma}(a, b, v_{n+1,b})$ above). Thus, in order to control the error proposed by theorem 4.3, it is necessary that the order of magnitude of the quantization error $\|\Theta_n - \widehat{\Theta}_n\|_p$ be at most b^{-1}.

4.7. Example

We now apply our numerical procedure to the repair workshop model presented in section 1.8.5. Suppose that the daily cost of maintenance is $q(x)$, while the daily cost of reparation is $p(x)$, with reasonably $p(x) > q(x)$, for a given setting x. Our aim is to find the value of the setting x that maximizes the expected total benefits $Bf(x)$, i.e. the discounted value (the interest rate is denoted by ρ) of production minus the costs related to maintenance and reparation over a period $T = 5$ years

$$Bf^* = \sup_{x \in [0;1]} Bf(x),$$

where

$$Bf(x) = \mathbb{E}_{(1,0,0)}\left[\int_0^T e^{-\rho t}(r(x)\mathbb{1}_{\{m_t=1\}} - p(x)\mathbb{1}_{\{m_t=2\}} - q(x)\mathbb{1}_{\{m_t=3\}})dt\right].$$

We use the following values $r(x) = x$, $p(x) = 100x^2$, $q(x) = 5$, $s = 7$ days, $\rho = 0.03/365$ and λ is the intensity of a Weibull distribution with parameters $\alpha = 2$ and $\beta = 600$. As we are dealing with time-dependent functionals, we need to use the time augmented process

$$\widetilde{X}_t = (m_t, \zeta_t, t),$$

where m is the mode (1 for working, 2 for repair and 3 for maintenance), ζ_t is the time since the last change of mode and t is the time since the beginning. Naturally, we set $\widetilde{X}_0 = (1, 0, 0)$. Our regularity assumptions clearly hold so that we may run our numerical method. We first need to find $N \in \mathbb{N}$ such that $\mathbb{P}_{(1,0,0)}(T_N < T)$ be small enough. Monte Carlo simulations lead to the value $N = 18$, corresponding to $\mathbb{P}_{(1,0,0)}(T_N < T) = 10^{-5}$. For a fixed $x \in [0;1]$, we will, therefore, compute $\widetilde{J}_N(\widetilde{l}, 0)(1,0,0)$ where

$$\widetilde{l}(m, \zeta, t) = \mathrm{e}^{-\rho t}\big(r(x)\mathbb{1}_{\{m=1\}} - p(x)\mathbb{1}_{\{m=2\}} - q(x)\mathbb{1}_{\{m=3\}}\big)\mathbb{1}_{\{t \leq T\}}.$$

Notice that we also could have chosen r, p and q slightly more general by allowing them to be time-dependent.

It is important to stress the fact that, once the Markov chain associated with the process is quantized, we are able to compute the approximation of $Bf(x)$ almost instantly for any $x \in [0;1]$ according to algorithm 4.1 because the same grids are used for every computation. Due to this flexibility, we are able to compute the values of $Bf(x)$ on a fine enough grid for x, and thus to solve the above optimization problem very easily. This is a very important advantage of our method. Indeed, if we computed $Bf(x)$ through standard methods such as Monte Carlo simulations, we would have to repeat the whole algorithm again and again for each value of x and solving the optimization problem would be much longer.

Figure 4.3 represents the approximation of the function Bf computed on a constant step grid of $[0;1]$ with step 10^{-2}. This leads to the solution of the earlier optimization problem. Indeed, we obtain $Bf^* = Bf(x^*) = 537.84$ where $x^* = 0.78$ is the value of the setting x that maximizes the benefits of the factory.

Number of points in the quantization grids	$\widehat{V}_0$	Relative error to 537.69
20 points	542.14	0.83%
50 points	539.57	0.35%
100 points	538.24	0.10%
500 points	537.84	0.03%

Table 4.1. *Values of $\widehat{V}_0$ for increasing number of points in the quantization grids*

Now let $x = x^* = 0.78$, Table 4.1 presents the values of $\widehat{V}_0$, which equal the approximations of $Bf(x)$, for an increasing number of points in the quantization grids. A reference value $Bf_{MC} = 537.69$ is obtained with 10^8 Monte Carlo simulations. Figure 4.4 shows a log-log plot of the approximation versus the number of points in the quantization grids. An estimation of the slope by a simple linear regression yields to a convergence rate of -1.08. The theoretical value from theorems 3.1 and 4.1 should be $-1/(d+1)$, where $d = 3$ is the dimension of the PDMP. However, the mode m is already discrete, after a jump $\zeta = 0$ and the running time T_n is related to the inter-jump times as $T_n = S_1 + \cdots + S_n$. Therefore, only one free variable really needs to be quantized, which explains why the empirical rate is close to -1 corresponding to a one-dimensional process.

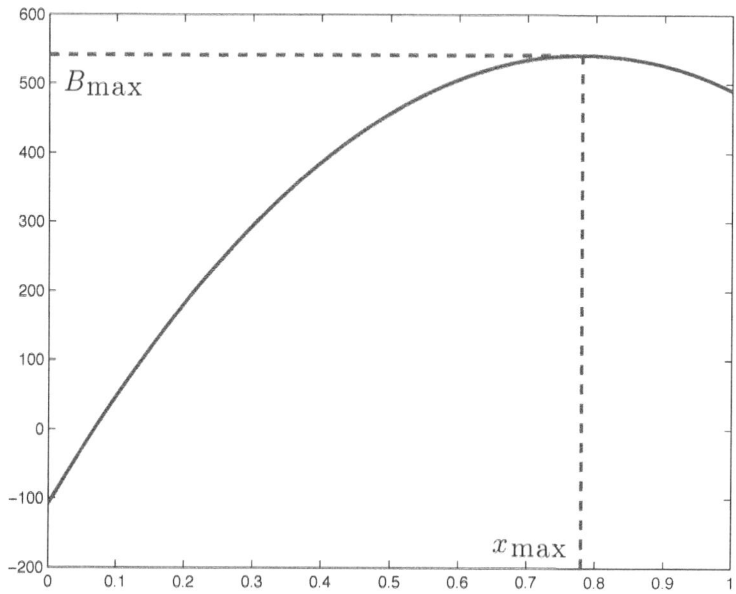

Figure 4.3. *The function Bf drawn with 500 points in the quantization grids*

From a computational time point of view, we already explained that the computation of large quantization grids is, by far, the most costly step since it may take up to several hours, whereas the approximation of the expectation that follows is then almost instantaneous. However, we may note, in Table 4.1, that grids containing only 50 points yield a quite accurate result with merely 0.35% error. Such grids only require a few minutes to be designed. The same grids may also serve several purposes as they are not dependent on the cost functions. For instance, we may compute the mean time spent by the machine in the workshop by taking $l(m,\zeta,t) = \mathbb{1}_{\{m\in\{2;3\}\}}$. Also see section 6.1 for another application of this procedure on the corrosion model.

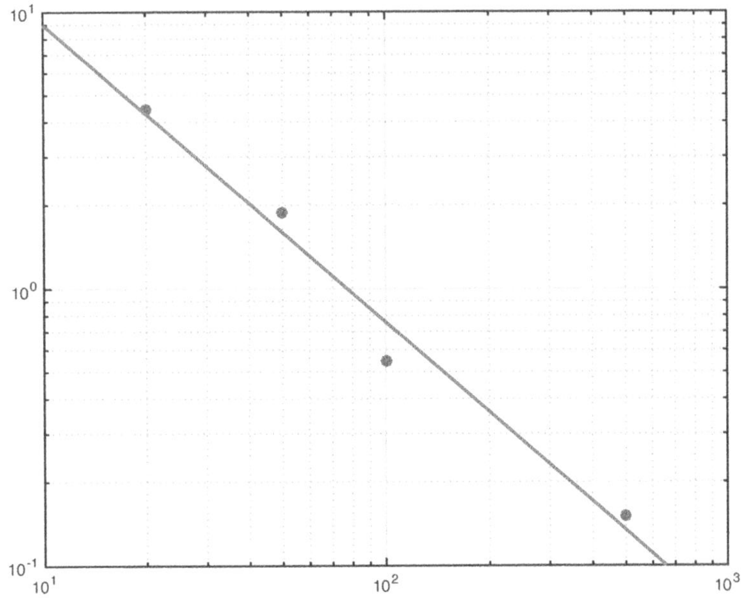

Figure 4.4. *Log-log plot of the approximation error versus the number of points in the quantization grids*

4.8. Conclusion

We have presented an efficient and easy-to-implement numerical method to approximate the expectations of functionals of PDMPs. We proved the convergence of our algorithm with bounds for the rate of convergence. Although our original method concerns time invariant functionals, we proved that we are able to tackle time-depending problems such as Lipschitz continuous time-depending functionals and deterministic time horizon expectations, due to the introduction of the time augmented process.

Our method is easy to implement because it merely requires us to be able to simulate the process. Furthermore, although the computation of the quantization grids may be quite time-consuming, it may be performed preliminarily because the grids only depend on the dynamics of the process and not on the cost functions l and c. Therefore, they may be stored offline and serve several purposes. As it is illustrated by the example, storing the grids provides to our approximation scheme efficiency and flexibility. Indeed, the computation of the expectation can be performed very quickly once the grids are available. Thus, if we decide, for instance, to modify the functional, the same grids may be used so that the new result is obtained very quickly. This flexibility is an important advantage over standard Monte Carlo simulations. The same quantization grids may also be used to compute other quantities of interest, as shown in Chapters 6 and 9.

5

Exit Time

5.1. Introduction

The aim of this chapter is to propose a practical numerical method for approximating the survival function and the moments of the exit time for a piecewise deteerministic Markov process (PDMP). Numerical computation of the moments of the exit time for a Markov process has been studied by K. Helmes, S. Röhl and R.H. Stockbridge in [HEL 01]. Starting from an assumption related to the generator of the process, they derive a system of linear equations satisfied by the moments. In addition to these equations, they include finitely many Hausdorff moment conditions that are also linear constraints. This optimization problem is a standard linear programming problem for which many pieces of efficient software are available. J.-B. Lasserre and T. Prieto-Rumeau introduced in [LAS 04] a similar method but they improved the efficiency of the algorithm by replacing the Hausdorff moment conditions with semi-definite positivity constraints of some moment matrices. Nevertheless, their approach cannot be applied to PDMPs because the assumption related to the generator of the process is generally not satisfied.

In [DAV 93, section 33], M.H.A. Davis gives an iterative method to compute the mean exit time for a PDMP but his approach involves solving a large set of Ordinary Differential Equations (ODEs) whose forms are very problem specific, depending on the behavior of the process at the boundary of the state space. The approach developed in the previous chapter 4 cannot be directly used as indicator functions are involved. However, they could be dealt with by using upper and lower piecewise linear Lipschitz continuous approximations as in sections 4.2 and 4.6.2. We do not choose to follow this approach here. The interested reader is referred to [BRA 12a] for more details.

In the context of applications to reliability, it seems important to study not only the expectation but also the distribution of the exit time. This leads us to a direct approach. We present approximation methods to compute the moments and the survival function of the exit time τ of a PDMP $(X_t)_{t\in\mathbb{R}}$ from a subset U of the state space E given the fact that it happens before the N-th jump T_N of the PDMP. Roughly speaking, we estimate the moments and the survival function for $\tau \wedge T_N$. In our approach, the first step consists of expressing the j-th moment (respectively, the survival function) as the last term of some sequence satisfying a recursion specifically built within this chapter. The second step is then to rewrite the recursions in terms of the underlying discrete-time Markov chain $(Z_n, T_n)_{n\in\mathbb{N}}$. Naturally, it can then be replaced by its quantized approximation to obtain a tractable numerical scheme. However, the proof of convergence is not straightforward as the recursion operators involve some discontinuities related to indicator functions. We do not choose to replace them by regular approximations. Instead, we show that the discontinuities happen with small enough probability. Another main difficulty arises from the recursion operators not being Lipschitz continuous. This difficulty is overcome by deriving new and important properties of the Markov chain $(Z_n, T_n)_{n\in\mathbb{N}}$, combined with a sharp feature of the optimal quantization. Thus, we are able to prove the convergence of the approximation scheme. In the case of the moments, we obtain bounds for the rate of convergence.

The main material from this chapter comes from [BRA 12c]. The chapter is organized as follows. In section 5.2, we define our exit time problem and derive its recursive formulation. We give our approximation schemes in section 5.3 and prove their convergence in section 5.4. Finally, we illustrate our method on a Poisson process with trend in section 5.5.

5.2. Problem setting

Let $(X_t)_{t\geq 0}$ be a PDMP on the state space $E = \{(m, \xi), m \in M, \xi \in E_m\}$ with mode set M. For all mode $m \in M$, let U_m be an open subset of E_m and set $U = \{(m, \xi), m \in M, \xi \in U_m\}$. We deal with the exit time τ of $(X_t)_{t\geq 0}$ from U

$$\tau = \inf\{s \geq 0 : X_s \notin U\}.$$

Denote μ as the distribution of the initial state of the process $X_0 = Z_0$. The following assumption appears natural to make our problem relevant.

ASSUMPTION 5.1.– The process starts in U and eventually leaves it almost surely i.e. the support of μ is included in U and $\mathbb{P}_\mu(\tau < +\infty) = 1$.

The aim of this chapter is to provide approximation schemes for the survival function and moments of τ. As in the previous chapter, our approach is based on a recursive computation using the underlying discrete-time Markov chain of the PDMP. Therefore, we will use a computation horizon that equals a jump time of the process: we will study $\tau \wedge T_N$ rather than τ itself. This is not too restrictive. Indeed, due to assumption 1.1, when N goes to infinity, we have

$$\tau \wedge T_N \to \tau \qquad \mathbb{P}_\mu\text{a.s.}$$

We may thus approximate τ by $\tau \wedge T_N$ if N is chosen such that $\mathbb{P}_\mu(\tau > T_N)$ is small enough because the evolution of the process beyond T_N will have little impact on the law or the moments of the exit time. See section 5.2.3 for a discussion of the choice of N. The rest of this section presents the two problems we are interested in and describes recursive methods to solve them.

DEFINITION 5.1.– *Let us define $u^*(x)$ for all $x \in U$ as the time for the flow starting from the point x to exit from U*

$$u^*(x) = \inf\{s \geq 0 : \Phi(x, s) \notin U\}.$$

We now introduce some technical assumptions that will be in force throughout this chapter. Here, u^* will play a similar role as t^* in the other chapters. This is why we require similar assumptions to assumptions 1.2 and 1.3 on u^*.

ASSUMPTION 5.2.– The function u^* is in $\mathbf{B}(E)$ with upper bound C_{u^*}.

ASSUMPTION 5.3.– The function u^* is in $\mathbf{L}(E)$ with Lipschitz constant L_{u^*}.

The next assumption concerns the shape of the set U. It will be used to ensure that when Z_n is in U, then its quantized approximation $\widehat{Z}_n$ is also in U.

ASSUMPTION 5.4.– For all $m \in M$, the set U_m is convex.

Next, we need to control the probability for the PDMP to jump in a neighborhood of the boundary of U.

ASSUMPTION 5.5.– For $\alpha > 0$, set

$$U^\alpha = \{x \in E : \min\{|x - y| : y \in \partial U\} \leq \alpha\}.$$

There exist $C_U > 0$ and $\beta > 0$ such that for all $0 \leq n \leq N, \mathbb{P}_\mu(Z_n \in U^\alpha) \leq C_U\alpha^\beta$.

This technical condition can be checked in most applications. We will see in section 6.2.1 how it can be derived quite generally when Z_n has a bounded density. It could be replaced by the stronger assumption that the probability to jump in a small enough neighborhood of the boundary ∂U is zero.

Finally, we assume that once the process has exited U, it cannot get back in U. This assumption can be made without loss of generality. Indeed, we are only interested in the process up to time τ, thus once it has exited τ, it can be killing process and put to a cemetery state never to enter U again. In this case, we have $t^* = u^*$.

ASSUMPTION 5.6.– The process cannot go back to U once it has left U: for all $z \notin U$, we have

$$\mathbb{P}_z(X_t \in U \text{ for some } t \geq 0) = 0.$$

5.2.1. *Distribution*

The first goal of this chapter is to compute an approximation of the distribution of the exit time τ. More precisely, we intend to approximate the survival function $\mathbb{P}_\mu(\tau > s \big| \tau \leq T_N)$ for $s > 0$. Our strategy is first to obtain a recursive formulation for this quantity, then express it in terms of the embedded Markov chain and finally quantize the Markov chain to obtain a computable approximation. We start with the recursive construction of the survival function.

DEFINITION 5.2.– *For all $s > 0$, define as follows the sequences $(p_n(s))_{n\in\mathbb{N}}$, $(q_n)_{n\in\mathbb{N}}$ and $(r_n(s))_{n\in\mathbb{N}}$*

$$\begin{aligned} p_n(s) &= \mathbb{P}_\mu(\tau > s \big| \tau \leq T_n), \\ q_n &= \mathbb{P}_\mu(\tau \leq T_n), \\ r_n(s) &= \mathbb{P}_\mu(\{\tau > s\} \cap \{T_n < \tau \leq T_{n+1}\}). \end{aligned}$$

The conditional probability $p_n(s)$ does not exist when $q_n = 0$. We then choose to extend the sequence by setting $p_n(s) = 0$ in this case. Our objective is to approximate $p_N(s)$ where N represents the computation horizon. The following proposition provides a recursion for the sequence $(p_n)_{n\leq N}$, pointing out that p_N may be computed as soon as the sequences $(q_n)_{n\leq N}$ and $(r_n)_{n\leq N-1}$ are known.

PROPOSITION 5.1.– Under assumption 5.1, for all $n \in \mathbb{N}$, $s > 0$, we have $p_0(s) = 0$ and

$$p_{n+1}(s) = \frac{p_n(s)q_n + r_n(s)}{q_{n+1}} \mathbb{1}_{\{q_{n+1} \neq 0\}}.$$

PROOF.– First, recall that $T_0 = 0$ so that we have $p_0 = 0$ since the process starts in U according to assumption 5.1. Then, let $n \in \mathbb{N}$ such that $q_{n+1} \neq 0$ and note that $\{\tau \leq T_{n+1}\} = \{\tau \leq T_n\} \cup \{T_n < \tau \leq T_{n+1}\}$. Thus, we have

$$\begin{aligned} p_{n+1}(s) &= \frac{\mathbb{P}_\mu(\{\tau > s\} \cap \{\tau \leq T_{n+1}\})}{\mathbb{P}_\mu(\tau \leq T_{n+1})} \\ &= \frac{\mathbb{P}_\mu(\{\tau > s\} \cap \{\tau \leq T_n\}) + \mathbb{P}_\mu(\{\tau > s\} \cap \{T_n < \tau \leq T_{n+1}\})}{q_{n+1}} \\ &= \frac{p_n(s)q_n + r_n(s)}{q_{n+1}}, \end{aligned}$$

showing the result. □

5.2.2. *Moments*

Our second goal is to approximate the moments of τ i.e. for all $j \in \mathbb{N}$, we are interested in $\mathbb{E}_\mu[\tau^j \mid \tau \leq T_N]$. This is a very classical problem and some results are already available, as mentioned in the introduction, although they do not apply to general PDMPs. It is also possible to use a Monte Carlo method and we will point out why the recursion-based method we propose is more efficient and flexible.

DEFINITION 5.3.– *For all $j \in \mathbb{N}$, let the sequences $(p_{n,j})_{n\in\mathbb{N}}$ and $(r_{n,j})_{n\in\mathbb{N}}$ be defined as follows*

$$p_{n,j} = \mathbb{E}_\mu[\tau^j | \tau \leq T_n], \quad r_{n,j} = \mathbb{E}_\mu[\tau^j \mathbb{1}_{\{T_n < \tau \leq T_{n+1}\}}].$$

Our objective is to approximate $p_{N,j}$ where N still represents the computation horizon. Similar to the previous section, the sequence $(p_{n,j})_{n\leq N}$ satisfies a recursion involving the sequences $(q_n)_{n\leq N}$, previously introduced, and $(r_{n,j})_{n\leq N-1}$.

PROPOSITION 5.2.– Under assumption 5.1, for all $n, j \in \mathbb{N}$, we have $p_{0,j} = 0$ and

$$p_{n+1,j} = \frac{p_{n,j}q_n + r_{n,j}}{q_{n+1}} \mathbb{1}_{\{q_{n+1} \neq 0\}}.$$

PROOF.– The proof is similar to that of proposition 5.1. □

5.2.3. *Computation horizon*

Before giving the details of our approximation scheme, let us discuss the choice of the computation horizon N. It is necessary to choose N large enough such that $\mathbb{P}_\mu(T_N < \tau)$ is small. It is difficult to estimate this probability for a general process because the links between τ and the jump times are largely problem-dependent. For instance, the geometry of U can be very complex. Therefore, N will generally be estimated through Monte Carlo simulations.

We introduce here another method to bound this probability that may prove useful in applications. First, note that, for any $T > 0$, we have

$$(T_N < \tau) \subset (T_N < T) \cup (\tau > T).$$

This implies that

$$\mathbb{P}_\mu(T_N < \tau) \leq \mathbb{P}_\mu(T_N < T) + \mathbb{P}_\mu(\tau > T).$$

This bound is especially useful whenever τ is bounded, which happens quite often in applications, because then there exists some T such that $\mathbb{P}_\mu(\tau > T) = 0$. On the contrary, when τ is not bounded, it remains however sometimes possible to obtain T such that $\mathbb{P}_\mu(\tau > T)$ is small.

Consider, for instance, the crack propagation model introduced in section 1.8.4. We are interested in the time τ before the crack size reaches a critical size y_c. It is possible to bound the exit time by considering the slowest flow. We clearly have for all $t \geq 0, Y_t \geq Y_0 e^{\alpha t}$ and thus we obtain

$$\mathbb{P}_\mu(\tau > \frac{1}{\alpha}\ln(\frac{y_c}{Y_0})) = 0$$

A bound for $\mathbb{P}_\mu(T_N < T)$ is given in section 1.8.4. Therefore it is possible, given the parameters, to find a theoretical value of N such that $\mathbb{P}_\mu(T_N < \tau)$ is smaller than a given precision ε.

5.3. Approximation schemes

In this section, we express our recursion in terms of the embedded Markov chain. We work with the post-jump locations and jump times chain $(Z_n, T_n)_{n\in\mathbb{N}}$ rather than the post-jump locations and inter-jump times one $(Z_n, S_n)_{n\in\mathbb{N}}$. Note that they play similar roles as $S_n = T_n - T_{n-1}$ for all

$n \geq 1$. As seen in the previous section, our recursions already involve (T_n), therefore it is natural to keep this sequence.

Concerning the distribution, we already emphasized that $p_N(s) = \mathbb{P}_\mu(\tau > s|\tau \leq T_N)$ may be computed as soon as the sequences $(q_n)_{n\leq N}$ and $(r_n)_{n\leq N-1}$ are known. Therefore, we will find expressions of these sequences depending on the Markov chain $(Z_n, T_n)_{n\leq N}$ that we will replace by the quantized process $(\widehat{Z}_n, \widehat{T}_n)_{n\leq N}$ in order to define their quantized approximations $(\widehat{q}_n)_{n\leq N}$ and $(\widehat{r}_n)_{n\leq N-1}$. First, note that

$$(T_n < \tau) = (Z_n \in U), \qquad (\tau \leq T_n) = (Z_n \notin U),$$

due to assumption 5.6. Moreover, on the event $(Z_n \in U, Z_{n+1} \notin U)$, we have

$$\tau = (T_n + u^*(Z_n)) \wedge T_{n+1},$$

as u^* is the deterministic exit time from U. Thus, we obtain

$$\begin{aligned} q_n &= \mathbb{E}_\mu[\mathbb{1}_{U^c}(Z_n)], \\ r_n(s) &= \mathbb{E}_\mu[\mathbb{1}_{\{(T_n+u^*(Z_n))\wedge T_{n+1}>s\}} \mathbb{1}_U(Z_n)\mathbb{1}_{U^c}(Z_{n+1})]. \end{aligned}$$

We now naturally define the quantized approximations of the previous sequences.

DEFINITION 5.4.– *For all $s > 0$, define $(\widehat{q}_n)_{n\leq N}$ and $(\widehat{r}_n)_{n\leq N-1}$ by*

$$\begin{aligned} \widehat{q}_n &= \mathbb{E}_\mu[\mathbb{1}_{U^c}(\widehat{Z}_n)], \\ \widehat{r}_n(s) &= \mathbb{E}_\mu[\mathbb{1}_{\{(\widehat{T}_n+u^*(\widehat{Z}_n))\wedge \widehat{T}_{n+1}>s\}} \mathbb{1}_U(\widehat{Z}_n)\mathbb{1}_{U^c}(\widehat{Z}_{n+1})]. \end{aligned}$$

It is important to note that both sequences $(\widehat{q}_n)$ and $(\widehat{r}_n(s))$ may be computed easily from the output of the quantization algorithm, see algorithm 5.1.

Recall from proposition 5.1 that the sequence $(p_n)_{n\leq N}$ satisfies a recursion depending on the two sequences $(q_n)_{n\leq N}$ and $(r_n)_{n\leq N-1}$, that we are now able to approximate. Hence, replacing them by their quantized approximations within the same recursion leads to a new sequence denoted by $(\widehat{p}_n)_{n\leq N}$.

DEFINITION 5.5.– *For all $s > 0$ and for all $n \leq N-1$, let $\widehat{p}_0(s) = 0$ and*

$$\widehat{p}_{n+1}(s) = \frac{\widehat{p}_n(s)\widehat{q}_n + \widehat{r}_n(s)}{\widehat{q}_{n+1}} \mathbb{1}_{\{\widehat{q}_{n+1}\neq 0\}}.$$

Algorithm 5.1: Approximation of (q_n) and $(r_n(s))$

input : Quantization grids $\Gamma_n, 0 \leq n \leq N$,
Weight vectors $\widehat{W}_n, 0 \leq n \leq N$, Transition matrices $\widehat{P}_n$, $1 \leq n \leq N$

output: $(\widehat{q}_n)$ and $(\widehat{r}_n(s))$

begin

  $\widehat{q}_0 \leftarrow 0$

  for $n \leftarrow 0$ **to** $N-1$ **do**

    $\widehat{q}_{n+1} \leftarrow \sum_{\substack{(z,t)\in\Gamma_{n+1} \\ z\notin U}} \widehat{W}_{n+1}(z,t)$

    $\widehat{r}_n(s) \leftarrow \sum_{\substack{(z,t)\in\Gamma_n \\ z\in U}} \sum_{\substack{(z',t')\in\Gamma_{n+1} \\ z'\notin U}} \mathbb{1}_{\{(t+u^*(z))\wedge t'>s\}} \widehat{W}_n(z,t)\widehat{P}_n((z,t),(z',t'))$

  end

  return: $(\widehat{q}_n), (\widehat{r}_n(s))$

end

Let us now consider the moments of the exit time τ. Of course they may be derived from the distribution but we present a method to approximate them directly. Similar to the case of the distribution, the moments $p_{N,j} = \mathbb{E}_\mu[\tau^j|\tau \leq T_N]$ may be computed as soon as the sequences $(q_n)_{n\leq N}$ and $(r_{n,j})_{n\leq N-1}$ are known. The first sequence has already been approximated in the previous section but we still need to find an expression of the second sequence depending on the Markov chain (Z_n, T_n). Due to assumption 5.6, we have

$$r_{n,j} = \mathbb{E}_\mu[((T_n + u^*(Z_n)) \wedge T_{n+1})^j \mathbb{1}_U(Z_n)\mathbb{1}_{U^c}(Z_{n+1})].$$

So that we may now naturally define the quantized approximation of the sequences $(r_{n,j})_{n\leq N-1}$ and $(p_{n,j})_{n\leq N}$.

DEFINITION 5.6.– *For all $j \in \mathbb{N}$, define the sequence $(\widehat{r}_{n,j})_{n\leq N-1}$ as*

$$\widehat{r}_{n,j} = \mathbb{E}_\mu\left[((\widehat{T}_n + u^*(\widehat{Z}_n)) \wedge \widehat{T}_{n+1})^j \mathbb{1}_U(\widehat{Z}_n)\mathbb{1}_{U^c}(\widehat{Z}_{n+1})\right],$$

and the sequence $(\widehat{p}_{n,j})_{n\leq N}$ by $\widehat{p}_{0,j} = 0$ and

$$\widehat{p}_{n+1,j} = \frac{\widehat{p}_{n,j}\widehat{q}_n + \widehat{r}_{n,j}}{\widehat{q}_{n+1}} \mathbb{1}_{\{\widehat{q}_{n+1}\neq 0\}}.$$

As in the previous section, $\widehat{r}_{n,j}$ may be computed easily from the output of the quantization algorithm, see algorithm 5.2.

Algorithm 5.2: Approximation of $(r_{n,j})$

---

input : Quantization grids $\Gamma_n, 0 \leq n \leq N$,
Weight vectors $\widehat{W}_n, 0 \leq n \leq N$,
Transition matrices $\widehat{P}_n, 1 \leq n \leq N$

output: $(\widehat{r}_{n,j}(s))$

begin

  for $n \leftarrow 0$ **to** $N-1$ **do**

    $$\widehat{r}_n \leftarrow \sum_{\substack{(z,t)\in\Gamma_n \\ z\in U}} \sum_{\substack{(z',t')\in\Gamma_{n+1} \\ z'\notin U}} ((t+u^*(z)) \wedge t')^j \widehat{W}_n(z,t)\widehat{P}_n((z,t),(z',t'))$$

  end

  return: $(\widehat{r}_{n,j})$

end

5.4. Convergence

We now prove that the approximation schemes defined in the previous section converge as the quantization error goes to zero.

5.4.1. *Distribution*

The convergence of $(\widehat{p}_n)_{n\leq N}$ toward $(p_n)_{n\leq N}$ is far from being trivial because on the one hand, the definitions of the sequences $(q_n)_{n\leq N}$ and $(r_n)_{n\leq N-1}$ contain many indicator functions that are not Lipschitz continuous and on the other hand, the recursive function giving p_{n+1} from p_n, q_n, q_{n+1} and r_n is not Lipschitz continuous either. We first establish the convergence of the sequences $(\widehat{q}_n)_{n\leq N}$ and $(\widehat{r}_n)_{n\leq N-1}$ toward $(q_n)_{n\leq N}$ and $(r_n)_{n\leq N-1}$.

PROPOSITION 5.3.– Under assumptions 5.1, 5.5 and 5.6, for all $n \leq N$, $\widehat{q}_n$ converges toward q_n when the quantization error goes to zero. More precisely, the error is bounded by

$$|q_n - \widehat{q}_n| \leq C_U^{\frac{p}{p+\beta}} \left(\left(\frac{\beta}{p}\right)^{\frac{p}{p+\beta}} + \left(\frac{p}{\beta}\right)^{\frac{\beta}{p+\beta}} \right) \|Z_n - \widehat{Z}_n\|_p^{\frac{p\beta}{p+\beta}},$$

where C_U and β are defined in assumption 5.5.

PROOF.– For all $n \leq N$, the definitions of q_n and $\widehat{q}_n$ yield

$$|q_n - \widehat{q}_n| = |\mathbb{E}_\mu[\mathbb{1}_U(Z_n) - \mathbb{1}_U(\widehat{Z}_n)]|.$$

The difference of the indicator functions is non-zero if and only if Z_n and $\widehat{Z}_n$ are on either side of ∂U. Therefore, in this case, for all $\alpha > 0$, if $|Z_n - \widehat{Z}_n| \leq \alpha$, then $\min\{|Z_n - y| : y \in \partial U\} \leq \alpha$. Hence, either $|Z_n - \widehat{Z}_n| > \alpha$ or $Z_n \in U^\alpha$. The Markov inequality and assumption 5.5 thus yield

$$\begin{aligned}\mathbb{E}_\mu[|\mathbb{1}_U(Z_n) - \mathbb{1}_U(\widehat{Z}_n)|] &\leq \mathbb{P}_\mu(|Z_n - \widehat{Z}_n| > \alpha) + \mathbb{P}_\mu(Z_n \in U^\alpha)\\ &\leq \frac{\|Z_n - \widehat{Z}_n\|_p^p}{\alpha^p} + C_U\alpha^\beta.\end{aligned}$$

This bound reaches a minimum when $\alpha = (p\|Z_n - \widehat{Z}_n\|_p^p(\beta C_U)^{-1})^{\frac{1}{p+\beta}}$ and the result follows. □

PROPOSITION 5.4.– Under assumptions 5.1, 5.3, 5.5 and 5.6, for all $n \leq N - 1$ and for almost every $s > 0$ w.r.t. the Lebesgue measure on $\mathbb{R}$, we have

$$\widehat{r}_n(s) \to r_n(s),$$

when the quantization error goes to zero.

PROOF.– Let $n \leq N - 1$ and $s > 0$, by definition, we have

$$|r_n(s) - \widehat{r}_n(s)| \leq A + B,$$

where

$$\begin{aligned}A =& |\mathbb{E}_\mu[(\mathbb{1}_{\{(T_n+u^*(Z_n))\wedge T_{n+1}>s\}} - \mathbb{1}_{\{(\widehat{T}_n+u^*(\widehat{Z}_n))\wedge \widehat{T}_{n+1}>s\}})\\ &\times \mathbb{1}_U(Z_n)\mathbb{1}_{U^c}(Z_{n+1})]|,\\ B =& |\mathbb{E}_\mu[\mathbb{1}_{\{(\widehat{T}_n+u^*(\widehat{Z}_n))\wedge \widehat{T}_{n+1}>s\}}(\mathbb{1}_U(Z_n)\\ &\mathbb{1}_{U^c}(Z_{n+1}) - \mathbb{1}_U(\widehat{Z}_n)\mathbb{1}_{U^c}(\widehat{Z}_{n+1}))]|.\end{aligned}$$

In the A term, we crudely bound $\mathbb{1}_U(Z_n)$ and $\mathbb{1}_{U^c}(Z_{n+1})$ by 1 and turn to the difference of the two indicator functions. This difference is non-zero if and only if $(T_n + u^*(Z_n)) \wedge T_{n+1}$ and $(\widehat{T}_n + u^*(\widehat{Z}_n)) \wedge \widehat{T}_{n+1}$ are on the either side of s yielding that they both belong to $[s - \eta; s + \eta]$ where

$$\eta = |(T_n + u^*(Z_n)) \wedge T_{n+1} - (\widehat{T}_n + u^*(\widehat{Z}_n)) \wedge \widehat{T}_{n+1}|.$$

Then, we have

$$\begin{aligned}&\left|\mathbb{1}_{\{(T_n+u^*(Z_n))\wedge T_{n+1}>s\}} - \mathbb{1}_{\{(\widehat{T}_n+u^*(\widehat{Z}_n))\wedge \widehat{T}_{n+1}>s\}}\right|,\\ &\qquad\qquad\leq \mathbb{1}_{\{|(T_n+u^*(Z_n))\wedge T_{n+1}-s|\leq \eta\}}\end{aligned}$$

so that we obtain

$$A \leq \mathbb{P}_\mu\big(\big|(T_n + u^*(Z_n)) \wedge T_{n+1} - s\big| \leq \eta\big).$$

The following discussion consists of noting that either η is small and so is the probability that $(T_n + u^*(Z_n)) \wedge T_{n+1}$ belongs to the interval $[s-\eta; s+\eta]$, or η is large but this happens with a small probability when the quantization error goes to zero. For all $\alpha > 0$, we have

$$\begin{aligned}A &\leq \mathbb{P}_\mu(|(T_n + u^*(Z_n)) \wedge T_{n+1} - s| \leq \eta, \eta \leq \alpha) + \mathbb{P}_\mu(\eta > \alpha)\\ &\leq \mathbb{P}_\mu(|(T_n + u^*(Z_n)) \wedge T_{n+1} - s| \leq \alpha) + \mathbb{P}_\mu(\eta > \alpha)\\ &\leq \big|\varphi_n(s+\alpha) - \varphi_n(s-\alpha)\big| + \|\eta\|_p^p \alpha^{-p},\end{aligned}$$

where φ_n denotes the distribution function of $(T_n + u^*(Z_n)) \wedge T_{n+1}$. Let $\epsilon > 0$ and assume that s is not an atom of this distribution so that there exists $\alpha_1 > 0$ such that $|\varphi_n(s+\alpha_1) - \varphi_n(s-\alpha_1)| \leq \epsilon$. Besides, due to the Lipschitz continuity of u^*, we have $\eta \leq |T_n - \widehat{T}_n| + L_{u^*}|Z_n - \widehat{Z}_n| + |T_{n+1} - \widehat{T}_{n+1}|$. Moreover, since the quantization error goes to 0, we may assume that $\|\eta\|_p \leq \alpha_1 \epsilon^{\frac{1}{p}}$. Setting $\alpha = \alpha_1$ in the previous computations yields

$$A \leq |\varphi_n(s+\alpha_1) - \varphi_n(s-\alpha_1)| + \|\eta\|_p^p \alpha_1^{-p} \leq 2\epsilon.$$

As the set of the atoms of the distribution function of $(T_n + u^*(Z_n)) \wedge T_{n+1}$ is at most countable, the previous discussion is true for almost every $s > 0$ w.r.t. the Lebesgue measure. Let us now bound the B term by

$$\begin{aligned}B &\leq \mathbb{E}_\mu|\mathbb{1}_U(Z_n)\mathbb{1}_{U^c}(Z_{n+1}) - \mathbb{1}_U(\widehat{Z}_n)\mathbb{1}_{U^c}(\widehat{Z}_{n+1})|\\ &\leq \mathbb{E}_\mu[\mathbb{1}_{U^c}(Z_{n+1})|1_U(Z_n) - 1_U(\widehat{Z}_n)|]\\ &\quad + \mathbb{E}_\mu[1_U(\widehat{Z}_n)|1_{U^c}(Z_{n+1}) - 1_{U^c}(\widehat{Z}_{n+1})|]\\ &\leq |q_n - \widehat{q}_n| + |q_{n+1} - \widehat{q}_{n+1}|,\end{aligned}$$

that goes to zero dues to proposition 5.3. □

Before turning to the convergence of our scheme, let us state a property of the sequence $(q_n)_{n\leq N}$ that will be important in the following proof. Indeed, the sequence (q_n) is non-decreasing since $\{\tau \le T_n\} \subset \{\tau \le T_{n+1}\}$ for all $n \le N-1$. Moreover, note that $q_0 = 0$ and $\lim_{n\to+\infty} q_n = 1$ due to assumption 5.1. Therefore, there exists an index denoted by $\widetilde{n} \ge 1$ such that

– for all $n < \widetilde{n}$, we have $q_n = 0$;

– for all $n \ge \widetilde{n}$, we have $q_n > 0$.

We denote $\widetilde{q} = q_{\widetilde{n}}$ as the first positive value of the sequence so that $q_n \ge \widetilde{q}$ for all $n \ge \widetilde{n}$. The convergence of the approximation scheme of the distribution of the exit time is now a straightforward consequence of the following proposition.

PROPOSITION 5.5.– Suppose assumptions 5.1, 5.3, 5.4, 5.5 and 5.6 hold. Let $(\sigma_n)_{n\le N-1}$ and $(\widehat{\sigma}_n)_{n\le N-1}$ be two sequences of $[0,1]$-valued real numbers. Let $(\pi_n)_{n\le N}$ and $(\widehat{\pi}_n)_{n\le N}$ be defined as follows, $\pi_0 = \widehat{\pi}_0 = 0$ and

$$\pi_{n+1} = \frac{\pi_n q_n + \sigma_n}{q_{n+1}} \mathbb{1}\{q_{n+1} \ne 0\}, \quad \widehat{\pi}_{n+1} = \frac{\widehat{\pi}_n \widehat{q}_n + \widehat{\sigma}_n}{\widehat{q}_{n+1}} \mathbb{1}\{\widehat{q}_{n+1} \ne 0\}.$$

For $0 \le n \le N$, if the quantization error is such that

$$C_U^{\frac{p}{p+\beta}} \left(\left(\frac{\beta}{p}\right)^{\frac{p}{p+\beta}} + \left(\frac{p}{\beta}\right)^{\frac{\beta}{p+\beta}} \right) \|Z_n - \widehat{Z}_k\|_p^{\frac{p\beta}{p+\beta}} \le \frac{\widetilde{q}}{2},$$

for all $k \le n$, then we have

$$\begin{aligned} |\pi_n - \widehat{\pi}_n| \le & \frac{2}{\widetilde{q}} (\overline{\pi} |q_{n-1} - \widehat{q}_{n-1}| + |\pi_{n-1} - \widehat{\pi}_{n-1}| + |\sigma_{n-1} - \widehat{\sigma}_{n-1}|) \\ & + \frac{2(\overline{\pi}+1)}{\widetilde{q}^2} |q_n - \widehat{q}_n|, \end{aligned}$$

where $\overline{\pi} = \max_{0\le n\le N} \pi_n$.

PROOF.– The difficulty of this proof lies in the fact that the recursive function giving π_{n+1} from π_n, q_n, q_{n+1} and σ_n is not Lipschitz continuous because of the division by q_{n+1}. To overcome this drawback, we will use the positive lower bound $\widetilde{q}$ for q_n described above. Moreover, a similar bound can be derived for the quantized values $\widehat{q}_n$ due to the convergence of $\widehat{q}_n$ toward q_n. We now prove by induction that $\widehat{\pi}_n$ converges toward π_n. First, we have $\widehat{\pi}_0 = \pi_0 = 0$. Set $1 \le n \le N$. If $n < \widetilde{n}$, then $q_n = 0$ which means that $Z_n \in U$ a.s. Since U is a convex set, $\widehat{Z}_n \in U$ a.s. too by proposition 3.1, yielding $\widehat{q}_n = 0$. Finally, from the definitions, we have $\pi_n = \widehat{\pi}_n = 0$. If $n \ge \widetilde{n}$, then $q_n \ge \widetilde{q} > 0$. Now we need to obtain the same kind of bound for $\widehat{q}_n$. This can be achieved due to proposition 5.3 giving the convergence of $\widehat{q}_n$

toward q_n. Indeed, assume from now on that the number of points in the quantization grids is large enough so that the quantization error is sufficiently small to ensure that for all $\widetilde{n} \leq n \leq N$, we have $|q_n - \widehat{q}_n| \leq \widetilde{q}/2$. Hence, the required lower bound is $\widehat{q}_n \geq \widetilde{q}/2 > 0$. Recall also that by construction $q_n \leq 1$ and $\widehat{q}_n \leq 1$ also. Therefore, we get

$$\begin{aligned}
|\pi_n - \widehat{\pi}_n| \leq & \left|\frac{\pi_{n-1}q_{n-1} + \sigma_{n-1}}{q_n} - \frac{\widehat{\pi}_{n-1}\widehat{q}_{n-1} + \widehat{\sigma}_{n-1}}{\widehat{q}_n}\right| \\
\leq & \frac{\pi_{n-1}}{\widehat{q}_n}|q_{n-1} - \widehat{q}_{n-1}| + \frac{\widehat{q}_{n-1}}{\widehat{q}_n}|\pi_{n-1} - \widehat{\pi}_{n-1}| \\
& + \frac{1}{\widehat{q}_n}|\sigma_{n-1} - \widehat{\sigma}_{n-1}| + |\pi_{n-1}q_{n-1} + \sigma_{n-1}|\frac{|q_n - \widehat{q}_n|}{q_n\widehat{q}_n} \\
\leq & \frac{\overline{\pi}}{\widehat{q}_n}|q_{n-1} - \widehat{q}_{n-1}| + \frac{1}{\widehat{q}_n}|\pi_{n-1} - \widehat{\pi}_{n-1}| \\
& + \frac{1}{\widehat{q}_n}|\sigma_{n-1} - \widehat{\sigma}_{n-1}| + (\overline{\pi} + 1)\frac{|q_n - \widehat{q}_n|}{q_n\widehat{q}_n} \\
\leq & \frac{2}{\widetilde{q}}\left(\overline{\pi}|q_{n-1} - \widehat{q}_{n-1}| + |\pi_{n-1} - \widehat{\pi}_{n-1}| + |\sigma_{n-1} - \widehat{\sigma}_{n-1}|\right) \\
& + \frac{2(\overline{\pi} + 1)}{\widetilde{q}^2}|q_n - \widehat{q}_n|,
\end{aligned}$$

hence the result. □

Note that a bound for the convergence rate of $\widehat{\pi}_n$ toward π_n may be obtained as soon as a bound for the convergence rate of $\widehat{\sigma}_n$ toward σ_n and an upper bound for the sequence $(\pi_n)_{0 \leq n \leq N}$ are available. Let us state one of our first main result, namely the convergence of the approximation scheme of the distribution of the exit time.

THEOREM 5.1.– Under assumptions 5.1, 5.3, 5.4, 5.5 and 5.6, for all $n \leq N$ and for almost every $s > 0$ w.r.t. the Lebesgue measure on $\mathbb{R}$, we have

$$\widehat{p}_n(s) \to p_n(s),$$

when the quantization error goes to zero.

PROOF.– Choose $s > 0$ such that $\widehat{r}_n(s)$ converges toward $r_n(s)$ and apply proposition 5.5 with $\sigma_n = r_n(s)$ and $\widehat{\sigma}_n = \widehat{r}_n(s)$ so that $\pi_n = p_n(s)$ and $\widehat{\pi}_n = \widehat{p}_n(s)$ for all n. Moreover, note that $(p_n(s))_n$ is bounded by 1. □

5.4.2. *Moments*

The following proposition proves the convergence of $\widehat{r}_{n,j}$ toward $r_{n,j}$.

PROPOSITION 5.6.– Under assumptions 5.1, 5.3, 5.4, 5.5 and 5.6, for all $n \leq N-1$ and for all $j \in \mathbb{N}$, $\widehat{r}_{n,j}$ converges toward $r_{n,j}$ when the quantization error goes to zero. More precisely, the error is bounded by

$$\begin{aligned}|r_{n,j} - \widehat{r}_{n,j}| \leq{} & j\big((n+1)C_{t^*}\big)^{j-1}\big(\|T_n - \widehat{T}_n\|_p + L_{u^*}\|Z_n - \widehat{Z}_n\|_p \\ & + \|T_{n+1} - \widehat{T}_{n+1}\|_p\big) + \big((n+1)C_{t^*}\big)^j\big(|q_n - \widehat{q}_n| \\ & + |q_{n+1} - \widehat{q}_{n+1}|\big).\end{aligned}$$

PROOF.– Let $n \leq N-1$ and $j \in \mathbb{N}$, we have

$$|r_{n,j} - \widehat{r}_{n,j}| \leq A + B.$$

where

$$\begin{aligned}A ={} & \big|\mathbb{E}_\mu\big[\big(((T_n + u^*(Z_n)) \wedge T_{n+1})^j - ((\widehat{T}_n + u^*(\widehat{Z}_n)) \wedge \widehat{T}_{n+1})^j\big) \\ & \times \mathbb{1}_U(Z_n)\mathbb{1}_{U^c}(Z_{n+1})\big]\big|, \\ B ={} & \big|\mathbb{E}_\mu\big[((\widehat{T}_n + u^*(\widehat{Z}_n)) \wedge \widehat{T}_{n+1})^j\big(\mathbb{1}_U(Z_n)\mathbb{1}_{U^c}(Z_{n+1}) \\ & - \mathbb{1}_U(\widehat{Z}_n)\mathbb{1}_{U^c}(\widehat{Z}_{n+1})\big)\big]\big|.\end{aligned}$$

Assumption 1.2 yields that the inter-jump times S_k are a.s. bounded by C_{t^*} so that $T_n \leq nC_{t^*}$ a.s. and $(T_n + u^*(Z_n)) \wedge T_{n+1} \leq (n+1)C_{t^*}$ a.s. These bounds are equally true when replacing T_n with the quantized approximation $\widehat{T}_n$ due to proposition 3.1, as the set $[0, C_{t^*}]$ is convex. Let us first consider the term A. We crudely bound the indicator functions by 1. Moreover, set

$$\eta = \big|(T_n + u^*(Z_n)) \wedge T_{n+1} - (\widehat{T}_n + u^*(\widehat{Z}_n)) \wedge \widehat{T}_{n+1}\big|,$$

and note that the function $t \mapsto t^j$ is Lipschitz continuous on any set $[0, M]$ with Lipschitz constant jM^{j-1}. Thus we have

$$A \leq \mathbb{E}_\mu\big[j\big((n+1)C_{t^*}\big)^{j-1}\eta\big] \leq j\big((n+1)C_{t^*}\big)^{j-1}\|\eta\|_p,$$

and due to the Lipschitz continuity of u^*, we have

$$A \leq j\big((n+1)C_{t^*}\big)^{j-1}\big(\|T_n - \widehat{T}_n\|_p + L_{u^*}\|Z_n - \widehat{Z}_n\|_p + \|T_{n+1} - \widehat{T}_{n+1}\|_p\big).$$

Moreover, the term B is bounded by

$$B \leq ((n+1)C_{t^*})^j \mathbb{E}_\mu[|\mathbb{1}_U(Z_n)\mathbb{1}_{U^c}(Z_{n+1}) - \mathbb{1}_U(\widehat{Z}_n)\mathbb{1}_{U^c}(\widehat{Z}_{n+1})|]$$
$$\leq ((n+1)C_{t^*})^j \left(|q_n - \widehat{q}_n| + |q_{n+1} - \widehat{q}_{n+1}|\right).$$

We conclude thanks to proposition 5.3. □

We may now state the other important result of this chapter namely the convergence of the approximation scheme of the moments of the exit time with a bound for the rate of convergence.

THEOREM 5.2.– Under assumption 5.1, 5.3, 5.4, 5.5 and 5.6, for all $n \leq N$ and for all $j \in \mathbb{N}$, $\widehat{p}_{n,j}$ converges toward $p_{n,j}$ when the quantization error goes to zero. More precisely, if the quantization error is such that for all $k \leq n$

$$C_U^{\frac{p}{p+\beta}}\left(\left(\frac{\beta}{p}\right)^{\frac{p}{p+\beta}} + \left(\frac{p}{\beta}\right)^{\frac{\beta}{p+\beta}}\right)\|Z_n - \widehat{Z}_k\|_p^{\frac{p\beta}{p+\beta}} \leq \frac{\widetilde{q}}{2},$$

then we have

$$|p_{n,j} - \widehat{p}_{n,j}| \leq \frac{2}{\widetilde{q}}\big((NC_{t^*})^j|q_{n-1} - \widehat{q}_{n-1}| + |p_{n-1,j} - \widehat{p}_{n-1,j}|$$
$$+|r_{n-1,j} - \widehat{r}_{n-1,j}|\big) + \frac{2((NC_{t^*})^j + 1)}{\widetilde{q}^2}|q_n - \widehat{q}_n|.$$

PROOF.– Let $j \in \mathbb{N}$ and apply Proposition 5.5 with $(\sigma_n) = (r_{n,j})$ and $(\widehat{\sigma}_n) = (\widehat{r}_{n,j})$ such that $(\pi_n) = (p_{n,j})$ and $(\widehat{\pi}_n)_k = (\widehat{p}_{n,j})$. A bound for the rate of convergence is obtained since the sequence $(p_{n,j})_{0 \leq n \leq N}$ is bounded by

$$p_{n,j} = \mathbb{E}_\mu[\tau^j | \tau \leq T_n] \leq \mathbb{E}_\mu[T_n^j | \tau \leq T_n] \leq (nC_{t^*})^j \leq (NC_{t^*})^j.$$

Hence, the result. □

The rate of convergence depends on the quantity $\widetilde{q}$ whose exact value might be unknown in some complex applications. In that case, it may still be approximated through Monte Carlo simulations. Nevertheless, theorems 5.1 and 5.2 prove the convergence of our approximation schemes regardless of the value of $\widetilde{q}$.

5.5. Example

We give a first application of our procedure on the Poisson process with trend presented in section 1.8.1. Also see Chapter 6 for an example in reliability related to the computation of a service time.

The study of this Poisson process with trend is especially interesting because it is possible to compute the exact value of its distribution function in order to compare it with the numerical value given by our approximation scheme. Recall that the process is defined as $X_t = t + N_t$ where (N_t) is a Poisson process with intensity 1. We are interested in the exit time τ from the set $U = [0, 10)$. We may choose the horizon $N = 10$ since $X_{T_N} = T_N + N_{T_N} = T_N + N \geq N$. Besides, it is clear that for all $x \in U$, $u^*(x) = 10 - x$. assumptions 5.3 and 5.4 are clearly satisfied and so is assumption 5.5 due to the following lemma.

LEMMA 5.1.– For all $\alpha > 0$ and for all $0 \leq n \leq N$, we have

$$\mathbb{P}_\mu(Z_n \in U^\alpha) \leq 2\alpha.$$

PROOF.– Since $Z_0 = 0$, we have

$$\mathbb{P}_\mu(Z_0 \in U^\alpha) = \mathbb{P}_\mu(Z_0 \in (10 - \alpha, 10 + \alpha)) = \mathbb{1}_{\{\alpha \geq 10\}} \leq \frac{1}{10}\alpha \leq 2\alpha.$$

Set now $1 \leq n \leq N$. Denote $f_{n,1}$ as the density of the Gamma distribution with parameters n and 1, and denote $C_n = \frac{1}{(n-1)!}\left(\frac{n-1}{\mathrm{e}}\right)^{n-1}$ as its upper bound. Since T_n has distribution $\Gamma(n, 1)$, $Z_n = n + T_n$ has density $f_{Z_n}(\cdot) = f_{n,1}(\cdot - n)$ that is also bounded by C_n. Finally, we have

$$\mathbb{P}_\mu(Z_n \in U^\alpha) = \mathbb{P}_\mu(Z_n \in [10 - \alpha, 10 + \alpha]) \leq 2C_n\alpha \leq 2\alpha.$$

Indeed, the sequence (C_n) is non increasing so that for all $1 \leq n \leq N$, we have $C_n \leq C_1 = 1$. □

Moreover, assumption 5.6 is satisfied since the process increases but assumption 1.2 is not, because $t^*(x) = +\infty$ for all $x \in E$. However, this can be solved by killing the process at time τ. Then $t^* = u^*$ is bounded by 10.

Table 5.1 presents the simulations results for the approximation of the mean exit time and the second moment. It includes for different number of points in the quantization grids the value of $\widehat{p}_{N,1}$ which approximates the mean exit time and $\widehat{p}_{N,2}$ which approximates the second moment. Reference values are obtained due to Monte Carlo method (10^8 simulations): $\mathbb{E}[\tau] = 5.125, \mathbb{E}[\tau^2] = 27.5$. For the first and second moment, the empirical convergence rate is presented in Figure 5.1. The slope is estimated through a regression model as -0.60 for the first moment and -0.59 for the second moment. Remark that there are roughly the same order as the rate of convergence of the optimal quantizer as here the effective dimension is 2.

Number of points in the quantization grids	Approx. mean $\widehat{p}_{N,1}$	Relative error to 5.125	Approx. moment $\widehat{p}_{N,2}$	Relative error to 27.5
20	5.050	0.75 %	26.66	1.68 %
50	5.096	0.46 %	27.20	0.98 %
100	5.095	0.24 %	27.21	0.34 %
200	5.118	0.20 %	27.43	0.44 %
500	5.123	0.12 %	27.49	0.15 %
1000	5.095	0.05 %	27.21	0.14 %
2000	5.118	0.05 %	27.43	0.10 %
5000	5.123	0.02 %	27.49	0.06 %

Table 5.1. *Approximation results for the mean exit time and the second moment.*

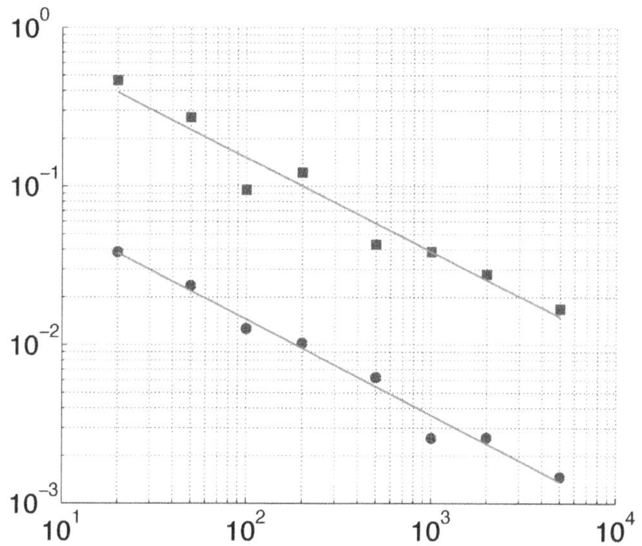

Figure 5.1. *Log-log plot of the error versus the number of points in the quantization grids for the first (lower curve) and second (upper curve) moment of the Poisson process*

We can obtain the exact value of the survival function of the exit time. Recall that the jump time T_n is distributed as $\Gamma(n, 1)$.

PROPOSITION 5.7.– Denote $\lfloor x \rfloor$ as the greatest integer smaller than or equal to x. For all $s \in \mathbb{R}^+$, we have

$$\mathbb{P}(\tau \geq s) = \mathbb{P}(T_{\lfloor 10-s \rfloor +1} > s)\mathbb{1}_{\{s \leq 10\}}.$$

PROOF.– Let $s > 0$. Notice that $X_s \geq s$, thus $\tau < s$ a.s. when $s > 10$. Assume now that $s \leq 10$, we have

$$\mathbb{P}(\tau \geq s) = \mathbb{P}(X_s \leq 10) = \mathbb{P}(N_s \leq \lfloor 10 - s \rfloor) = \mathbb{P}(T_{\lfloor 10-s \rfloor +1} \geq s).$$

Hence, the result. □

Figure 5.2 represents both the exact survival function of the exit time and its quantized approximation with 500 points. The functions appear indistinguishable. Table 5.2 contains the supremum of the empirical error between the two functions, the supreme being taken among the continuity points of the survival function, i.e. non-integer points. For the survival function, the empirical convergence rate is presented in Figure 5.3. It is estimated through a regression model as -0.67. It is again roughly of the same order as the rate of convergence of the optimal quantizer in dimension 2.

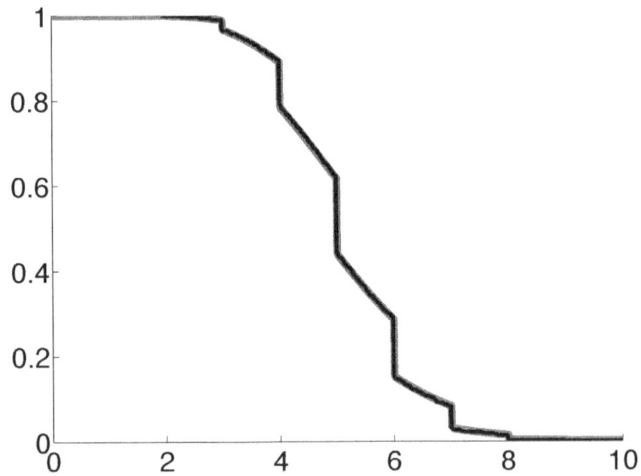

Figure 5.2. *Exact survival function of τ (gray) and its quantized approximation (black) with 500 points in the quantization grids*

Number of points in the quantization grids	Error $\max_s \lvert p_N(s) - \widehat{p}_N(s) \rvert$
20	0.595
50	0.0409
100	0.0261
200	0.0143
500	0.0085
1000	0.0062
2000	0.0033
5000	0.0014

Table 5.2. *Simulations results for the distribution: maximum of the error between the exact and approximate survival function for given number of points in the quantization grids*

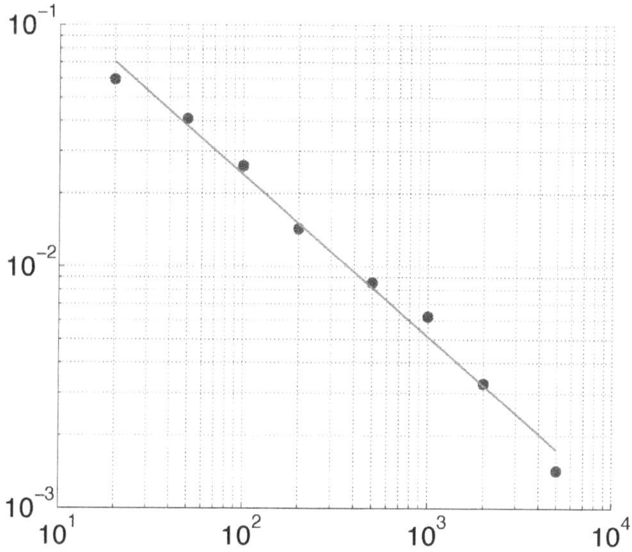

Figure 5.3. *Log–log plot of the error versus the number of points in the quantization grids for the survival function of the Poisson process*

Note from the proof of proposition 5.4 that a bound for the rate of convergence for the exit time distribution may be obtained as soon as for all $n \leq N-1$, the survival function of $(T_n + u^*(Z_{k-n})) \wedge T_{n+1}$ denoted by φ_n

is piecewise Lipschitz continuous. Although it is difficult to state general assumptions under which this is true, the following proposition proves that this condition is fulfilled for the Poisson process.

PROPOSITION 5.8.– For all $n \leq N-1$, the survival function φ_n of $(T_n + u^*(Z_n)) \wedge T_{n+1}$ is Lipschitz continuous on $(0; 10-n)$ and on $(10-n; +\infty)$ with Lipschitz constant $L_{\varphi_n} \leq 1$.

PROOF.– Set $n = 0$ and $s > 0$, we have

$$\varphi_0(s) = \mathbb{P}_\mu((T_0 + u^*(Z_0)) \wedge T_1 > s) = \mathbb{P}_\mu(10 \wedge T_1 > s) = \mathbb{1}_{\{10>s\}} e^{-s},$$

since T_1 has exponential distribution with parameter 1. Therefore, φ_0 equals zero on $[10; +\infty)$ and is Lipschitz continuous with Lipschitz constant 1 on $(0; 10)$.

Set $n \geq 1$, $s > 0$ and remember that the random variables $(S_k)_{k\in\mathbb{N}}$ are independent and all have exponential distribution with parameter 1 so that, in particular, T_n and S_{n+1} are independent and T_n has distribution $\Gamma(n, 1)$. Moreover, recall that $Z_n = n + T_n$ and that $u^*(x) = 10 - x$. We have

$$\begin{aligned} \varphi_n(s) &= \mathbb{P}_\mu((T_n + u^*(Z_n)) \wedge T_{n+1} > s) \\ &= \int_{(\mathbb{R}^+)^2} \mathbb{1}_{\{(t+(10-n-t))\wedge u > s\}} f_{n,1}(t) f_{n+1,1}(u) dt du, \end{aligned}$$

where $f_{k,1}$ denotes the density function of the distribution $\Gamma(k, 1)$. Set $s' > s > 0$, we have

$$\begin{aligned} |\varphi_n(s') - \varphi_n(s)| &\leq \int_{(\mathbb{R}^+)^2} |\mathbb{1}_{\{(10-n)\wedge u > s'\}} - \mathbb{1}_{\{(10-n)\wedge u > s\}}| \\ &\quad \times f_{n,1}(t) f_{n+1,1}(u) dt du \\ &\leq \int_{(\mathbb{R}^+)^2} \mathbb{1}_{\{(10-n)\wedge u \in (s;s']\}} f_{n,1}(t) f_{n+1,1}(u) dt du \\ &\leq \int_{(\mathbb{R}^+)^2} \left(\mathbb{1}_{\{10-n \in (s;s']\}} + \mathbb{1}_{\{u \in]s;s']\}}\right) \\ &\quad \times f_{n,1}(t) f_{n+1,1}(u) dt du \\ &\leq \mathbb{1}_{\{10-n \in [s;s']\}} + C_{f_{n+1,1}} |s' - s| \\ &\leq \mathbb{1}_{\{10-n \in [s;s']\}} + |s' - s|, \end{aligned}$$

since $C_{f_{n+1},1} \leq 1$. Finally if s and s' both belong to $(0; 10 - n)$ or if they both belong to $(10 - n; +\infty)$, we have $|\varphi_n(s') - \varphi_n(s)| \leq |s' - s|$. □

Consequently, in this example, we are now able to give a bound for the rate of convergence of the exit time distribution approximation scheme.

PROPOSITION 5.9.– For all $n \leq N - 1$, let $s > 0$ and if the quantization error is small enough to ensure that

$$\left(\frac{p}{2}\right)^{\frac{1}{p+1}} \left(\|T_n - \widehat{T}_n\|_p + \|Z_n - \widehat{Z}_n\|_p + \|T_{n+1} - \widehat{T}_{n+1}\|_p\right)^{\frac{p}{p+1}} < |10 - n - s|,$$

we have then

$$|r_n(s) - \widehat{r}_n(s)| \leq 2\left(\frac{p}{2}\right)^{\frac{1}{p+1}} \left(\frac{1}{p} + 1\right)\left(\|T_n - \widehat{T}_n\|_p + \|Z_n - \widehat{Z}_n\|_p + \|T_{n+1} - \widehat{T}_{n+1}\|_p\right)^{\frac{p}{p+1}} + \left|q_n - \widehat{q}_n\right| + \left|q_{n+1} - \widehat{q}_{n+1}\right|.$$

Moreover, for all $n \leq N$, if the quantization error is such that for all $k \leq n$

$$2\left(\frac{p}{2}\right)^{\frac{1}{p+1}} \left(\frac{1}{p} + 1\right)\|Z_k - \widehat{Z}_k\|_p^{\frac{p}{p+1}} \leq \frac{\widetilde{q}}{2},$$

we have then

$$|p_n(s) - \widehat{p}_n(s)| \leq \frac{2}{\widetilde{q}}\left(|q_{n-1} - \widehat{q}_{n-1}| + |p_{n-1}(s) - \widehat{p}_{n-1}(s)| + |r_{n-1}(s) - \widehat{r}_{n-1}(s)|\right) + \frac{4}{\widetilde{q}^2}|q_n - \widehat{q}_n|.$$

PROOF.– The proof derives directly from the proofs of proposition 5.4 and theorem 5.1. The A term may be bounded due to the piecewise Lipschitz continuity of the functions φ_n on $(0; 10 - n)$ and on $(10 - n; +\infty)$. Let $s > 0$, $s \neq 10 - n$, and let $\alpha > 0$ such that $10 - n \notin [s - \alpha; s + \alpha]$ i.e. $\alpha < |10 - n - s|$, we have

$$A \leq \left|\varphi_n(s + \alpha) - \varphi_n(s - \alpha)\right| + \frac{\|\eta\|_p^p}{\alpha^p} \leq 2L_{\varphi_n}\alpha + \frac{\|\eta\|_p^p}{\alpha^p},$$

that reaches a minimum when $\alpha = \left(\frac{p\|\eta\|_p^p}{2L_{\varphi_n}}\right)^{\frac{1}{p+1}}$. Notice besides that $L_{\varphi_n} = 1$ and $L_{u^*} = 1$. □

We can also calculate the exact value of $\widetilde{q}$ that is the first non-negative value of the sequence $(\mathbb{P}_\mu(Z_n \notin U))$. We have $\widetilde{q} = \mathbb{P}_\mu(Z_1 \notin (0; 10)) = \mathbb{P}_\mu(T_1 \geq 9) = e^{-9}$ because T_1 has exponential distribution with parameter 1.

5.6. Conclusion

We have presented an efficient and easy to implement numerical method to approximate the survival function and moments of exit times for PDMPs. We proved the convergence of our algorithm with bounds for the rate of convergence for the moments. Let us describe the practical interest of our approach:

– the quantizations grids only have to be computed once and can be used for several purposes. Moreover, once they are obtained, the procedures leading to $\widehat{p}_N(s)$ and to $\widehat{p}_{N,j}$ can be achieved very simply since we only have to compute finite sums;

– concerning the distribution, since $\widehat{p}_N(s)$ can be computed almost instantly for any value of s, the whole survival function can be obtained very quickly. Similarly, concerning the moments, $\widehat{p}_{N,j}$ can be computed very quickly for any j, so that any moment is almost instantly available;

– furthermore, in both cases, we may decide to change the set U and consider the exit time τ' from a new set U'. This will yield new sequences $(\widehat{q}_n)$, $(\widehat{r}_{n,j})$ and $(\widehat{p}_{n,j})$ in the case of the j-th moment approximation or new sequences $(\widehat{q}_n)$, $(\widehat{r}_n(s))$ and $(\widehat{p}_n(s))$ if we are interested in the distribution. These new sequences are obtained quickly and easily since the quantized process remains the same and we only have to compute finite sums. Of course, the set U' satisfy the same assumptions as U and be such that $\mathbb{P}_\mu(T_N < \tau')$ remains small without changing the computation horizon N. This last condition is fulfilled if, for instance, $U' \subset U$. This flexibility is an important advantage of our method over, for instance, a Monte Carlo method.

Another application of this procedure is given in Chapter 6.

6

Example in Reliability: Service Time

Classically, we characterize the performance of a system by several indicators: availability, quality, maintainability, safety, etc. Evaluating all of these indicators is crucial from a practical point of view. It makes possible the calculation of a cost function measuring the overall performance of the system. However, in the context of dynamic reliability for complex and large systems, these indicators are quite complicated to evaluate. In the framework of Piecewise Deterministic Markov Processes (PDMPs), these performance indices are defined as expectations of functionals of the process or as exit times from a given area of the state space. In this chapter, we illustrate the numerical methods developed in Chapters 4 and 5 on the corrosion example presented in section 2.2. We evaluate two indicators of performance: the mean thickness loss in section 6.1 as well as the service time of the structure in section 6.2. In particular, we will see how the same quantization grids can be used to compute many different quantities of interest. We also discuss the validity of the assumptions made in Chapters 4 and 5.

6.1. Mean thickness loss

We want to compute the mean thickness loss in environment 2 for a horizon of $T = 18$ years for the corrosion process. As we deal with a deterministic horizon, we will use the time-augmented process. The running cost function corresponding to this problem is $l \in \mathbf{B}(\widetilde{E})$ defined as follows. For modes $m \in \{1, 2, 3\} \times \{0\}$, there is no thickness loss so $l(m, \zeta, t) = 0$ for all $\zeta \in E_m$

and $t \geq 0$. For modes $m = (i, 1) \in \{1, 2, 3\} \times \{1\}$, we define l along the flow as

$$l(\widetilde{\Phi}(\xi, u)) = \frac{d}{du}\big(d_i(\rho, a + u)\big)\mathbb{1}_{\{i=2\}} = \rho_i(1 - \mathrm{e}^{-(a+u)/\eta_i})\mathbb{1}_{\{i=2\}},$$

for all $\xi = (m, \zeta, t)$, $\zeta = (d, \rho, a)$, $t \geq 0$ and $u \geq 0$. Thus, l is clearly in $\mathbf{L}_{\widetilde{\Phi}}(\widetilde{E})$. We then define $\widetilde{l}(\widetilde{\Phi}(\xi, u)) = l(\widetilde{\Phi}(\xi, u))\mathbb{1}_{\{t+u \leq T\}}$ so that, we have

$$\begin{aligned} L(\xi, s) &= \int_0^s \widetilde{l}(\widetilde{\Phi}(\xi, u))du \\ &= \int_0^{s \wedge (T-t)^+} l(\widetilde{\Phi}(\xi, u))du \\ &= \big(d_m(\rho, a + s \wedge (T - t)^+) - d_m(\rho, a)\big)\mathbb{1}_{\{m=(2,1)\}}, \end{aligned}$$

this is indeed the thickness lost in mode 2 from the point $\xi = (m, d, \rho, a, t)$ during a time $u \wedge (T - t)^+$.

We now turn to the assumptions that the process must satisfy in order to apply our numerical procedure. Assumption 1.3 on the deterministic time to the boundary holds true as $\widetilde{t}^*(m, d, \rho, a, t) = T - t$ if $m \in \{1, 2, 3\} \times \{1\}$ and $\widetilde{t}^*(m, \gamma, a, t) = \gamma \wedge (T - t)$ otherwise. Assumptions 1.4 and 1.5 on the jump rates λ_i clearly hold as the rates are constant in each environment. We now turn to assumption 1.7 on the Markov kernel $\widetilde{Q}$. This assumption does not hold for any function w in $\mathbf{L}^u_{\widetilde{\Phi}}(\widetilde{E})$. However, we will prove that it holds true for a sufficiently large subset of $\mathbf{L}^u_{\widetilde{\Phi}}(\widetilde{E})$.

DEFINITION 6.1.– A function w in $\mathbf{L}^u_{\widetilde{\Phi}}(\widetilde{E})$ is said to vanish after T if for all $\xi = (x, t) \in \widetilde{E}$ we have $t \geq T \implies w(\xi) = 0$.

As for all $\xi = (x, t) \in \widetilde{E}$ and $0 \leq n \leq N$, by construction we have

$$v_{N-n}(x) = \mathbb{E}_x\Big[\int_0^{T_n} l(\widetilde{\Phi}(\xi, u))\mathbb{1}_{\{t+u \leq T\}}du\Big],$$

the sequence of functions (v_n) vanishing after T. Hence, it is sufficient to prove that assumption 1.7 holds for this class of functions.

LEMMA 6.1.– *The Markov kernel $\widetilde{Q}$ is locally Lipschitz in the following sense: there exists $[\widetilde{Q}] \in \mathbb{R}^+$ such that for all $u \geq 0$ and for all functions $w \in \mathbf{L}^u_{\Phi}(E)$ that vanishes after T,*

1) for all $\xi, \xi' \in \widetilde{E}$ *and* $s \in [0, t^*(\xi) \wedge t^*(\xi') \wedge u)$*, we have*

$$|\widetilde{Q}w(\widetilde{\Phi}(\xi, s)) - \widetilde{Q}w(\widetilde{\Phi}(\xi', s))| \leq [\widetilde{Q}][w]_1^u|\xi - \xi'|,$$

2) for all $\xi, \xi' \in \widetilde{E}$ *such that* $\widetilde{t^*}(\xi) \vee \widetilde{t^*}(\xi') \leq u$*, we have*

$$|\widetilde{Q}w(\widetilde{\Phi}(\xi, \widetilde{t^*}(\xi))) - \widetilde{Q}w(\widetilde{\Phi}(\xi', \widetilde{t^*}(\xi')))| \leq [\widetilde{Q}]\big([w]_*^u + [w]_1^u\big)|\xi - \xi'|.$$

PROOF.– Set $\xi = (m, \zeta, t)$ and $\xi' = (m', \zeta', t')$ in $\widetilde{E}$ and $s \in [0, t^*(\xi) \wedge t^*(\xi') \wedge u)$. As $|\xi - \xi'| = +\infty$ if $m \neq m'$, we only need to consider the case $m = m'$. Suppose that $m = (i, 0)$, $\zeta = (\gamma, a)$ and $\zeta' = (\gamma', a')$. Set $j = i + 1[3]$. Then, we have

$$\begin{aligned}
&|\widetilde{Q}w(\widetilde{\Phi}(\xi, s)) - \widetilde{Q}w(\widetilde{\Phi}(\xi', s))| \\
&\quad = |w(j, 0, \gamma - s, 0, t + s) - w(j, 0, \gamma' - s, 0, t' + s)| \\
&\quad = |w(\widetilde{\Phi}((j, 0, \gamma - s, 0, t + s), 0)) \\
&\quad - w(\widetilde{\Phi}((j, 0, \gamma' - s, 0, t' + s), 0))| \\
&\quad \leq [w]_1|\xi - \xi'|,
\end{aligned}$$

as $w \in \mathbf{L}_\Phi^u(E)$ and by assumption $\gamma - s > 0$ and $\gamma' - s > 0$. Still suppose that $m = (i, 0)$, $\zeta = (\gamma, a)$ and $\zeta' = (\gamma', a')$ and that the boundary is reached in both cases when the protection against corrosion is gone (i.e. $\gamma < T - t$ and $\gamma' < T - t'$). In this case, we have

$$\begin{aligned}
&|\widetilde{Q}w(\widetilde{\Phi}(\xi, \widetilde{t^*}(\xi))) - \widetilde{Q}w(\widetilde{\Phi}(\xi', \widetilde{t^*}(\xi')))| \\
&\quad \leq \frac{1}{\rho_i^+ - \rho_i^-} \int_{\rho_i^-}^{\rho_i^+} |w(i, 1, 0, \rho, 0, t + \gamma) \\
&\quad - w(i, 1, 0, \rho, 0, t' + \gamma')|d\rho \\
&\quad \leq [w]_1|\xi - \xi'|,
\end{aligned}$$

In the case when the boundary is reached in both cases when the horizon T is reached (i.e. $\gamma \geq T - t$ and $\gamma' \geq T - t'$), we use the fact that w vanishes after T. Thus, the difference $|\widetilde{Q}w(\widetilde{\Phi}(\xi, \widetilde{t^*}(\xi))) - \widetilde{Q}w(\widetilde{\Phi}(\xi', \widetilde{t^*}(\xi')))|$ also vanishes. As regards the case where say ξ reaches the boundary when γ is reached ($\gamma < T - t$) and ξ' when T is reached ($\gamma' < T - t'$), we may not have to consider it if we add an extra mode when T is reached, so that $|\xi - \xi'| = +\infty$.

Suppose now that $m = (i, 0)$, $\zeta = (d, \rho, a)$ and $\zeta' = (d', \rho', a')$. Set $j = i + 1[3]$. Then, we have

$$\begin{aligned}
&|\widetilde{Q}w(\widetilde{\Phi}(\xi, s)) - \widetilde{Q}w(\widetilde{\Phi}(\xi', s))| \\
&\quad =|w(j, 0, \gamma - s, 0, t + s) - w(j, 0, \gamma' - s, 0, t' + s)| \\
&\quad \leq \frac{1}{\rho_j^+ - \rho_j^-} \int_{\rho_j^-}^{\rho_j^+} |w(j, 1, d + d_i(\rho, a + s) - d_i(\rho, a), \widetilde{\rho}, 0, t + s) \\
&\qquad - w(j, 1, d' + d_i(\rho', a' + s) - d_i(\rho', a'), \widetilde{\rho}, 0, t' + s)| d\widetilde{\rho} \\
&\quad \leq [w]_1 (1 + 2L_d) |\xi - \xi'|,
\end{aligned}$$

where L_d is the maximum of the Lipschitz constants of the d_i, and

$$\begin{aligned}
|d_i(\rho, a) - d_i(\rho', a')| &\leq |\rho - \rho'| (a + \eta_i (\mathrm{e}^{-a/\eta_i} - 1)) + 2\rho' |a - a'| \\
&\leq (2 \max_{1 \leq i \leq 3} \{\rho_i^+\} \vee T) |\xi - \xi'| \\
&\leq T |\xi - \xi'|,
\end{aligned}$$

using the value from Table 2.1, so that $L_d = T = 18$. Here again, the difference $|\widetilde{Q}w(\widetilde{\Phi}(\xi, \widetilde{t^*}(\xi))) - \widetilde{Q}w(\widetilde{\Phi}(\xi', \widetilde{t^*}(\xi')))|$ vanishes as w vanishes after T. So, we obtain the expected result for $[\widetilde{Q}] \leq 1 + 2L_d$. □

Table 6.1 presents the values of the loss of thickness in environment 2 obtained through our approximation scheme. The reference value for the relative error is 0.01374, obtained with 10^8 Monte Carlo simulations. Figure 6.1 gives a log-log representation of the error versus the number of points in the quantization grids. The convergence rate, estimated through a regression model, is -1.135. This rate is much faster than the rate of convergence of the optimal quantizer given in theorem 3.1 since here the effective dimension is 3 (indeed, m is deterministic and $a = 0$ immediately after a jump so that we only quantize the variables ρ, d and t), and the rate thus should be around $1/3$. In this very special case, our upper bounds for the convergence rate are, therefore, not sharp at all. Recall that they are very general and valid for a wide class of PDMPs.

6.2. Service time

We now consider the service time of the structure. When the loss of thickness reaches 0.2 mm, the structure is said to be unusable, this will be the exit criterion. Therefore, set

$$U_m = [0, 0.2) \times [\rho_i^-, \rho_i^+] \times \mathbb{R}^+,$$

for modes $m = (i, 1) \in \{1, 2, 3\} \times \{1\}, U_m = E_m$ for modes $m \in \{1, 2, 3\} \times \{0\}$. The set

$$U = \bigcup_{m \in M} \{m\} \times U_m$$

Number of points in the quantization grids	$\widehat{V_0}$ (10^{-2})	Relative error
10	1.32596	3.510 %
20	1.32937	3.262 %
50	1.33322	2.982 %
100	1.36455	0.702 %
200	1.36513	0.629 %
500	1.37792	0.271 %
1000	1.37502	0.060 %
2000	1.37400	0.014 %
4000	1.37410	0.007 %
6000	1.37413	0.005 %

Table 6.1. *Approximation of the mean loss of thickness (in mm) in environment 2 for different numbers of points in the quantization grids*

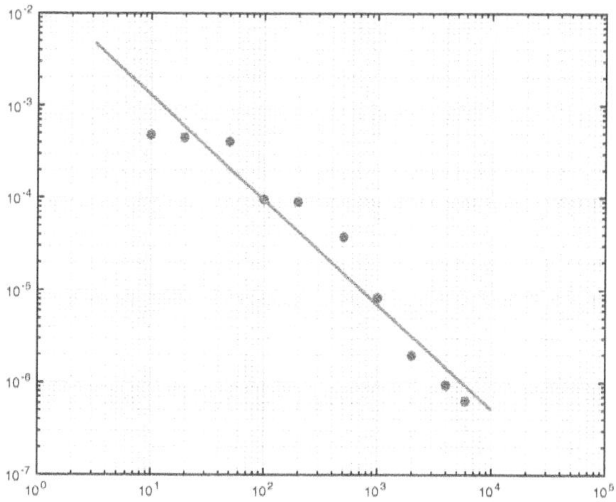

Figure 6.1. *Log-log plot of the error when approximating the loss of thickness in environment 2 w.r.t. the number of points in the quantization grids*

is clearly convex, so that assumption 5.4 is satisfied. Let τ be the exit time of the PDMP from U. Note that τ corresponds to the service time of the structure. In this case, we will consider the process killed when it exits U. We will denote by Δ the cemetery state.

6.2.1. *Mean service time*

Let us study the first moment of τ, i.e. the mean service time of the structure. Recall that the process starts in mode $(1,0)$ in the workshop environment with an anticorrosion protection that will hold for a time γ drawn according to a Weibull distribution. Thus, we have

$$\tau = \gamma + \tau',$$

where τ' is the exit time from U for a process starting in mode $m = (i,1)$ with a suitable law for i at position $\zeta = (0, \rho, 0)$ with ρ uniformly distributed on $[\rho_i^-; \rho_i^+]$. In other words, we only need to study the process after the corrosion actually starts attacking the structure. Therefore, it is sufficient to check whether τ' satisfies the required assumptions.

Let $K = \inf\{n \geq 0 : T_n = \gamma\}$, the number of the jump occurring when the protection against corrosion ends. We will only study the process after the K-th jump, i.e. in modes $m = (i,1) \in \{1,2,3\} \times \{1\}$. Assumption 5.2 holds as in modes $m = (i,1)$, we have $\rho_i \geq 10^{-7}$ and $\eta_i \leq 200000$ so that $d_t \geq 10^{-7}(t - 200000)$ and leading to $u^* \leq 0.2 \cdot 10^7 + 200000 = 2.2 \cdot 10^6$ h. Denote by C_{u^*} this bound. Assumption 5.6 holds as the thickness loss is increasing. Assumption 1.2 does not hold as explained in section 2.2.2, but this can be easily overcome by killing the process when it exits U. Thus, $t^* = u^*$ is bounded. We now prove that assumption 5.5 holds.

LEMMA 6.2.– *For all $\alpha > 0$ and for all $K \leq n \leq N$, we have*

$$\mathbb{P}(Z_n \in U^\alpha) \leq 4\alpha.$$

PROOF.– As $n \geq K$, the mode of Z_n is $m = (M_n, 1)$. Thus, we can rewrite Z_n as $Z_n = ((M_n, 1), D_n, R_n, 0)$ where D_n is the thickness loss at time T_n and R_n is the corrosion rate at time T_n. Thus, we have

$$\mathbb{P}(Z_n \in U^\alpha) = \mathbb{P}(|D_n - 0.2| \leq \alpha).$$

We, therefore, study more precisely the distribution of D_n. Denote $F_i(s) = s + \eta_i(\mathrm{e}^{-s\eta_i} - 1)$. We have

$$D_n = D_{n-1} + R_{n-1} F_{M_{n-1}}(S_n).$$

Let us now prove that the random variable $R_nF_{M_n}(S_{n+1})$ has a bounded density. Recall that R_n has a uniform distribution on $[\rho^-_{M_n};\rho^+_{M_n}] \subset [10^{-7};10^{-5}]$ and S_{n+1} has an exponential distribution with parameter λ_{M_n}. Let h be a real bounded measurable function in $\mathbf{B}(\mathbb{R}^+)$, we have

$$\mathbb{E}[h(R_nF_{M_n}(S_{n+1})) \mid M_n = i] = \int_0^{+\infty} \int_{\rho_i^-}^{\rho_i^+} h(\rho F_i(s)) \frac{\lambda_i e^{-\lambda_i s}}{\rho_i^+ - \rho_i^-} d\rho ds.$$

Introduce the change of variable $v = \rho F_i(s)$ to obtain

$$\begin{aligned}&\mathbb{E}_\mu[h(R_nF_{M_n}(S_{n+1})) \mid M_n = i] \\ &\quad = \int_0^{+\infty} h(v)\Big(\int_{\rho_i^-}^{\rho_i^+} \frac{\lambda_i e^{-\lambda_i F_i^{-1}(v\rho^{-1})}(F_i^{-1})'(v\rho^{-1})}{(\rho_i^+ - \rho_i^-)u} d\rho\Big) dv.\end{aligned}$$

Integration by parts yields

$$\begin{aligned}&\int_{\rho_i^-}^{\rho_i^+} \frac{\lambda_i e^{-\lambda_i F_i^{-1}(v\rho^{-1})}(F_i^{-1})'(v\rho^{-1})}{(\rho_i^+ - \rho_i^-)u} d\rho \\ &\quad = \int_{\rho_i^-}^{\rho_i^+} u \frac{\lambda_i e^{-\lambda_i F_i^{-1}(v\rho^{-1})}(F_i^{-1})'(v\rho^{-1})}{u^2(\rho_i^+ - \rho_i^-)} d\rho \\ &\quad = \frac{1}{\rho_i^+ - \rho_i^-}\Big(\big[\rho e^{-\lambda_i F_i^{-1}(v\rho^{-1})}\big]_{\rho_i^-}^{\rho_i^+} - \int_{\rho_i^-}^{\rho_i^+} e^{-\lambda_i F_i^{-1}(v\rho^{-1})} d\rho\Big).\end{aligned}$$

Finally, the density of the random variable $R_nF_{M_n}(S_{n+1})$ is bounded by

$$\Big|\int_{\rho_i^-}^{\rho_i^+} \frac{\lambda_i e^{-\lambda_i F_i^{-1}(v\rho^{-1})}(F_i^{-1})'(v\rho^{-1})}{(\rho_i^+ - \rho_i^-)u} d\rho\Big| \leq \frac{\rho_i^- + \rho_i^+}{\rho_i^+ - \rho_i^-} + 1 \leq \frac{2\rho_i^+}{\rho_i^+ - \rho_i^-} \leq 2.$$

Set $j \in \mathbb{N}$, we now study the distribution of the random variables $(D_n)_{n\geq K}$ conditionally to the event $(K = j)$. An induction argument provides that, conditionally to the event $(K = j)$, the random variable D_n has a density ψ_n bounded by 2 for $n \geq j$. Therefore, for $n \geq j$, we have

$$\mathbb{P}(|D_n - 0.2| \leq \alpha | K = j) = \int_{0.2-\alpha}^{0.2+\alpha} \psi_n(v) dv \leq 4\alpha,$$

since $\psi_n \leq 2$. Finally, we obtain

$$\begin{aligned}\mathbb{P}(Z_n \in U^\alpha) &= \mathbb{P}(|D_n - 0.2| \leq \alpha) \\ &= \sum_{j\in\mathbb{N}} \mathbb{P}(|D_n - 0.2| \leq \alpha | K = j)\mathbb{P}(K = j)\end{aligned}$$

The result follows. □

Regarding assumption 5.3, we will replace it by a weaker local Lipschitz property by reworking the proof of propositions 5.4 and 5.6.

LEMMA 6.3.– *For all $n \in \mathbb{N}$, set*

$$\eta_n = \left|\left((T_n + u^*(Z_n)) \wedge T_{n+1}\right) - \left((\widehat{T}_n + u^*(\widehat{Z}_n)) \wedge \widehat{T}_{n+1}\right)\right|.$$

We have for all $\alpha > 0$,

$$\begin{aligned}\|\eta_n\|_p \leq \|T_n - \widehat{T}_n\|_p + 2\|T_{n+1} - \widehat{T}_{n+1}\|_p \\ + \left([u^*]_{\frac{\alpha}{2}} + \frac{4C_{u^*}}{\alpha}\right)\|Z_n - \widehat{Z}_n\|_p + 10C_{u^*}\alpha^{1/p},\end{aligned}$$

where $[u^*]_\alpha = (1 + C_{u^*} + 4 \times 10^5)10^7(1 - e^{-\alpha/2})^{-1}$.

PROOF.– Set $\alpha > 0$ and $m = (i, 1)$ for $1 \leq i \leq 3$. We will prove that the function $(d, \rho, a) \mapsto u^*(m, d, \rho, a)$ is Lipschitz continuous on $\widetilde{U}_\alpha = U \backslash U_\alpha$. The scalar $u = u^*(m, d, \rho, 0)$ satisfies the following equivalent equations

$$d + d_i(u) = 0.2 \qquad \Longleftrightarrow \qquad d + \rho(u + \eta_i(e^{-u/\eta_i} - 1)) = 0.2.$$

The implicit equation satisfied by u yields $u \geq \alpha/\rho_{max} = (\rho_i^+)^{-1}\alpha \geq 10^5\alpha$ on the set $(U_m)_\alpha$. This lower bound is crucial to prove the Lipschitz continuity. Let $d, d' \leq 0.2 - \alpha$ and denote $u = u^*(d, \rho, 0)$ and $u' = u^*(d', \rho, 0)$. Note that $d + d_i(u) = d' + d_i(u')$ because they both equal 0.2. Consequently, we have $|d_i(u) - d_i(u')| = |d' - d|$ and, noting that $\eta_i \leq 2 \cdot 10^5$ we have

$$\begin{aligned}|d - d'| &= \rho|u - u' + \eta_i(e^{-u/\eta_i} - e^{-u'/\eta_i})| \\ &\geq \rho(1 - e^{-u\wedge u'/\eta_i})|u - u'| \\ &\geq 10^{-7}(1 - e^{-\alpha/2})|u - u'|\end{aligned}$$

that proves the Lipschitz continuity of u^* w.r.t. d on $\widetilde{U}_\alpha$.

Similarly, let $\rho, \rho' \in [\rho_i^-; \rho_i^+]$ and denote $u = u^*(d, \rho, 0)$ and $u' = u^*(d, \rho', 0)$. Note that $d + \rho(u + \eta_i(e^{-u/\eta_i} - 1)) =$

$d + \rho'(u' + \eta_i(\mathrm{e}^{-u'/\eta_i} - 1))$ because they both equal 0.2. Subtracting $d + \rho(u' + \eta_i(\mathrm{e}^{-u'/\eta_i} - 1))$ in both terms yields

$$\rho|u - u' + \eta_i(\mathrm{e}^{-u/\eta_i} - \mathrm{e}^{-u'/\eta_i})| = |\rho - \rho'||u' + \eta(\mathrm{e}^{-u'/\eta_i} - 1)|.$$

A lower bound for the left-hand side term has already been computed earlier, while the right hand-side is easily bounded by $(C_{u^*} + 4 \cdot 10^5)|\rho - \rho'|$, since $\eta_i \leq 2 \cdot 10^5$, so that we have

$$(C_{u^*} + 4 \cdot 10^5)|\rho - \rho'| \geq 10^{-7}(1 - e^{-\alpha/2})|u - u'|,$$

that proves the Lipschitz continuity of u^* w.r.t. ρ on $\widetilde{U}_\alpha$. Finally, for all $\alpha > 0$, the function u^* is Lipschitz continuous on $\widetilde{U}_\alpha$ with Lipschitz constant $[u^*]_\alpha = (1 + C_{u^*} + 4 \cdot 10^5)10^7(1 - e^{-\frac{\alpha}{2}})^{-1}$.

Let $n \in \mathbb{N}$, we now intend to bound $\|\eta_n\|_p$. Recall that K is the number of the jump ending the protection against corrosion. First, note that, on the event $(n \leq K)$, i.e. when the protection from corrosion is still active, we have $u^* = +\infty$ so that

$$\|\eta_n \mathbb{1}_{(n \leq K)}\|_p = \|(T_{n+1} - \widehat{T}_{n+1})\mathbb{1}_{(n \leq K)}\|_p \leq \|T_{n+1} - \widehat{T}_{n+1}\|_p.$$

Second, if $Z_n = \Delta$ where Δ denotes the cemetery state, then $\widehat{Z}_n = \Delta$ too and we have $\eta_n = 0$ so that we obtain

$$\begin{aligned}\|\eta_n \mathbb{1}_{\{n>K\}}\|_p \leq &\|T_n - \widehat{T}_n\|_p + \|T_{n+1} - \widehat{T}_{n+1}\|_p \\ &+ \|(u^*(Z_n) - u^*(\widehat{Z}_n))\mathbb{1}_{\{n>K\}}\mathbb{1}_{\{Z_n \neq \Delta\}}\|_p.\end{aligned}$$

We need to bound the last term of the previous sum. Consider, therefore, the event $(n > K) \cap (Z_n \neq \Delta)$. As U is convex, we have $Z_n \in U$ and $\widehat{Z}_n \in U$. Set $\alpha > 0$. We have

$$\|(u^*(Z_n) - u^*(\widehat{Z}_n))\mathbb{1}_{\{n \geq K\}}\mathbb{1}_{\{Z_n \neq \Delta\}}\|_p \leq A + B,$$

where

$$\begin{aligned}A &= \|(u^*(Z_n) - u^*(\widehat{Z}_n))\mathbb{1}_{\{Z_n \in \widetilde{U}_\alpha\}}\mathbb{1}_{\{n \geq K\}}\|_p, \\ B &= \|(u^*(Z_n) - u^*(\widehat{Z}_n))\mathbb{1}_{\{Z_n \in U^\alpha\}}\mathbb{1}_{\{n \geq K\}}\|_p.\end{aligned}$$

The term B is bounded by $2C_{u^*}\mathbb{P}(Z_n \in U^\alpha)^{1/p} \leq 10C_{u^*}\alpha^{1/p}$. We now turn to the term A and use the Lipschitz continuity of u^* on $\widetilde{U}_\beta$ for any $\beta > 0$. We have

$$\begin{aligned}A \leq &\|\big(u^*(Z_n) - u^*(\widehat{Z}_n)\big)\mathbb{1}_{\{Z_n\in\widetilde{U}_\alpha\}}\mathbb{1}_{\{\widehat{Z}_n\in\widetilde{U}_{\frac{\alpha}{2}}\}}\mathbb{1}_{\{n\geq K\}}\|_p\\ &+\|\big(u^*(Z_n) - u^*(\widehat{Z}_n)\big)\mathbb{1}_{\{Z_n\in\widetilde{U}_\alpha\}}\mathbb{1}_{\{\widehat{Z}_n\notin\widetilde{U}_{\frac{\alpha}{2}}\}}\mathbb{1}_{\{n\geq K\}}\|_p\\ \leq &[u^*]_{\frac{\alpha}{2}}\|Z_n - \widehat{Z}_n\|_p + 2C_{u^*}\|\mathbb{1}_{\{Z_n\in\widetilde{U}_\alpha\}}\mathbb{1}_{\{\widehat{Z}_n\notin\widetilde{U}_{\frac{\alpha}{2}}\}}\|_p.\end{aligned}$$

Note now that $\mathbb{1}_{\{Z_n\in\widetilde{U}_\alpha\}}\mathbb{1}_{\{\widehat{Z}_n\notin\widetilde{U}_{\frac{\alpha}{2}}\}} \leq \mathbb{1}_{\{|Z_n-\widehat{Z}_n|\geq\frac{\alpha}{2}\}}$ so that finally we have

$$\begin{aligned}A &\leq [u^*]_{\frac{\alpha}{2}}\|Z_n - \widehat{Z}_n\|_p + 2C_{u^*}\Big(\mathbb{P}(|Z_n - \widehat{Z}_n| \geq \frac{\alpha}{2})\Big)^{1/p}\\ &\leq [u^*]_{\frac{\alpha}{2}}\|Z_n - \widehat{Z}_n\|_p + 4C_{u^*}\alpha^{-1}\|Z_n - \widehat{Z}_n\|_p\end{aligned}$$

and the result follows. □

Simulation results for the approximation of the mean exit time are given in Table 6.2. In order to have a value of reference, a Monte Carlo method (10^6 simulations) yields the value $E[\tau] = 493.03 \times 10^3$ h. For the first moment, the empirical convergence rate is presented in Figure 6.2. It is estimated through a regression model at -1.28. Remark that it is roughly the same order as for the expectation in section 6.1 (even though this mean time is computed using another algorithm, the quantity approximated is of the same nature), and again much sharper than the predicted quantization error.

Number of points in the quantization grids	Approximation $\widehat{p}_{N,1}$ ($\times 10^3$ h)	Relative error to 493.03×10^3h
20	528.37	7.16%
50	495.79	0.56%
100	491.40	0.33%
200	494.22	0.24%
500	493.40	0.08%

Table 6.2. *Approximation results for the mean exit time*

6.2.2. *Distribution of the service time*

Considering the approximation scheme for the exit time distribution, we may note that the quantized value $\widehat{p}_N(s)$ is not necessary smaller than 1.

Therefore, it appears natural to replace $\widehat{p}_N(s)$ by $\widehat{p}_N(s) \wedge 1$. This does not change the convergence theorem and can only improve the approximation error.

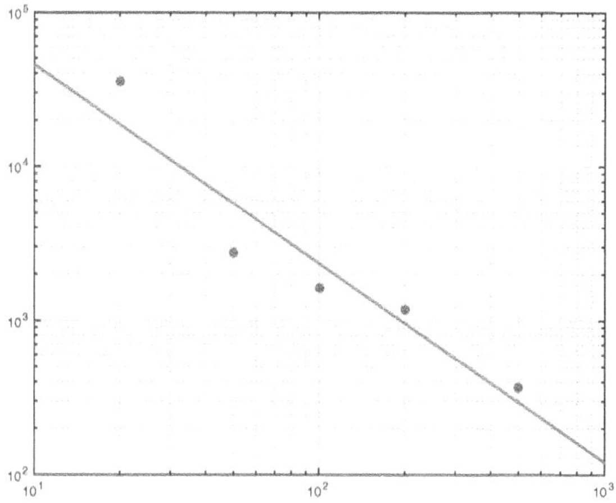

Figure 6.2. *Logarithm of the error w.r.t. the logarithm of the number of points in the quantization grids for the first moment of the service time of the corrosion process*

Figure 6.3 presents the survival function of τ obtained through Monte Carlo simulations (the dashed line), through our approximation scheme (the solid line) and the error. Table 6.3 contains the empirical error for different numbers of points in the quantization grids. For the survival function, the empirical convergence rate is presented in Figure 6.4. It is estimated through a regression model as -0.66. Remark that unlike in the previous cases, we retrieve here the order of magnitude of the quantization error. Note that for this example, by using Monte Carlo simulations, we can estimate the value of $\widetilde{q}$ as approximately $\widetilde{q} = 0.0187$ (for 10^7 histories).

Number of points in the quantization grids	Approximation $\max_s \lvert p_N(s) - \widehat{p}_N(s) \rvert$
20	0.145
50	0.119
100	0.040
200	0.039
500	0.020

Table 6.3. *Approximation results for the distribution of the exit time*

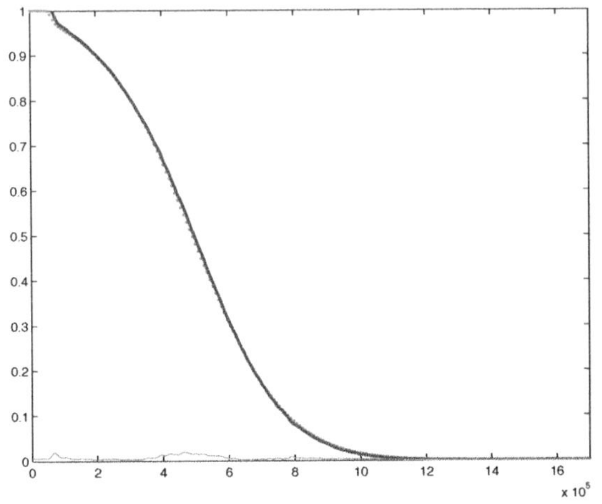

Figure 6.3. *Survival function of the service time of the corrosion process obtained through Monte Carlo simulations (dashed), quantized approximation (solid) and the error with 2,000 points in the quantization grids*

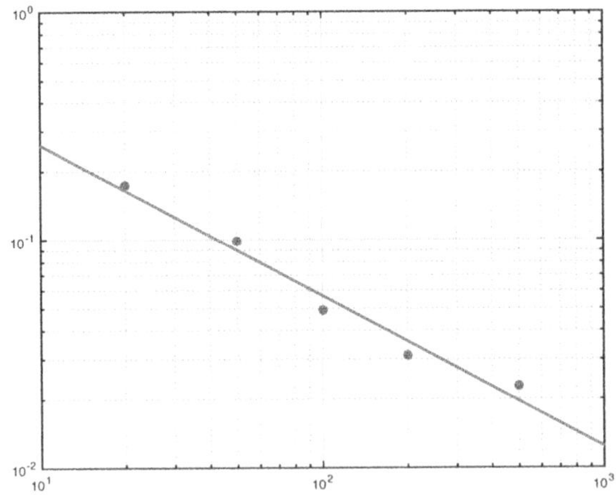

Figure 6.4. *Logarithm of the error w.r.t. the logarithm of the number of points in the quantization grids for the survival of the corrosion process*

6.3. Conclusion

To conclude, we have proved the applicability of our methodology on the corrosion example. Although it does not quite satisfy the required assumptions, we proved rigorously that we can still apply our procedure with a guaranteed convergence. The relative errors of our approximation are small enough for grids with as few as 50 or 100 points. Such grids are very fast to construct. We can obtain more precise results with more grid points. The same grids can be used to compute very different quantities, as will also be further illustrated in Chapter 9.

PART 3

Optimization

7

Optimal Stopping

7.1. Introduction

The aim of this chapter is to propose a computational method for the optimal stopping of a piecewise deterministic Markov process (PDMP) (X_t). Roughly speaking, an optimal stopping problem consists of choosing the best admissible time to stop the process (X_t) in order to maximize some performance. In this chapter, an admissible time is a stopping time lower than the N-th jump time of the process, i.e. a stopping time τ in the set $\mathcal{M}_N$ defined in section 1.7. The performance is the expectation of a function g, called *reward function*, of the process at the chosen stopping time. The *value function* v is the best possible performance. It depends on the (deterministic) starting point x_0 of the process

$$v(x_0) = \sup_{\tau \in \mathcal{M}_N} \mathbb{E}_{x_0}[g(X_\tau)]. \qquad [7.1]$$

Optimal stopping problems have been studied for PDMPs in [COS 88, COS 00, DAV 93, GAT 91, GUG 86, LEN 85]. In [GUG 86], the author defines an operator related to the first jump time of the process, and shows that the value function of the optimal stopping problem is a fixed point for this operator. The basic assumption in this case is that the final cost function is continuous along trajectories, and it is shown that the value function will also have the same property. In [GAT 91, LEN 85], the authors adopt some stronger continuity assumptions and boundary conditions to show that the value function of the optimal stopping problem satisfies some variational inequalities, related to integro-differential equations. In [DAV 93], M.H.A. Davis assumes that the value function is bounded and locally

Lipschitz along trajectories to show that the variational inequalities are necessary and sufficient to characterize the value function of the optimal stopping problem. In [COS 00], the authors weakened the continuity assumptions of [DAV 93, GAT 91, LEN 85]. A paper related to our work is [COS 88] by O.L.V. Costa and M.H.A. Davis. It is the only one presenting a computational technique for solving the optimal stopping problem for a PDMP based on a discretization of the state space similar to the one proposed by H. J. Kushner in [KUS 77]. In particular, the authors in [COS 88] derive a convergence result for the approximation scheme but no estimation of the rate of convergence is derived.

Powerful and interesting methods based on the quantization techniques presented in Chapter 3 have been developed in [BAL 03, BAL 05, PAG 04b] for computing the Snell-envelope associated with discrete-time Markov chains and diffusion processes. Roughly speaking, the approach proposed in [BAL 03, BAL 05, PAG 04b] for studying the optimal stopping problem for a continuous-time diffusion process (Y_t) is based on a time-discretization scheme to obtain a discrete-time Markov chain $(\overline{Y}_n)$. It is shown that the original continuous-time optimization problem can be converted to an auxiliary optimal stopping problem associated with the discrete-time Markov chain $(\overline{Y}_n)$. Under some suitable assumptions, a rate of convergence of the auxiliary value function to the original one can be derived. Then, in order to address the optimal stopping problem of the discrete-time Markov chain, a twofold computational method is proposed. The first step consists of approximating the Markov chain by a quantized process. The second step is to approximate the conditional expectations which are used to compute the backward dynamic programming formula by the conditional expectation related to the quantized process. This procedure leads to a tractable formula called a *quantization tree algorithm* (see proposition 4 in [BAL 03] or section 4.1 in [PAG 04b]). Providing the cost function and the Markov kernel are Lipschitz, some bounds and rates of convergence are obtained, see for example section 2.2.2 in [BAL 03].

Regarding PDMPs, it was shown in [GUG 86] that the value function of the optimal stopping problem can be calculated by iterating a functional operator, labeled as L (see equation [7.6] for its definition), which is related to a continuous-time maximization and a discrete-time dynamic programming formula. Thus, in order to approximate the value function of the optimal stopping problem of a PDMP (X_t), a natural approach would have been to follow the same lines as in [BAL 03, BAL 05, PAG 04b]. However their method cannot be directly applied to our problem for two main reasons related to the specificities of PDMPs.

First, PDMPs are in essence discontinuous at random times. Therefore, as pointed out in [GUG 86], it will be problematic to convert the original optimization problem into an optimal stopping problem associated with a time discretization of (X_t) with nice convergence properties. In particular, it appears ill-advised to propose as in [BAL 03] a fixed-step time-discretization scheme $(X_{n\Delta})$ of the original process (X_t). Besides, another important intricacy concerns the transition semi-group $(P_t)_{t\in\mathbb{R}^+}$ of (X_t). On the one hand, it cannot be explicitly calculated from the local characteristics (Φ, λ, Q) of the PDMP, see [COS 08, DUF 99]. Consequently, it will be complicated to express the Markov kernel P_Δ associated with the Markov chain $(X_{n\Delta})$. On the other hand, the Markov chain $(X_{n\Delta})$ is in general not even a Feller chain (see [DAV 93, pages 76-77]), therefore it will be hard to ensure it is Lipschitz in the sense of definition 1 in [BAL 03].

Second, the other main difference arises from the fact that the function appearing in the backward dynamic programming formula associated with L and the reward function g is not continuous even if some strong regularity assumptions are made on g. Consequently, the approach developed in [BAL 03, BAL 05, PAG 04b] has to be refined since it can only handle conditional expectations of Lipschitz-continuous functions.

However, by using the special structure of PDMPs, we are able to overcome both these difficulties. Indeed, as mentioned in section 1.6, there exists, associated with the PDMP (X_t), a natural embedded discrete-time Markov chain (Θ_n) with $\Theta_n = (Z_n, S_n)$. The main operator L can be expressed using the Markov chain (Θ_n) and a continuous-time maximization. We first convert the continuous-time maximization of operator L into a discrete-time maximization by using a path-dependent time-discretization scheme. This enables us to approximate the value function by the solution of a backward dynamic programming equation in discrete-time involving conditional expectation of the Markov chain (Θ_n). Then, a natural approximation of this optimization problem is obtained by replacing (Θ_n) by its quantized approximation as described in Chapter 3. It must be pointed out that this optimization problem is related to the calculation of conditional expectations of indicator functions of the Markov chain (Θ_n). As said above, it is not straightforward to obtain convergence results as in [BAL 03, BAL 05, PAG 04b]. We deal successfully with indicator functions by showing that the event on which the discontinuity actually occurs is of small enough probability. This enables us to provide a rate of convergence for the approximation scheme.

In addition and more importantly, this numerical approximation scheme enables us to propose a computable stopping rule which also is an ϵ-optimal stopping time of the original stopping problem. Indeed, for any $\epsilon > 0$ we can construct a stopping time, labeled as τ, such that

$$v(x_0) - \epsilon \leq \mathbb{E}_{x_0}\Big[g(X_\tau)\Big] \leq v(x_0).$$

Our computational approach is attractive in the sense that it does not require any additional calculations. Moreover, we can characterize how far it is from optimal in terms of the value function. In [BAL 03, section 2.2.3, Proposition 6], another criteria for the approximation of the optimal stopping time has been proposed. In the context of PDMPs, it must be noticed that an optimal stopping time does not generally exists as shown in [GUG 86, section 2].

The main material from this chapter comes from [SAP 10b]. This chapter is organized as follows. In section 7.2, we state the dynamic programming equation satisfied by the value function of our optimal stopping problem. In section 7.3, we build an approximation of the value function. Technical results related to Lipschitz continuity are presented in section 7.4. In section 7.5, we evaluate the error between the approximate value function and the real value function. In section 7.6 we propose a computable ϵ-optimal stopping time and evaluate its sharpness. Finally our procedure is illustrated on the Tansmission control Protocol (TCP) process (see section 1.8.2) in section 7.7. More involved applications related to maintenance optimization will be presented in Chapter 9.

7.2. Dynamic programming equation

Assume that the distribution of X_0 is given by δ_{x_0} for a fixed state $x_0 \in E$. We suppose throughout this chapter that assumptions 1.3, 1.4, 1.5 and 1.6 hold and moreover, the reward function g also has a suitable Lipschitz regularity.

ASSUMPTION 7.1.– The reward function g is in $\mathbf{L}_\Phi(E)$.

The optimal stopping problem [7.1] has been studied by U.S. Gugerli in [GUG 86]. In particular, the author showed that the value function v can be recursively constructed by the following procedure. Set

$$v_N = g, \quad v_n = L(v_{n+1}, g), \qquad [7.2]$$

$0 \leq n \leq N$. The last term v_0 of this recurrence is then exactly the value function

$$v = v_0.$$

In the dynamic programming equation [7.2], L is an operator from $\mathbf{L}_\Phi(E) \times \mathbf{L}_\Phi(E)$ onto $\mathbf{B}(E)$ defined by

$$L(w, w')(x) = \sup_{t \geq 0} J(w, w')(x, t) \vee Kw(x),$$

and operators $H : \mathbf{L}_\Phi(E) \to \mathbf{B}(E \times \mathbb{R}^+)$, $I : \mathbf{L}_\Phi(E) \to \mathbf{B}(E \times \mathbb{R}^+)$, $J : \mathbf{L}_\Phi(E) \times \mathbf{L}_\Phi(E) \to \mathbf{B}(E \times \mathbb{R}^+)$, $K : \mathbf{L}_\Phi(E) \to \mathbf{B}(E)$ are defined by

$$\begin{aligned}
Hw(x,t) &= w\big(\Phi(x, t \wedge t^*(x))\big)\mathrm{e}^{-\Lambda(x, t\wedge t^*(x))},\\
Iw(x,t) &= \int_0^{t\wedge t^*(x)} \lambda Q w(\Phi(x,s))\mathrm{e}^{-\Lambda(x,s)} ds,\\
J(w,w')(x,t) &= Iw(x,t) + Hw'(x,t),\\
Kw(x) &= \int_0^{t^*(x)} \lambda Q w(\Phi(x,s))\mathrm{e}^{-\Lambda(x,s)} ds && [7.3]\\
&\quad + Qw(\Phi(x, t^*(x)))\mathrm{e}^{-\Lambda(x,t^*(x))}, && [7.4]
\end{aligned}$$

As in the previous part, the key to our approximation procedure is to rewrite the operators in terms of the embedded Markov chain. It is easy to derive a probabilistic interpretation of operators H, I, K and L involving $(Z_n, S_n)_{n\in\mathbb{N}}$.

LEMMA 7.1.– For all $x \in E$, $w \in \mathbf{L}_\Phi(E)$, $w' \in \mathbf{L}_\Phi(E)$, and $t \geq 0$, one has

$$\begin{aligned}
Hw(x,t) &= w\big(\Phi(x, t\wedge t^*(x))\big)\mathbb{P}_x\big(S_1 \geq t \wedge t^*(x)\big),\\
Iw(x,t) &= \mathbb{E}_x\Big[w(Z_1)\mathbb{1}_{\{S_1 < t\wedge t^*(x)\}}\Big],\\
Kw(x) &= \mathbb{E}_x\big[w(Z_1)\big], && [7.5]\\
L(w,w')(x) &= \sup_{u \leq t^*(x)} \Big\{\mathbb{E}_x\big[w(Z_1)\mathbb{1}_{\{S_1<u\}}\big]\\
&\quad + w'\big(\Phi(x,u)\big)\mathbb{P}_x\big(S_1 \geq u\big)\Big\} \vee \mathbb{E}_x\big[w(Z_1)\big]. && [7.6]
\end{aligned}$$

Therefore, the value function at point x_0 is solution of the following dynamic programming equation involving only the embedded Markov chain: $v(x_0) = V_0 = v(Z_0)$ where the sequence of random variables (V_n) is recursively defined as follows.

DEFINITION 7.1.– *Introduce the random variables* $(V_n)_{0\leq n\leq N}$ *by*

$$V_n = v_n(Z_n),$$

or equivalently

$$\begin{aligned} V_n = &\sup_{u\leq t^*(Z_n)} \Big\{\mathbb{E}_{x_0}\Big[v_{n+1}(Z_{n+1})\mathbb{1}_{\{S_{n+1}<u\}} \\ &+ g(\Phi(Z_n,u))\mathbb{1}_{\{S_{n+1}\geq u\}}\big|Z_n\Big]\Big\} \vee \mathbb{E}_{x_0}\Big[v_{n+1}(Z_{n+1})\big|Z_n\Big] \\ = &\sup_{u\leq t^*(Z_n)} \Big\{\mathbb{E}_{x_0}\Big[V_{n+1}\mathbb{1}_{\{S_{n+1}<u\}} + g(\Phi(Z_n,u))\mathbb{1}_{\{S_{n+1}\geq u\}}\big|Z_n\Big]\Big\} \\ &\vee \mathbb{E}_{x_0}\Big[V_{n+1}\big|Z_n\Big]. \end{aligned} \quad [7.7]$$

It is a highly remarkable feature of this problem that the dynamic programming equation involving a sequence of function (v_n) can simply be rewritten as a self-contained dynamic programming equation involving only the random variables (V_n). As random variables are much more tractable numerically than functions, this is a critical step toward a numerical approximation of the value function $v(x_0)$.

7.3. Approximation of the value function

The aim of this chapter is to provide a numerical approximation of the sequence (V_n), to be able to approximate the value function v at point x_0. To approximate the sequence of value functions (V_n), we proceed in two steps. First, the continuous-time maximization of operator L is converted into a discrete-time maximization by using a path-dependent time-discretization scheme to give a new operator L^d. In particular, it is important to remark that these time-discretization grids depend on the the post-jump locations (Z_n) of the PDMP (see definition 7.2 and the following comment). Second, the conditional expectations of the Markov chain (Θ_n) in the definition of L^d are replaced by the conditional expectations of its quantized approximation $(\widehat{\Theta}_n)$ to define an operator $\widehat{L}^d$.

First, we define the path-adapted discretization grids as follows.

DEFINITION 7.2.– *For* $z \in E$, *set* $\Delta(z) \in (0, t^*(z))$. *Define* $n(z) = \lfloor \frac{t^*(z)}{\Delta(z)} \rfloor - 1$, *where* $\lfloor x \rfloor$ *denotes the greatest integer smaller than or equal to* x. *The set* *of points* $(t_i)_{0 \leq i \leq n(z)\}}$ *with* $t_i = i\Delta(z)$ *is denoted by* $G(z)$. *This is the grid associated with the time interval* $[0, t^*(z)]$.

It is important to note that, for all $z \in E$, not only we have $t^*(z) \notin G(z)$, but also $\max G(z) = t_{n(z)} \leq t^*(z) - \Delta(z)$. This property is crucial for the sequel.

DEFINITION 7.3.– *Consider for* $w \in \mathbf{L}_\Phi(E)$ *and* $z \in E$,

$$L^d(w, g)(z) = \max_{s \in G(z)} \Big\{ \mathbb{E}_z \big[w(Z_1) \mathbb{1}_{\{S_1 < s\}} + g\big(\Phi(z, s)\big) \mathbb{1}_{\{S_1 \geq s\}} \big] \Big\} \\ \vee \mathbb{E}_z \big[w(Z_1) \big].$$

Now let us turn to the quantization of (Θ_n). As explained in section 3.4, the quantization algorithm will provide us with a finite grid $\Gamma_n^\Theta \subset E \times \mathbb{R}^+$ at each time $0 \leq n \leq N$ as well as weights for each point of the grid. Set $p \geq 1$ such that Θ_n has finite moments at least up to the order p and let $p_n = proj_{\Gamma_n^\Theta}$ be the closest-neighbor projection from $E \times \mathbb{R}^+$ onto Γ_n^Θ. Recall that the quantization of Θ_n is defined by

$$\widehat{\Theta}_n = (\widehat{Z}_n, \widehat{S}_n) = p_n(Z_n, S_n).$$

Also recall that Γ_n^Z is the projection of Γ_n^Θ on E and Γ_n^S the projection of Γ_n^Θ on $\mathbb{R}^+$.

In practice, we will first compute the quantization grids and weights using the competitive learning vector quantization (CLVQ) algorithm 3.3, and then compute a path-adapted time-grid for each $z \in \Gamma_n^Z$, for all $0 \leq n \leq N - 1$. Hence, there is only a finite number time grids to compute, and like the quantization grids, they can be computed and stored off-line.

The definition of the discretized operators now naturally follows the characterization given in lemma 7.1.

DEFINITION 7.4.– *For* $k \in \{1, \ldots, N\}$, $w \in \mathbf{B}(\Gamma_k^Z)$, $z \in \Gamma_{k-1}^Z$, *and* $s \in \mathbb{R}^+$, *set*

$$\widehat{J}_k(w, g)(z, s) = \mathbb{E}_{x_0} \Big[w(\widehat{Z}_k) \mathbb{1}_{\{\widehat{S}_k < s\}} + g\big(\Phi(z, s)\big) \mathbb{1}_{\{\widehat{S}_k \geq s\}} \big| \widehat{Z}_{k-1} = z \Big],$$

$$\widehat{K}_k(w)(z) = \mathbb{E}_{x_0} \Big[w(\widehat{Z}_k) \big| \widehat{Z}_{k-1} = z \Big],$$

$$\widehat{L}_k^d(w, g)(z) = \max_{s \in G(z)} \Big\{ \widehat{J}_k(w, g)(z, s) \Big\} \vee \widehat{K}_k(w)(z).$$

Note that $\widehat{\Theta}_n$ is a random variable taking finitely many values, hence the expectations above actually are finite sums, the probability of each atom being given by its weight on the quantization grid. We can now give the complete construction of the sequence approximating (V_n).

DEFINITION 7.5.– *Consider* $\widehat{v}_N(z) = g(z)$ *for* $z \in \Gamma_N^Z$ *and*

$$\widehat{v}_{k-1}(z) = \widehat{L}_k^d(\widehat{v}_k, g)(z), \qquad [7.8]$$

for $1 \leq k \leq N$ *and* $z \in \Gamma_{k-1}^Z$.

DEFINITION 7.6.– *The approximation of* V_k *is denoted by*

$$\widehat{V}_k = \widehat{v}_k(\widehat{Z}_k), \qquad [7.9]$$

for $0 \leq k \leq N$.

Our approximation procedure is detailed in algorithm 7.1. Note that the fixed time discretization step Δ can be replaced (for instance) by a fixed number of grid points N_G by replacing Line 8 by **for** $u \leftarrow 0$ **to** $t^*(z) - \Delta$ **by step**: $(t^*(z) - \Delta)/N_G$. Note also that the time discretization grids can be either stored or computed online as presented here.

One of our main objectives is to show that $\widehat{V}_0$ is a good approximation of the value function $v(x_0)$. Our strategy to get this result is first to prove that $\widehat{v}_k$ is a good approximation of v_k for $0 \leq k \leq N$ on the quantization grids, and thus that $\widehat{V}_k$ is a good approximation of V_k for $0 \leq k \leq N$.

7.4. Lipschitz continuity properties

In order to evaluate the sharpness of our approximation procedure, we need to derive some technical preliminary results related to the Lipschitz continuity properties of the operators and value functions. This section can be skipped at first reading.

7.4.1. *Lipschitz properties of* J *and* K

In this section, we derive useful Lipschitz-type properties of operators J and K. The first result is straightforward. It is a generalization of lemma 4.5.

Algorithm 7.1: Approximation of the value function $v(x_0)$

input : Quantization grids $(\Gamma_n^\Theta)_{0\leq n\leq N}$, Transition matrices $(\widehat{P}_n^\Theta)_{1\leq n\leq N}$, cost function g, time discretization step Δ

output : $\widehat{V}_0$

begin

  for $z \in \Gamma_N^Z$ **do**

    $\widehat{v}_N(z) \leftarrow g(z)$

  end

  for $n \leftarrow N-1$ **to** 0 **do**

    for $z \in \Gamma_n^Z$ **do**

      $\widehat{K}\widehat{v}_{n+1}(z) \leftarrow \sum_{(z',s')\in\Gamma_{n+1}^\Theta} \widehat{P}_{n+1}(z,(z',s'))v_{n+1}(z')$

      for $u \leftarrow 0$ **to** $t^*(z)-\Delta$ **by step**: Δ

      do

        $\widehat{J}\widehat{v}_{n+1}(z,u) \leftarrow \sum_{(z',s')\in\Gamma_{n+1}^\Theta} \widehat{P}_{n+1}(z,(z',s'))\big(v_{n+1}(z')\mathbb{1}_{\{s'<u\}}$

        $+g(\Phi(z,u))\mathbb{1}_{\{s'\geq u\}}\big)$

      end

      $\widehat{v}_n(z) \leftarrow \max_u\{\widehat{J}\widehat{v}_{n+1}(z,u)\} \vee \widehat{K}\widehat{v}_{n+1}(z)$

    end

  end

  return : $\widehat{v}_0$

end

LEMMA 7.2.– Let $w \in \mathbf{L}_\Phi(E)$. Then for all $(x,y)\in E^2$ and $(t,u)\in(\mathbb{R}^+)^2$, we have

$$\Big|w\big(\Phi(x,t\wedge t^*(x))\big)\mathrm{e}^{-\Lambda(x,t\wedge t^*(x))} - w\big(\Phi(y,u\wedge t^*(y))\big)\mathrm{e}^{-\Lambda(y,u\wedge t^*(y))}\Big| \leq D_1(w)|x-y| + D_2(w)|t-u|,$$

where

– if $t < t^*(x)$ and $u < t^*(y)$,

$$D_1(w) = [w]_1 + C_w C_{t^*}[\lambda]_1, \quad D_2(w) = [w]_2 + C_w C_\lambda,$$

– if $t = t^*(x)$ and $u = t^*(y)$,

$$D_1(w) = [w]_* + C_w C_{t^*}[\lambda]_1 + C_w C_\lambda L_{t^*}, \quad D_2(w) = 0,$$

– otherwise,

$$D_1(w) = [w]_1 + C_w C_{t^*}[\lambda]_1 + [w]_2 L_{t^*} + C_w C_\lambda L_{t^*}, \; D_2(w) = [w]_2 + C_w C_\lambda.$$

LEMMA 7.3.– Let $w \in \mathbf{L}_\Phi(E)$. Then for all $x \in E$, $(t, u) \in (\mathbb{R}^+)^2$, we have

$$\Big| J(w,g)(x,t) - J(w,g)(x,u)\Big| \leq \big(C_w C_\lambda + [g]_2 + C_g C_\lambda\big)|t-u|.$$

PROOF.– By definition of J, we obtain

$$\begin{aligned}&\Big| J(w,g)(x,t) - J(w,g)(x,u)\Big|\\ &\quad\leq \Big| g(\Phi(x, t\wedge t^*(x)))e^{-\Lambda(x,t\wedge t^*(x))} - g(\Phi(x,u\wedge t^*(x)))e^{-\Lambda(x,u\wedge t^*(x))}\Big|\\ &\quad+ \Big|\int_{t\wedge t^*(x)}^{u\wedge t^*(x)} \lambda Q w(\Phi(x,s))e^{-\Lambda(x,s)}ds\Big|.\end{aligned}$$

Applying lemma 7.2 to g, the result follows. □

LEMMA 7.4.– Let $w \in \mathbf{L}_\Phi(E)$. Then for all $(x, y) \in E^2$, $t \in \mathbb{R}^+$,

$$\Big| J(w,g)(x,t) - J(w,g)(y,t)\Big| \leq \big(C_w E_1 + [w]_1 E_2 + E_3\big)|x-y|,$$

and

$$\Big| Iw(x,t) - Iw(y,t)\Big| \leq \big(C_w E_1 + [w]_1 E_2\big)|x-y|,$$

where

$$\begin{aligned}E_1 &= C_\lambda L_{t^*} + C_{t^*}[\lambda]_1\big(1 + C_{t^*}C_\lambda\big),\\ E_2 &= C_{t^*}C_\lambda[Q],\\ E_3 &= [g]_1 + [g]_2 L_{t^*} + C_g\big\{C_{t^*}[\lambda]_1 + C_\lambda L_{t^*}\big\}.\end{aligned}$$

PROOF.– again by definition, we obtain

$$\begin{aligned}&\Big| J(w,g)(x,t) - J(w,g)(y,t)\Big|\\ &\quad\leq \Big|\int_0^{t\wedge t^*(x)} \lambda Q w(\Phi(x,s))e^{-\Lambda(x,s)}ds\\ &\qquad- \int_0^{t\wedge t^*(y)} \lambda Q w(\Phi(y,s))e^{-\Lambda(y,s)}ds\Big|\\ &\quad+ \Big| g(\Phi(x, t\wedge t^*(x)))e^{-\Lambda(x,t\wedge t^*(x))} - g(\Phi(y,t\wedge t^*(y)))e^{-\Lambda(y,t\wedge t^*(y))}\Big|.\end{aligned}$$

Without loss of generality it can be assumed that $t^*(x) \le t^*(y)$. From lemma 7.2 for g and using the fact that $\big|t \wedge t^*(x) - t \wedge t^*(y)\big| \le \big|t^*(x) - t^*(y)\big|$, we get

$$\Big|J(w,g)(x,t) - J(w,g)(y,t))\Big|$$
$$\le \int_0^{t\wedge t^*(x)} \Big|\lambda Qw(\Phi(x,s))e^{-\Lambda(x,s)} - \lambda Qw(\Phi(y,s))e^{-\Lambda(y,s)}\Big| ds$$
$$+ \big(C_w C_\lambda L_{t^*} + E_3\big)|x-y|.$$

By using a similar result as lemma 7.2 for λQw, we obtain the first inequality. The second inequality is just a straightforward consequence of the previous one by setting $g = 0$. □

LEMMA 7.5.– Let $w \in \mathbf{L}_\Phi(E)$. Then for all $(x,y) \in E^2$,

$$\big|Kw(x) - Kw(y)\big| \le \big(C_w E_4 + [w]_1 E_2 + [w]_*[Q]\big)|x-y|,$$

where $E_4 = 2C_\lambda L_{t^*} + C_{t^*}[\lambda]_1\big(2 + C_{t^*}C_\lambda\big)$.

PROOF.– The proof is similar to the previous ones and therefore omitted. □

7.4.2. *Lipschitz properties of the value functions*

Now we turn to the Lipschitz continuity of the sequence of value functions (v_n). Namely, we prove that under our assumptions, v_n belongs to $\mathbf{L}_\Phi(E)$ for all $0 \le n \le N$. We also compute the Lipschitz constant of v_n on $\overline{E}$ as it is much sharper in this case than $[v_n]_1$, see section 1.4.1.

We start with proving sharper results on operator J.

LEMMA 7.6.– Let $w \in \mathbf{L}_\Phi(E)$. Then for all $x \in E$ and $(s,t) \in (\mathbb{R}^+)^2$,

$$\Big|\sup_{u\ge t} J(w,g)(x,u) - \sup_{u\ge s} J(w,g)(x,u)\Big| \le \big(C_w C_\lambda + [g]_2 + C_g C_\lambda\big)|t-s|.$$

PROOF.– Without loss of generality, it can be assumed that $t \le s$. Therefore, we have

$$\Big|\sup_{u\ge t} J(w,g)(x,u) - \sup_{u\ge s} J(w,g)(x,u)\Big|$$
$$= \sup_{u\ge t} J(w,g)(x,u) - \sup_{u\ge s} J(w,g)(x,u). \qquad [7.10]$$

Remark that $\sup_{u \geq t} J(w,g)(x,u) = J(w,g)(x,\bar{t})$ for some $\bar{t} \in [t \wedge t^*(x), t^*(x)]$. Consequently, if $\bar{t} \geq s$ then we have $\left| \sup_{u \geq t} J(w,g)(x,u) - \sup_{u \geq s} J(w,g)(x,u) \right| = 0$.
Now if $\bar{t} \in [t \wedge t^*(x), s[$, then we have

$$\sup_{u \geq t} J(w,g)(x,u) - \sup_{u \geq s} J(w,g)(x,u) \leq J(w,g)(x,\bar{t}) - J(w,g)(x,s).$$

From lemma 7.3, we obtain the following inequality

$$\sup_{u \geq t} J(w,g)(x,u) - \sup_{u \geq s} J(w,g)(x,u) \leq \big(C_w C_\lambda + [g]_2 + C_g C_\lambda\big)|\bar{t} - s|. \quad [7.11]$$

Combining [7.10], [7.11] and the fact that $|\bar{t} - s| \leq |t - s|$ the result follows. □

Similarly, we obtain the following result.

LEMMA 7.7.– Let $w \in \mathbf{L}_\Phi(E)$. Then for all $(x,y) \in E^2$,

$$\left| \sup_{t \leq t^*(x)} J(w,g)(x,t) - \sup_{t \leq t^*(y)} J(w,g)(y,t) \right| \leq \big(C_w E_5 + [w]_1 E_2 + E_6\big)|x - y|,$$

where $E_5 = E_1 + C_\lambda L_{t^*}$ and $E_6 = E_3 + \big([g]_2 + C_g C_\lambda\big) L_{t^*}$.

Now we turn to (v_n). Recall that for all $0 \leq n \leq N$, (v_n) is bounded with $C_{v_n} = C_g$.

PROPOSITION 7.1.– *For all* $0 \leq n \leq N$, $v_n \in \mathbf{L}_\Phi(E)$ *and*

$$[v_n]_1 \leq e^{C_\lambda C_{t^*}} \big(2[v_{n+1}]_1 E_2 + C_g E_1 + C_g E_4 + C_g C_{t^*}[\lambda]_1 (1 + C_\lambda C_{t^*})\big) + e^{C_\lambda C_{t^*}} \Big(\big([g]_1 + [g]_2 L_{t^*}\big) \vee \big([v_{n+1}]_* [Q]\big)\Big), \quad [7.12]$$

$$[v_n]_2 \leq e^{C_\lambda C_{t^*}} \Big(C_g C_\lambda \big(4 + C_\lambda C_{t^*}\big) + [g]_2\Big), \quad [7.13]$$

$$[v_n]_* \leq [v_n]_1 + [v_n]_2 L_{t^*},$$

$$L_{v_n} \leq [v_{n+1}]_1 E_2 + C_g E_5 + \Big(E_6 \vee \big([v_{n+1}]_* [Q] + C_g C_{t^*}[\lambda]_1\big)\Big).$$

PROOF.– Clearly, $v_N = g$ is in $\mathbf{L}_\Phi(E)$. Assume that v_{n+1} is in $\mathbf{L}_\Phi(E)$, then by using the semi-group property of the drift Φ it can be shown that for any $x \in E, t \in [0, t^*(x)]$, we have (see [GUG 86, Eq. (8)])

$$v_n(\Phi(x,t)) = e^{\Lambda(x,t)}\Big(\big(\sup_{u\geq t} J(v_{n+1}, g)(x,u) \vee Kv_{n+1}(x)\big) - Iv_{n+1}(x,t)\Big). \qquad [7.14]$$

Remark that for $x \in E, t \in \mathbb{R}^+$, we have

$$\sup_{u\geq t} J(v_{n+1}, g)(x,u) \vee Kv_{n+1}(x) \leq \sup_{u} J(v_{n+1}, g)(x,u) \vee Kv_{n+1}(x) = v_n(x). \qquad [7.15]$$

Set $(x,y) \in E^2$ and $t \in [0, t^*(x) \wedge t^*(y)]$. It is easy to show that

$$\left|e^{\Lambda(x,t)} - e^{\Lambda(y,t)}\right| \leq e^{C_\lambda C_{t^*}}[\lambda]_1 C_{t^*}|x-y|, \qquad [7.16]$$

$$\left|Iv_{n+1}(x,t) - Iv_{n+1}(y,t)\right| \leq \Big(C_{v_{n+1}}E_1 + [v_{n+1}]_1 E_2\Big)|x-y|. \qquad [7.17]$$

Then, equations [7.14]–[7.17] yield

$$\begin{aligned}
&\left|v_n(\Phi(x,t)) - v_n(\Phi(y,t))\right| \\
&\leq \Big\{\left|v_n(x)\right| + \left|Iv_{n+1}(x,t)\right|\Big\} e^{C_\lambda C_{t^*}}[\lambda]_1 C_{t^*}|x-y| \\
&\quad + e^{\Lambda(y,t)}\Big\{\sup_{u\geq t}\left|J(v_{n+1},g)(x,u) - J(v_{n+1},g)(y,u)\right| \\
&\quad \vee \left|Kv_{n+1}(x) - Kv_{n+1}(y)\right|\Big\} \\
&\quad + e^{\Lambda(y,t)}\Big(C_{v_{n+1}}E_1 + [v_{n+1}]_1 E_2\Big)|x-y|. \qquad [7.18]
\end{aligned}$$

For $x \in E, t \in [0, t^*(x)]$ and $n \in \mathbb{N}$, note that

$$e^{\Lambda(x,t)} \leq e^{C_\lambda C_{t^*}}, \quad \left|Iv_{n+1}(x,t)\right| \leq C_\lambda C_{v_{n+1}} C_{t^*}, \; |v_{n+1}(x)| \leq C_g. \quad [7.19]$$

Therefore, we obtain inequality [7.12] by using equations [7.18], [7.19] and lemma 7.4, 7.6, and the fact that $C_g E_1 + E_3 = C_g E_4 + [g]_1 + [g]_2 L_{t^*}$.

Now, set $x \in E$ and $t, s \in [0, t^*(x)]$. Similarly, we have

$$\left| e^{\Lambda(x,t)} - e^{\Lambda(x,s)} \right| \leq e^{C_\lambda C_{t^*}} C_\lambda |t-s|, \qquad [7.20]$$

$$\left| Iv_{n+1}(x,t) - Iv_{n+1}(x,s) \right| \leq C_\lambda C_{v_{n+1}} |t-s|. \qquad [7.21]$$

Combining equations [7.14], [7.15], [7.20] and [7.21], it yields

$$\begin{aligned} \left| v_n(\Phi(x,t)) - v_n(\Phi(x,s)) \right| &\leq \Big\{ \left| v_n(x) \right| \\ &+ \left| Iv_{n+1}(x,t) \right| \Big\} e^{C_\lambda C_{t^*}} C_\lambda |t-s| \\ + e^{\Lambda(x,t)} \Big\{ \Big| &\sup_{u \geq t} J(v_{n+1}, g)(x,u) \\ &- \sup_{u \geq s} J(v_{n+1}, g)(x,u) \Big| + C_\lambda C_{v_{n+1}} |t-s| \Big\}. \end{aligned} \qquad [7.22]$$

Finally, inequality [7.13] follows from equations [7.19], [7.22] and lemma 7.5.

We clearly have $[v_n]_* \leq [v_n]_1 + [v_n]_2 L_{t^*}$. Finally, set $(x,y) \in \overline{E}^2$. By definition, we have

$$\begin{aligned} \left| v_n(x) - v_n(y) \right| \leq \left| Kv_{n+1}(x) - Kv_{n+1}(y) \right| \\ \vee \Big| \sup_{u \leq t^*(x)} J(v_{n+1}, g)(x,u) \\ - \sup_{u \leq t^*(y)} J(v_{n+1}, g)(y,u) \Big|, \end{aligned}$$

and since $E_4 = E_5 + C_{t^*}[\lambda]_1$ we conclude by using lemmas 7.7 and 7.5. □

REMARK 7.1.– *Note that L_{v_n} is much sharper than $[v_n]_1$. If in addition to our assumptions, the drift Φ is Lipschitz-continuous in both variables, then with obvious notation, we have $[v_n]_i \leq L_{v_n}[\Phi]_i$ for $i \in \{1, 2, *\}$, which should yield better constants, see the example in section 7.7.*

7.5. Error estimation for the value function

We are now able to state our main result, namely the convergence of our approximation scheme with an upper bound for the rate of convergence.

THEOREM 7.1.– Set $n \in \{0, \dots, N-1\}$, and suppose that $\Delta(z)$ satisfies

$$\min_{z \in \Gamma_n^z} \{\Delta(z)\} > C_\lambda^{-1/2} \big(4L_{t^*} \|Z_n - \widehat{Z}_n\|_p + 2\|S_{n+1} - \widehat{S}_{n+1}\|_p\big)^{1/2}, \qquad [7.23]$$

for any $z \in \Gamma_n^z$. Then the discretization error for V_n is no greater than the following:

$$\begin{aligned} \|V_n - \widehat{V}_n\|_p \leq & \left\|V_{n+1} - \widehat{V}_{n+1}\right\|_p + \alpha \|\Delta(\widehat{Z}_n)\|_p + \beta_n \|\widehat{Z}_n - Z_n\|_p \\ & + 2L_{v_{n+1}} \|\widehat{Z}_{n+1} - Z_{n+1}\|_p \\ & + \gamma \big(L_{t^*} \|\widehat{Z}_n - Z_n\|_p + \|S_{n+1} - \widehat{S}_{n+1}\|_p\big)^{1/2}, \end{aligned}$$

where $\beta_n = L_{v_n} + C_g E_4 + [v_{n+1}]_1 E_2 + \big([\lambda]_1 C_{t^*}^2 C_\lambda\big) \vee \big([v_{n+1}]_* [Q]\big)$, $\alpha = 2([g]_2 + 2C_g C_\lambda)$, $\gamma = 4C_g C_\lambda^{1/2}$, and E_2 and E_4 are defined in section 7.4.

Observe that $\|V_N - \widehat{V}_N\|_p \leq [g] \left\|\widehat{Z}_N - Z_N\right\|_p$ by recalling that $V_N = g(Z_N)$ and $\widehat{V}_N = g(\widehat{Z}_N)$. Moreover, the quantization error $\|\Theta_n - \widehat{\Theta}_n\|_p$ goes to zero as the number of points in the grids goes to infinity, see theorem 3.1. Hence $|V_0 - \widehat{V}_0|$ can be made arbitrarily small by an adequate choice of the discretization parameters. Note that the square root in the last error term is the price to pay for integrating non-continuous functions, see the definition of operator J with the indicator functions, and the introduction of section 7.5.2.

Note that both the left-hand side and right-hand side of condition [7.23] depend on the quantization grids, therefore it may not be very easy to select parameters satisfying this condition. One general case when such a condition holds is, for instance, when assumption 1.8 holds, and the support of the distribution of Z_n is uniformly far from the boundary of the state space. We can thus first select a value of $\Delta(z)$ independent from z and then adjust the number of points in the quantization grids for the error terms on the right-hand side of condition [7.23] to be small enough for the condition to hold true.

To prove theorem 7.1, we split the left-hand-side difference into four terms:

$$\|V_n - \widehat{V}_n\|_p \leq \sum_{i=1}^{4} \Xi_i,$$

where

$$\Xi_1 = \|v_n(Z_n) - v_n(\widehat{Z}_n)\|_p,$$

$$\Xi_2 = \|L(v_{n+1}, g)(\widehat{Z}_n) - L^d(v_{n+1}, g)(\widehat{Z}_n)\|_p,$$

$$\Xi_3 = \|L^d(v_{n+1}, g)(\widehat{Z}_n) - \widehat{L}^d_{n+1}(v_{n+1}, g)(\widehat{Z}_n)\|_p,$$

$$\Xi_4 = \|\widehat{L}^d_{n+1}(v_{n+1}, g)(\widehat{Z}_n) - \widehat{L}^d_{n+1}(\widehat{v}_{n+1}, g)(\widehat{Z}_n)\|_p.$$

The first term is easy enough to handle due to proposition 7.1 in section 7.4.2.

LEMMA 7.8.– An upper bound for Ξ_1 is

$$\|v_n(Z_n) - v_n(\widehat{Z}_n)\|_p \leq L_{v_n}\|Z_n - \widehat{Z}_n\|_p.$$

We are going to study the other terms one by one in the following sections.

7.5.1. *Second term*

In this part, we study the error induced by the replacement of the supremum over all non-negative t smaller than or equal to $t^*(z)$ by the maximum over the finite grid $G(z)$ in the definition of operator L.

LEMMA 7.9.– Let $w \in \mathbf{L}_\Phi(E)$. Then for all $z \in E$,

$$\Big|\sup_{t\leq t^*(z)} J(w,g)(z,t) - \max_{s\in G(z)} J(w,g)(z,s)\Big| \leq 2(C_w C_\lambda + [g]_2 + C_g C_\lambda)\Delta(z).$$

PROOF.– From lemma 7.3, it follows that for any $z \in E$, $J(w,g)(z,\cdot)$ is continuous on $[0, t^*(z)]$ and so, there exists $\bar{t} \in [0, t^*(z)]$ such that $\sup_{t\leq t^*(z)} J(w,g)(z,t) = J(w,g)(z,\bar{t})$, and there exists $0 \leq i \leq n(z)$ such that $\bar{t} \in [t_i, t_{i+1}]$ (with $t_{n(z)+1} = t^*(z)$). Consequently, lemma 7.3 yields

$$\begin{aligned}
0 \leq & \sup_{t\leq t^*(z)} J(w,g)(z,t) - \max_{s\in G(z)} J(w,g)(z,s) \\
\leq & J(w,g)(z,\bar{t}) - J(w,g)(z,t_i) \\
\leq & (C_w C_\lambda + [g]_2 + C_g C_\lambda)|\bar{t} - t_i| \\
\leq & (C_w C_\lambda + [g]_2 + C_g C_\lambda)|t_{i+1} - t_i|.
\end{aligned}$$

However, $|t_{i+1} - t_i| \leq \Delta(z)$, for any $i < n(z)$ and $|t_{n(z)+1} - t_{n(z)}| \leq 2\Delta(z)$ implying the result. □

Turning back to the second error term, we get the following bound.

LEMMA 7.10.– An upper bound for Ξ_2 is

$$\|L(v_{n+1}, g)(\widehat{Z}_n) - L^d(v_{n+1}, g)(\widehat{Z}_n)\|_p \leq 2\big([g]_2 + 2C_g C_\lambda\big)\|\Delta(\widehat{Z}_n)\|_p.$$

PROOF.– From the definition of L and L^d, we readily obtain

$$\|L(v_{n+1}, g)(\widehat{Z}_n) - L^d(v_{n+1}, g)(\widehat{Z}_n)\|_p$$
$$\leq \Big\| \sup_{t \leq t^*(\widehat{Z}_n)} J(v_{n+1}, g)(\widehat{Z}_n, t) - \max_{s \in G(\widehat{Z}_n)} J(v_{n+1}, g)(\widehat{Z}_n, s)\Big\|_p.$$

Now in view of the previous lemma, we have

$$\|L(v_{n+1}, g)(\widehat{Z}_n) - L^d(v_{n+1}, g)(\widehat{Z}_n)\|_p$$
$$\leq 2\big(C_{v_{n+1}} C_\lambda + [g]_2 + C_g C_\lambda\big)\|\Delta(\widehat{Z}_n)\|_p.$$

Finally, recalling that $C_{v_{n+1}} = C_g$, this completes the proof. □

7.5.2. *Third term*

This is the crucial part of our derivation, where we need to compare conditional expectations relative to the real Markov chain (Z_n, S_n) and its quantized approximation $(\widehat{Z}_n, \widehat{S}_n)$. The main difficulty arises from the fact that some functions inside the expectations are indicator functions and in particular they are not Lipschitz-continuous. We manage to overcome this difficulty by proving that the event on which the discontinuity actually occurs is of small enough probability, this is the aim of the following two lemmas.

LEMMA 7.11.– For all $n \in \{0, \ldots, N\}$ and $\eta > 0$,

$$\Big\|\mathbb{1}_{\{t^*(Z_n) < t^*(\widehat{Z}_n) - \eta\}}\Big\|_p \leq \frac{L_{t^*}\|Z_n - \widehat{Z}_n\|_p}{\eta}.$$

PROOF.– We use Chebychev's inequality. We clearly have

$$\mathbb{E}_{x_0}\Big[\Big|\mathbb{1}_{\{t^*(Z_n) < t^*(\widehat{Z}_n) - \eta\}}\Big|^p\Big] = \mathbb{P}\Big(t^*(Z_n) < t^*(\widehat{Z}_n) - \eta\Big)$$
$$\leq \mathbb{P}\Big(\big|t^*(Z_n) - t^*(\widehat{Z}_n)\big| > \eta\Big) \leq \frac{L^p_{t^*}\|Z_n - \widehat{Z}_n\|^p_p}{\eta^p},$$

showing the result. □

LEMMA 7.12.– For all $n \in \{0, \dots, N-1\}$ and $0 < \eta < \min_{z \in \Gamma_n^Z} \{\Delta(z)\}$,

$$\Big\| \max_{s \in G(\widehat{Z}_n)} \mathbb{E}_{x_0}\big[|\mathbb{1}_{\{S_{n+1}<s\}} - \mathbb{1}_{\{\widehat{S}_{n+1}<s\}}| \big| \widehat{Z}_n\big] \Big\|_p$$
$$\leq \frac{2}{\eta}\|S_{n+1} - \widehat{S}_{n+1}\|_p + C_\lambda \eta + \frac{2L_{t^*}\|Z_n - \widehat{Z}_n\|_p}{\eta}.$$

PROOF.– Set $0 < \eta < \min_{z \in \Gamma_n^Z} \{\Delta(z)\}$. Remark that the difference of indicator functions is non-zero if and only if S_{n+1} and $\widehat{S}_{n+1}$ are on either side of s. Hence, we have

$$\big|\mathbb{1}_{\{S_{n+1}<s\}} - \mathbb{1}_{\{\widehat{S}_{n+1}<s\}}\big| \leq \mathbb{1}_{\{|S_{n+1}-\widehat{S}_{n+1}|>\frac{\eta}{2}\}} + \mathbb{1}_{\{|S_{n+1}-s|\leq\frac{\eta}{2}\}}.$$

This yields

$$\| \max_{s \in G(\widehat{Z}_n)} \mathbb{E}_{x_0}\big[|\mathbb{1}_{\{S_{n+1}<s\}} - \mathbb{1}_{\{\widehat{S}_{n+1}<s\}}| \big| \widehat{Z}_n\big] \|_p$$
$$\leq \big\|\mathbb{1}_{\{|S_{n+1}-\widehat{S}_{n+1}|>\frac{\eta}{2}\}}\big\|_p + \| \max_{s \in G(\widehat{Z}_n)} \mathbb{E}_{x_0}\big[\mathbb{1}_{\{s-\frac{\eta}{2}\leq S_{n+1}\leq s+\frac{\eta}{2}\}}\big|\widehat{Z}_n\big] \|_p. \quad [7.24]$$

On the one hand, Chebychev's inequality yields

$$\big\|\mathbb{1}_{\{|S_{n+1}-\widehat{S}_{n+1}|>\frac{\eta}{2}\}}\big\|_p^p = \mathbb{P}(|S_{n+1}-\widehat{S}_{n+1}| > \frac{\eta}{2}) \leq \frac{2^p\big\|S_{n+1} - \widehat{S}_{n+1}\big\|_p^p}{\eta^p}. \quad [7.25]$$

On the other hand, as $s \in G(\widehat{Z}_n)$ and by definition of η, we have $s + \eta < t^*(\widehat{Z}_n)$, see the comment after definition 7.2. Thus, we have

$$\mathbb{E}_{x_0}\big[\mathbb{1}_{\{s-\frac{\eta}{2}\leq S_{n+1}\leq s+\frac{\eta}{2}\}}\big|\widehat{Z}_n\big] = \mathbb{E}_{x_0}\Big[\mathbb{E}_{x_0}\big[\mathbb{1}_{\{s-\frac{\eta}{2}\leq S_{n+1}\leq s+\frac{\eta}{2}\}}|Z_n\big]\Big|\widehat{Z}_n\Big]$$
$$\leq \mathbb{E}_{x_0}\Big[\int_{s-\frac{\eta}{2}}^{s+\frac{\eta}{2}} \lambda(\Phi(Z_n,u))du\Big|\widehat{Z}_n\Big] + \mathbb{E}_{x_0}\Big[\mathbb{1}_{\{t^*(Z_n)\leq s+\frac{\eta}{2}\}}\Big|\widehat{Z}_n\Big]$$
$$\leq \eta C_\lambda + \mathbb{E}_{x_0}\Big[\mathbb{1}_{\{t^*(Z_n)<t^*(\widehat{Z}_n)-\frac{\eta}{2}\}}\Big|\widehat{Z}_n\Big]. \quad [7.26]$$

Combining equations [7.24]–[7.26] and lemma 7.11 the result follows. □

Now we turn to the consequences of replacing the Markov chain (Z_n, S_n) by its quantized approximation $(\widehat{Z}_n, \widehat{S}_n)$ in the conditional expectations.

LEMMA 7.13.– Let $w \in \mathbf{L}_\Phi(E)$, then we have

$$\begin{aligned}\Big|\mathbb{E}_{x_0}\big[w(Z_{n+1})\big|Z_n = \widehat{Z}_n\big] - \mathbb{E}_{x_0}\big[w(\widehat{Z}_{n+1})\big|\widehat{Z}_n\big]\Big|& \\ \leq (C_w E_4 + [w]_1 E_2 + [w]_*[Q])\mathbb{E}_{x_0}\big[|Z_n - \widehat{Z}_n| \mid \widehat{Z}_n\big]& \\ + L_w \mathbb{E}_{x_0}\big[|Z_{n+1} - \widehat{Z}_{n+1}|\big|\widehat{Z}_n\big].&\end{aligned}$$

PROOF.– First, note that

$$\begin{aligned}\mathbb{E}_{x_0}\big[w(Z_{n+1})\big|Z_n = \widehat{Z}_n\big] - \mathbb{E}_{x_0}\big[w(\widehat{Z}_{n+1})\big|\widehat{Z}_n\big]& \\ = \mathbb{E}_{x_0}\big[w(Z_{n+1})\big|Z_n = \widehat{Z}_n\big] - \mathbb{E}_{x_0}\big[w(Z_{n+1})\big|\widehat{Z}_n\big]& \\ + \mathbb{E}_{x_0}\big[w(Z_{n+1})\big|\widehat{Z}_n\big] - \mathbb{E}_{x_0}\big[w(\widehat{Z}_{n+1})\big|\widehat{Z}_n\big].&\end{aligned}$$

On the one hand, section 1.4.1 yields

$$\Big|\mathbb{E}_{x_0}\big[w(Z_{n+1})\big|\widehat{Z}_n\big] - \mathbb{E}_{x_0}\big[w(\widehat{Z}_{n+1})\big|\widehat{Z}_n\big]\Big| \leq L_w \mathbb{E}_{x_0}\big[|Z_{n+1} - \widehat{Z}_{n+1}|\big|\widehat{Z}_n\big].$$

On the other hand, recall that we have $\big(\widehat{Z}_n, \widehat{S}_n\big) = p_n\big(Z_n, S_n\big)$. Hence, we have the following property: $\sigma\{\widehat{Z}_n\} \subset \sigma\{Z_n, S_n\}$. By using the special structure of the PDMP (X_t), we have $\sigma\{Z_n, S_n\} \subset \mathcal{F}_{T_n}$. Now, by using the Markov property of the process (X_t), it follows

$$\begin{aligned}\mathbb{E}_{x_0}\big[w(Z_{n+1})\big|\widehat{Z}_n\big] &= \mathbb{E}_{x_0}\Big[\mathbb{E}_{x_0}\big[w(Z_{n+1})\big|\mathcal{F}_{T_n}\big]\Big|\widehat{Z}_n\Big] \\ &= \mathbb{E}_{x_0}\Big[\mathbb{E}_{x_0}\big[w(Z_{n+1})\big|Z_n\big]\Big|\widehat{Z}_n\Big].\end{aligned}$$

Equation [7.5] thus yields

$$\begin{aligned}&\mathbb{E}_{x_0}\big[w(Z_{n+1})\big|Z_n = \widehat{Z}_n\big] - \mathbb{E}_{x_0}\big[w(Z_{n+1})\big|\widehat{Z}_n\big] \\ &= \mathbb{E}_{x_0}\Big[\mathbb{E}_{x_0}\big[w(Z_{n+1})\big|Z_n = \widehat{Z}_n\big] - \mathbb{E}_{x_0}\big[w(Z_{n+1})\big|Z_n\big]\Big|\widehat{Z}_n\Big] \\ &= \mathbb{E}_{x_0}\Big[Kw(\widehat{Z}_n) - Kw(Z_n)\Big|\widehat{Z}_n\Big].\end{aligned}$$

Now we use lemma 7.5 to conclude. □

Now we combine the preceding lemmas to derive the third error term.

LEMMA 7.14.– For all $0 < \eta < \min_{z \in \Gamma_n^Z} \{\Delta(z)\}$, an upper bound for Ξ_3 is

$$
\begin{aligned}
&\left\| L^d(v_{n+1}, g)(\widehat{Z}_n) - \widehat{L}^d_{n+1}(v_{n+1}, g)(\widehat{Z}_n) \right\|_p \\
&\leq L_{v_{n+1}} \| Z_{n+1} - \widehat{Z}_{n+1} \|_p \\
&+ \Big[C_g E_4 + [v_{n+1}]_1 E_2 + 8 C_g \frac{L_{t^*}}{\eta} + \big([\lambda]_1 C_{t^*}^2 C_\lambda \big) \\
&\vee \big([v_{n+1}]_* [Q] \big) \Big] \| Z_n - \widehat{Z}_n \|_p + \frac{4}{\eta} C_g \| S_{n+1} - \widehat{S}_{n+1} \|_p + 2 C_g C_\lambda \eta.
\end{aligned}
$$

PROOF.– From the definition of L^d and $\widehat{L}^d_{n+1}$, we readily obtain

$$
\begin{aligned}
&\left| L^d(v_{n+1}, g)(\widehat{Z}_n) - \widehat{L}^d_{n+1}(v_{n+1}, g)(\widehat{Z}_n) \right| \\
&\quad \leq \left| \mathbb{E}_{x_0} \big[v_{n+1}(Z_{n+1}) \big| Z_n = \widehat{Z}_n \big] - \mathbb{E}_{x_0} \big[v_{n+1}(\widehat{Z}_{n+1}) \big| \widehat{Z}_n \big] \right| \\
&\quad \vee \max_{s \in G(\widehat{Z}_n)} \Big[\Big| \mathbb{E}_{x_0} \big[\Psi(Z_{n+1}, S_{n+1}, s) \big| Z_n = \widehat{Z}_n \big] \\
&\qquad\qquad - \mathbb{E}_{x_0} \big[\Psi(\widehat{Z}_{n+1}, \widehat{S}_{n+1}, s) \big| \widehat{Z}_n \big] \Big| \\
&\qquad + \big| g(\Phi(\widehat{Z}_n, s)) \big| \Big| \mathbb{E}_{x_0} \big[\mathbb{1}_{\{S_{n+1} \geq s\}} \big| Z_n = \widehat{Z}_n \big] - \mathbb{E}_{x_0} \big[\mathbb{1}_{\{\widehat{S}_{n+1} \geq s\}} \big| \widehat{Z}_n \big] \Big| \Big],
\end{aligned}
\qquad [7.27]
$$

where $\Psi(y, t, s) = v_{n+1}(y) \mathbb{1}_{\{t < s\}}$. On the one hand, combining lemma 7.13 and the fact that v_{n+1} is in $\mathbf{L}_\Phi(E)$ (see proposition 7.1), we obtain

$$
\begin{aligned}
&\left| \mathbb{E}_{x_0} \big[v_{n+1}(Z_{n+1}) \big| Z_n = \widehat{Z}_n \big] - \mathbb{E}_{x_0} \big[v_{n+1}(\widehat{Z}_{n+1}) \big| \widehat{Z}_n \big] \right| \\
&\qquad \leq L_{v_{n+1}} \mathbb{E}_{x_0} \big[|Z_{n+1} - \widehat{Z}_{n+1} \, \big| \widehat{Z}_n \big] \\
&\qquad\quad + \big(C_g E_4 + [v_{n+1}]_1 E_2 + [v_{n+1}]_* [Q] \big) \mathbb{E}_{x_0} \big[|Z_n - \widehat{Z}_n| \, \big| \, \widehat{Z}_n \big].
\end{aligned}
\qquad [7.28]
$$

On the other hand, similar arguments as in the proof of lemma 7.13 yield

$$
\begin{aligned}
\mathbb{E}_{x_0} \big[\Psi(Z_{n+1}, S_{n+1}, s) \big| Z_n = \widehat{Z}_n \big] - \mathbb{E}_{x_0} \big[\Psi(\widehat{Z}_{n+1}, \widehat{S}_{n+1}, s) \big| \widehat{Z}_n \big] \\
= \Upsilon_1 + \Upsilon_2
\end{aligned}
\qquad [7.29]
$$

where

$$\Upsilon_1 = \mathbb{E}_{x_0}\Big[\mathbb{E}_{x_0}\big[\Psi(Z_{n+1}, S_{n+1}, s)\big|Z_n = \widehat{Z}_n\big]$$
$$- \mathbb{E}_{x_0}\big[\Psi(Z_{n+1}, S_{n+1}, s)\big|Z_n\big]\Big|\widehat{Z}_n\Big]$$

$$\Upsilon_1 = \mathbb{E}_{x_0}\big[\Psi(Z_{n+1}, S_{n+1}, s)\big|\widehat{Z}_n\big] - \mathbb{E}_{x_0}\big[\Psi(\widehat{Z}_{n+1}, \widehat{S}_{n+1}, s)\big|\widehat{Z}_n\big]$$

Regarding the first difference of the right hand side of equation [7.29], labeled as Υ_1. A difficulty is that we know by construction that $s < t^*(\widehat{Z}_n)$ and so,

$$\mathbb{E}_{x_0}\big[\Psi(Z_{n+1}, S_{n+1}, s)\big|Z_n = \widehat{Z}_n\big] = Iv_{n+1}(\widehat{Z}_n, s).$$

However, we know nothing about the relative positions of s and $t^*(Z_n)$. On the event where $s \leq t^*(Z_n)$ as well, we recognize the operator I inside the expectation, that is

$$\mathbb{E}_{x_0}\big[\Psi(Z_{n+1}, S_{n+1}, s)\big|Z_n\big]\mathbb{1}_{\{s \leq t^*(Z_n)\}} = Iv_{n+1}(Z_n, s)\mathbb{1}_{\{s \leq t^*(Z_n)\}}.$$

On the opposite event $s > t^*(Z_n)$, we crudely bound Ψ by C_g. Hence, we obtain

$$|\Upsilon_1| \leq \mathbb{E}_{x_0}\Big[\big|Iv_{n+1}(\widehat{Z}_n, s) - Iv_{n+1}(Z_n, s)\big|\mathbb{1}_{\{s \leq t^*(Z_n)\}}\Big|\widehat{Z}_n\Big]$$
$$+ 2C_g\mathbb{E}_{x_0}\big[\mathbb{1}_{\{t^*(Z_n) < s\}}\big|\widehat{Z}_n\big].$$

Now lemma 7.4 gives an upper bound for the first term. As for the indicator function, by definition of $G(\widehat{Z}_n)$ and our choice of η, we have $s < t^*(\widehat{Z}_n) - \eta$. Thus, we have

$$|\Upsilon_1| \leq (C_g E_1 + [v_{n+1}]_1 E_2)\mathbb{E}_{x_0}\big[|\widehat{Z}_n - Z_n|\big|\widehat{Z}_n\big]$$
$$+ 2C_g\mathbb{E}_{x_0}\big[\mathbb{1}_{\{t^*(Z_n) < t^*(\widehat{Z}_n) - \eta\}}\big|\widehat{Z}_n\big]. \qquad [7.30]$$

Let us now turn to the second difference of the right hand side of equation [7.29], labeled as Υ_2. It satisfies

$$|\Upsilon_2| \leq L_{v_{n+1}}\mathbb{E}_{x_0}\big[|\widehat{Z}_{n+1} - Z_{n+1}|\big|\widehat{Z}_n\big]$$
$$+ C_g\mathbb{E}_{x_0}\big[|\mathbb{1}_{\{S_{n+1} < s\}} - \mathbb{1}_{\{\widehat{S}_{n+1} < s\}}|\big|\widehat{Z}_n\big]. \qquad [7.31]$$

Combining equation [7.29]–[7.31], it follows that

$$\Big|\mathbb{E}_{x0}\big[\Psi(Z_{n+1},S_{n+1},s)\big|Z_n=\widehat{Z}_n\big]-\mathbb{E}_{x0}\big[\Psi(\widehat{Z}_{n+1},\widehat{S}_{n+1},s)\big|\widehat{Z}_n\big]\Big|$$

$$\leq \big(C_gE_1+[v_{n+1}]_1E_2\big)\mathbb{E}_{x0}\big[|\widehat{Z}_n-Z_n|\big|\widehat{Z}_n\big]$$
$$+2C_g\mathbb{E}_{x0}\big[\mathbb{1}_{\{t^*(Z_n)<t^*(\widehat{Z}_n)-\eta\}}\big|\widehat{Z}_n\big]$$
$$+L_{v_{n+1}}\mathbb{E}_{x0}\big[|\widehat{Z}_{n+1}-Z_{n+1}|\big|\widehat{Z}_n\big]$$
$$+C_g\mathbb{E}_{x0}\big[|\mathbb{1}_{\{S_{n+1}<s\}}-\mathbb{1}_{\{\widehat{S}_{n+1}<s\}}|\big|\widehat{Z}_n\big]. \qquad [7.32]$$

Now, observe that

$$\Big|\mathbb{E}_{x0}\big[\mathbb{1}_{\{S_{n+1}\geq s\}}\big|Z_n=\widehat{Z}_n\big]-\mathbb{E}_{x0}\big[\mathbb{1}_{\{\widehat{S}_{n+1}\geq s\}}\big|\widehat{Z}_n\big]\Big|$$
$$=\Big|\mathbb{E}_{x0}\big[\mathbb{1}_{\{S_{n+1}<s\}}\big|Z_n=\widehat{Z}_n\big]-\mathbb{E}_{x0}\big[\mathbb{1}_{\{\widehat{S}_{n+1}<s\}}\big|\widehat{Z}_n\big]\Big|$$

and a similar calculation as before (choose $\Psi(y,t,s)=\mathbb{1}_{\{t<s\}}$) shows that

$$\Big|\mathbb{E}_{x0}\big[\mathbb{1}_{\{S_{n+1}\geq s\}}\big|Z_n=\widehat{Z}_n\big]-\mathbb{E}_{x0}\big[\mathbb{1}_{\{\widehat{S}_{n+1}\geq s\}}\big|\widehat{Z}_n\big]\Big|$$
$$\leq E_1\mathbb{E}_{x0}\big[|\widehat{Z}_n-Z_n|\big|\widehat{Z}_n\big]+2\mathbb{E}_{x0}\big[\mathbb{1}_{\{t^*(Z_n)<t^*(\widehat{Z}_n)-\eta\}}\big|\widehat{Z}_n\big]$$
$$+\mathbb{E}_{x0}\big[|\mathbb{1}_{\{S_{n+1}<s\}}-\mathbb{1}_{\{\widehat{S}_{n+1}<s\}}|\big|\widehat{Z}_n\big]. \qquad [7.33]$$

Now, combining [7.27], [7.28], [7.32] and [7.33]

$$\big|L^d(v_{n+1},g)(\widehat{Z}_n)-\widehat{L}^d_{n+1}(v_{n+1},g)(\widehat{Z}_n)\big|\leq\Big[L_{v_{n+1}}\mathbb{E}_{x0}\big[|Z_{n+1}-\widehat{Z}_{n+1}|\big|\widehat{Z}_n\big]$$
$$+\big(C_gE_4+[v_{n+1}]_1E_2+[v_{n+1}]_*[Q]\big)\mathbb{E}_{x0}\big[|Z_n-\widehat{Z}_n|\;\big|\;\widehat{Z}_n\big]\Big]$$
$$\vee\Big[\big(2C_gE_1+[v_{n+1}]_1E_2\big)\mathbb{E}_{x0}\big[|Z_n-\widehat{Z}_n|\big|\widehat{Z}_n\big]$$
$$+L_{v_{n+1}}\mathbb{E}_{x0}\big[|\widehat{Z}_{n+1}-Z_{n+1}|\big|\widehat{Z}_n\big]+4C_g\mathbb{E}_{x0}\big[\mathbb{1}_{\{t^*(Z_n)<t^*(\widehat{Z}_n)-\eta\}}\big|\widehat{Z}_n\big]$$
$$+2C_g\mathbb{E}_{x0}\big[|\mathbb{1}_{\{S_{n+1}<s\}}-\mathbb{1}_{\{\widehat{S}_{n+1}<s\}}|\big|\widehat{Z}_n\big]\Big].$$

Note that $2E_1 = E_4 + [\lambda]_1 C_{t^*}^2 C_\lambda$ and so, we obtain

$$\left|L^d(v_{n+1},g)(\widehat{Z}_n) - \widehat{L}_{n+1}^d(v_{n+1},g)(\widehat{Z}_n)\right| \leq L_{v_{n+1}} \mathbb{E}_{x_0}\left[|Z_{n+1} - \widehat{Z}_{n+1}|\big|\widehat{Z}_n\right]$$

$$+ \Big[C_g E_4 + [v_{n+1}]_1 E_2 + \big([\lambda]_1 C_{t^*}^2 C_\lambda\big) \vee \big([v_{n+1}]_*[Q]\big)\Big] \mathbb{E}_{x_0}\left[|Z_n - \widehat{Z}_n| \mid \widehat{Z}_n\right]$$

$$+ 4C_g \mathbb{E}_{x_0}\left[\mathbb{1}_{\{t^*(Z_n) < t^*(\widehat{Z}_n) - \eta\}}\big|\widehat{Z}_n\right] + 2C_g \mathbb{E}_{x_0}\left[|\mathbb{1}_{\{S_{n+1} < s\}} - \mathbb{1}_{\{\widehat{S}_{n+1} < s\}}|\big|\widehat{Z}_n\right].$$

Finally, by using lemmas 7.11 and 7.12 and taking the L_p norm on both sides, we obtain

$$\left\|L^d(v_{n+1},g)(\widehat{Z}_n) - \widehat{L}_{n+1}^d(v_{n+1},g)(\widehat{Z}_n)\right\|_p \leq L_{v_{n+1}} \|Z_{n+1} - \widehat{Z}_{n+1}\|_p$$

$$+ \Big[C_g E_4 + [v_{n+1}]_1 E_2 + \big([\lambda]_1 C_{t^*}^2 C_\lambda\big) \vee \big([v_{n+1}]_*[Q]\big)\Big] \|Z_n - \widehat{Z}_n\|_p$$

$$+ 4C_g \frac{L_{t^*}\|Z_n - \widehat{Z}_n\|_p}{\eta} + 2C_g \left[\frac{2}{\eta}\|S_{n+1} - \widehat{S}_{n+1}\|_p\right.$$

$$\left. + C_\lambda \eta + \frac{2L_{t^*}\|Z_n - \widehat{Z}_n\|_p}{\eta}\right].$$

showing the result. □

7.5.3. *Fourth term*

The last error term is a mere comparison of two finite sums.

LEMMA 7.15.– An upper bound for Ξ_4 is

$$\|\widehat{L}_{n+1}^d(v_{n+1},g)(\widehat{Z}_n) - \widehat{L}_{n+1}^d(\widehat{v}_{n+1},g)(\widehat{Z}_n)\|_p$$
$$\leq L_{v_{n+1}} \left\|\widehat{Z}_{n+1} - Z_{n+1}\right\|_p + \left\|V_{n+1} - \widehat{V}_{n+1}\right\|_p.$$

PROOF.– By definition of operator $\widehat{L}_n^d$, we have

$$\|\widehat{L}_{n+1}^d(v_{n+1},g)(\widehat{Z}_n) - \widehat{L}_{n+1}^d(\widehat{v}_{n+1},g)(\widehat{Z}_n)\|_p$$

$$= \Bigg\| \max_{s \in G(\widehat{Z}_n)} \left\{\mathbb{E}_{x_0}\left[v_{n+1}(\widehat{Z}_{n+1})\mathbb{1}_{\{\widehat{S}_{n+1} < s\}} + g(\Phi(\widehat{Z}_n,s))\mathbb{1}_{\{\widehat{S}_{n+1} \geq s\}}\big|\widehat{Z}_n\right]\right\}$$

$$\vee \mathbb{E}_{x_0}\left[v_{n+1}(\widehat{Z}_{n+1})\big|\widehat{Z}_n\right]$$

$$- \max_{s \in G(\widehat{Z}_n)} \left\{\mathbb{E}_{x_0}\left[\widehat{v}_{n+1}(\widehat{Z}_{n+1})\mathbb{1}_{\{\widehat{S}_{n+1} < s\}} + g(\Phi(\widehat{Z}_n,s))\mathbb{1}_{\{\widehat{S}_{n+1} \geq s\}}\big|\widehat{Z}_n\right]\right\}$$

$$\vee \mathbb{E}_{x_0}\left[\widehat{v}_{n+1}(\widehat{Z}_{n+1})\big|\widehat{Z}_n\right]\Bigg\|_p$$

$$\leq \left\|\mathbb{E}_{x_0}[v_{n+1}(\widehat{Z}_{n+1}) - \widehat{v}_{n+1}(\widehat{Z}_{n+1})|\widehat{Z}_n]\right\|_p$$

$$\leq \left\|v_{n+1}(\widehat{Z}_{n+1}) - v_{n+1}(Z_{n+1})\right\|_p + \left\|v_{n+1}(Z_{n+1}) - \widehat{v}_{n+1}(\widehat{Z}_{n+1})\right\|_p.$$

We conclude using the fact that $v_{n+1} \in \mathbf{L}_\Phi(E)$ (see proposition 7.1) and the definitions of V_{n+1} and $\widehat{V}_{n+1}$. □

7.5.4. *Proof of theorem 7.1*

We can finally turn to the proof of theorem 7.1. lemmas 7.8, 7.10, 7.14 and 7.15 from the preceding sections directly yield, for all $0 < \eta < \min_{z \in \Gamma_n^z}\{\Delta(z)\}$,

$$\begin{aligned}\|V_n - \widehat{V}_n\|_p \leq\ & L_{v_n}\|\widehat{Z}_n - Z_n\|_p + 2([g]_2 + 2C_gC_\lambda)\|\Delta(\widehat{Z}_n)\|_p \\ & + L_{v_{n+1}}\|Z_{n+1} - \widehat{Z}_{n+1}\|_p + \Big[C_gE_4 + [v_{n+1}]_1E_2 + 8C_g\frac{L_{t^*}}{\eta} \\ & + ([\lambda]_1C_{t^*}^2C_\lambda) \vee ([v_{n+1}]_*[Q])\Big]\|Z_n - \widehat{Z}_n\|_p \\ & + \frac{4}{\eta}C_g\|S_{n+1} - \widehat{S}_{n+1}\|_p + 2C_gC_\lambda\eta \\ & + L_{v_{n+1}}\left\|\widehat{Z}_{n+1} - Z_{n+1}\right\|_p + \left\|V_{n+1} - \widehat{V}_{n+1}\right\|_p.\end{aligned}$$

The optimal choice for η clearly satisfies

$$C_\lambda\eta = \frac{1}{\eta}(4L_{t^*}\|\widehat{Z}_n - Z_n\|_p + 2\|S_{n+1} - \widehat{S}_{n+1}\|_p),$$

providing it also satisfies the condition $0 < \eta < \min_{z \in \Gamma_n^z}\{\Delta(z)\}$. Hence, rearranging the terms above, we get the expected result:

$$\begin{aligned}\|V_n - \widehat{V}_n\|_p \leq\ & \left\|V_{n+1} - \widehat{V}_{n+1}\right\|_p + 2([g]_2 + 2C_gC_\lambda)\|\Delta(\widehat{Z}_n)\|_p \\ & + \Big\{L_{v_n} + C_gE_4 + [v_{n+1}]_1E_2 + ([\lambda]_1C_{t^*}^2C_\lambda) \vee ([v_{n+1}]_*[Q])\Big\}\|\widehat{Z}_n - Z_n\|_p \\ & + 4C_g\Big(C_\lambda\{4L_{t^*}\|\widehat{Z}_n - Z_n\|_p + 2\|S_{n+1} - \widehat{S}_{n+1}\|_p\}\Big)^{1/2} \\ & + 2L_{v_{n+1}}\|Z_{n+1} - \widehat{Z}_{n+1}\|_p.\end{aligned}$$

7.6. Numerical construction of an ϵ-optimal stopping time

In [GUG 86, Theorem 1], U.S. Gugerli defined an ϵ-optimal stopping time for the original problem. Roughly speaking, this stopping time depends on the embedded Markov chain (Θ_n), and on the optimal value function. Therefore, a natural candidate for an ϵ-optimal stopping time should be obtained by replacing the Markov chain (Θ_n) and the optimal value function by their *quantized approximations*. However, this leads to intractable comparisons between some quantities involving (Θ_n) and its quantized approximation. It is then far from obvious to show that this method would provide a computable ϵ-optimal stopping rule. Nonetheless, by modifying the approach of U.S. Gugerli [GUG 86], we are able to propose a numerical construction of an ϵ-optimal stopping time of the original stopping problem.

Here is how we proceed. First, recall that p_n be the closest-neighbor projection from $E \times \mathbb{R}^+$ onto Γ_n^{Θ}, and for all $(z, s) \in E \times \mathbb{R}^+$ define $(\widehat{z}_n, \widehat{s}_n) = p_n(z, s)$. Note that $\widehat{z}_n$ and $\widehat{s}_n$ depend on both z and s. Now, for $n \in \{1, \ldots, N\}$, define

$$
\begin{aligned}
&s_n^*(z, s) \\
&= \min\left\{t \in G(\widehat{z}_{n-1}) \,\middle|\, \widehat{J}_n(\widehat{v}_n, g)(\widehat{z}_{n-1}, t) = \max_{u \in G(\widehat{z}_{n-1})} \widehat{J}_n(\widehat{v}_n, g)(\widehat{z}_{n-1}, u)\right\},
\end{aligned}
$$

and

$$
\begin{aligned}
&r_{n,\beta}(z, s) \\
&= \begin{cases} t^*(z) & \text{if } \widehat{K}_n \widehat{v}_n(\widehat{z}_{n-1}) > \displaystyle\max_{u \in G(\widehat{z}_{n-1})} \widehat{J}_n(\widehat{v}_n, g)(\widehat{z}_{n-1}, u), \\ s_n^*(z, s) \mathbb{1}_{\{s_n^*(z,s) < t^*(z)\}} + (t^*(z) - \beta) \mathbb{1}_{\{s^*(z,s) \geq t^*(z)\}} & \text{otherwise.} \end{cases}
\end{aligned}
$$

Note the use of both the real jump time horizon $t^*(z)$ and the quantized approximations of K, J and (z, s). Set for $n \in \{0, \cdots, N-1\}$

$$
\tau_{N-n} = \begin{cases} T_n + r_{n+1,\beta}(Z_n, S_n) & \text{if } r_{n+1,\beta}(Z_n, S_n) < S_{n+1}, \\ T_{n+1} + r_{n+2,\beta}(Z_{n+1}, S_{n+1}) & \text{if } r_{n+2,\beta}(Z_{n+1}, S_{n+1}) < S_{n+2}, \\ \vdots & \\ T_{N-1} + r_{N,\beta}(Z_{N-1}, S_{N-1}) & \text{if } r_{N,\beta}(Z_{N-1}, S_{N-1}) < S_N, \\ T_N & \text{otherwise.} \end{cases}
$$

and so we have clearly

$$
\begin{aligned}
\tau_{N-n} = &\left(T_n + r_{n+1,\beta}(Z_n, S_n)\right) \mathbb{1}_{\{r_{n+1,\beta}(Z_n,S_n) < S_{n+1}\}} \\
&+ \tau_{N-n-1} \mathbb{1}_{\{r_{n+1,\beta}(Z_n,S_n) \geq S_{n+1}\}}.
\end{aligned}
$$

The stopping rule τ_N is our candidate to provide a computable and ϵ-optimal stopping time.

This procedure is especially appealing because it requires no more calculation: we have already computed the values of $\widehat{K}_n$ and $\widehat{J}_n$ on the grids. We just have to store the point where the maximum of $\widehat{J}_n$ is reached. To achieve this aim, algorithm 7.1 is modified as detailed in algorithm 7.2.

Algorithm 7.2: Approximation of the value function $v(x_0)$

---

input : Quantization grids $(\Gamma_n^{\Theta})_{0\leq n\leq N}$, Transition matrices $(\widehat{P}_n^{\Theta})_{1\leq n\leq N}$, cost function g, time discretization step Δ
output: $\widehat{v}_n, s_n^*, i_n, 1 \leq n \leq N$

begin
  for $z \in \Gamma_N^Z$ **do**
    $\widehat{v}_N(z) \leftarrow g(z)$
  end
  for $n \leftarrow N-1$ **to** 0 **do**
    for $z \in \Gamma_n^Z$ **do**
      $\widehat{K}\widehat{v}_{n+1}(z) \leftarrow \sum_{(z',s')\in\Gamma_{n+1}^{\Theta}} \widehat{P}_{n+1}(z,(z',s'))v_{n+1}(z')$
      for $u \leftarrow 0$ **to** $t^*(z) - \Delta$ **by step**: Δ
      do
        $\widehat{J}\widehat{v}_{n+1}(z,u) \leftarrow \sum_{(z',s')\in\Gamma_{n+1}^{\Theta}} \widehat{P}_{n+1}(z,(z',s'))\big(v_{n+1}(z')\mathbb{1}_{\{s'<u\}}$
        $+g(\Phi(z,u))\mathbb{1}_{\{s'\geq u\}}()$
      end
      $\widehat{v}_n(z) \leftarrow \max_u\{\widehat{J}\widehat{v}_{n+1}(z,u)\} \vee \widehat{K}\widehat{v}_{n+1}(z)$
      $s_{n+1}^*(z) \leftarrow \min\big\{u^*; \widehat{J}\widehat{v}_{n+1}(z,u^*) = \max_u\{\widehat{J}\widehat{v}_{n+1}(z,u)\}\big\}$
      if $\widehat{K}\widehat{v}_{n+1}(z) > \max_u\{\widehat{J}\widehat{v}_{n+1}(z,u)\}$ **then**
        $i_{n+1}(z) \leftarrow 1$
      else
        $i_{n+1}(z) \leftarrow 0$
      end
    end
  end
  return : $\widehat{v}_0$
end

Algorithm 7.3: Computation of an ϵ-optimal stopping time

input : Quantization grids $(\Gamma_n^{\Theta})_{0\leq n\leq N}$, Outputs of Algorithm 7.2: s_n^*, i_n, $1 \leq n \leq N$, positive real number β, simulator of trajectories of the PDMP, projection operators p_n onto the quantization grids

output: Stopping date τ

```
begin
    n ← 0
    τ ← NaN
    while n < N do
        simulate (Z_n, S_n, S_{n+1})
        (z, s) ← p_n(Z_n, S_n)
        if i_{n+1}(z) = 1 then
            r ← t*(Z_n)
        else
            r ←
            s_{n+1}(z) 1_{s_{n+1}(z) < t*(Z_n)} + (t*(Z_n) − β) 1_{s_{n+1}(z) ≥ t*(Z_n)}
        end
        if r < S_{n+1} then
            τ ← T_n + r
            n ← N
        else
            n ← n + 1
        end
    end
    if n = N and τ = NaN then
        τ ← T_N
    end
    return: τ
end
```

The algorithm to compute the ϵ-optimal stopping time on a given trajectory is now detailed in algorithm 7.3.

Note that algortihm 7.3 can be run online in real time as it does not involve any heavy calculation. It can also be applied to real trajectories instead of simulated trajectories. At each time step, we just measure the values of (Z_n, S_n) and compare the candidate stopping time $T_n + r$ with the next jump time whenever it occurs, or stop at T_N if no jump occurred before that date.

We now turn to the investigation of the properties of our stopping rule. First, let us show that τ_N is a stopping time.

LEMMA 7.16.– For any $n \in \{0, \cdots, N-1\}, \tau_{N-n}$ is an $(\mathcal{F}_t)$-stopping time.

PROOF.– We use the notation of section 1.7. Let $n \in \{0, \cdots, N-1\}$. Set $U_k = r_{k,\beta}(Z_{k+1}, S_{k+1})\mathbb{1}_{\{U_{k-1} \geq S_{k-1}\}}$ for $n+2 \leq k \leq N-1$ and $U_{n+1} = r_{n+1,\beta}(Z_n, S_n)$. We then clearly have

$$\tau_{N-n} = \sum_{k=1}^{n} C_{t^*} \wedge S_k + \sum_{k=n+1}^{N} U_k \wedge S_k$$

which is an $(\mathcal{F}_t)$-stopping time by proposition 1.2. □

Now let us show that stopping at τ_N provides a good approximation of the value function V_0. Namely, for any $z \in E$, $s \in \mathbb{R}^+$ and $n \in \{0, \ldots, N-1\}$ set

$$\begin{aligned}\overline{v}_n(z, s) &= \mathbb{E}_{x_0}\left[g(X_{\tau_{N-n}})\big|Z_n = z, S_n = s\right],\\ \overline{V}_n &= \overline{v}_n(Z_n, S_n).\end{aligned}$$

The comparison between V_0 and $\overline{V}_0$ is provided by the next two theorems. However, we need first the following technical lemmas.

LEMMA 7.17.– The following inequality holds

$$\begin{aligned}&\left\|Kv_{n+1}(Z_n) - \widehat{K}_{n+1}\widehat{v}_{n+1}(\widehat{Z}_n)\right\|_p\\ &\leq 2\big(C_g E_4 + [v_{n+1}]_1 E_2 + [v_{n+1}]_*[Q]\big)\left\|Z_n - \widehat{Z}_n\right\|_p\\ &\quad + 2L_{v_{n+1}}\left\|Z_{n+1} - \widehat{Z}_{n+1}\right\|_p + \left\|V_{n+1} - \widehat{V}_{n+1}\right\|_p.\end{aligned}$$

PROOF.– Observe that

$$\begin{aligned}&\left\|Kv_{n+1}(Z_n) - \widehat{K}_{n+1}\widehat{v}_{n+1}(\widehat{Z}_n)\right\|_p\\ &\leq \left\|\mathbb{E}_{x_0}\left[v_{n+1}(Z_{n+1})\big|Z_n\right] - \mathbb{E}_{x_0}\left[v_{n+1}(Z_{n+1})\big|Z_n = \widehat{Z}_n\right]\right\|_p\\ &\quad + \left\|\mathbb{E}_{x_0}\left[v_{n+1}(Z_{n+1})\big|Z_n = \widehat{Z}_n\right] - \mathbb{E}_{x_0}\left[v_{n+1}(\widehat{Z}_{n+1})\big|\widehat{Z}_n\right]\right\|_p\\ &\quad + \left\|\mathbb{E}_{x_0}\left[v_{n+1}(\widehat{Z}_{n+1})\big|\widehat{Z}_n\right] - \mathbb{E}_{x_0}\left[\widehat{v}_{n+1}(\widehat{Z}_{n+1})\big|\widehat{Z}_n\right]\right\|_p.\end{aligned}$$

Lemma 7.13 gives

$$\begin{aligned}
&\left\|Kv_{n+1}(Z_n) - \widehat{K}_{n+1}\widehat{v}_{n+1}(\widehat{Z}_n)\right\|_p \\
&\quad \leq \left\|Kv_{n+1}(Z_n) - Kv_{n+1}(\widehat{Z}_n)\right\|_p \\
&\quad\quad + \big(C_g E_4 + [v_{n+1}]_1 E_2 + [v_{n+1}]_*[Q]\big)\left\|Z_n - \widehat{Z}_n\right\|_p \\
&\quad\quad + L_{v_{n+1}}\left\|Z_{n+1} - \widehat{Z}_{n+1}\right\|_p \\
&\quad\quad + \left\|v_{n+1}(Z_{n+1}) - v_{n+1}(\widehat{Z}_{n+1})\right\|_p \\
&\quad\quad + \left\|v_{n+1}(Z_{n+1}) - \widehat{v}_{n+1}(\widehat{Z}_{n+1})\right\|_p.
\end{aligned}$$

Consequently, from lemma 7.5, it follows

$$\begin{aligned}
&\left\|Kv_{n+1}(Z_n) - \widehat{K}_{n+1}\widehat{v}_{n+1}(\widehat{Z}_n)\right\|_p \\
&\quad \leq \big(C_g E_4 + [v_{n+1}]_1 E_2 + [v_{n+1}]_*[Q]\big)\left\|Z_n - \widehat{Z}_n\right\|_p \\
&\quad\quad + \big(C_g E_4 + [v_{n+1}]_1 E_2 + [v_{n+1}]_*[Q]\big)\left\|Z_n - \widehat{Z}_n\right\|_p \\
&\quad\quad + L_{v_{n+1}}\left\|Z_{n+1} - \widehat{Z}_{n+1}\right\|_p \\
&\quad\quad + L_{v_{n+1}}\left\|Z_{n+1} - \widehat{Z}_{n+1}\right\|_p + \left\|V_{n+1} - \widehat{V}_{n+1}\right\|_p,
\end{aligned}$$

showing the result. □

LEMMA 7.18.– For any $n \in \{0, \ldots, N-1\}$ and $0 < \eta < \min_{z \in \Gamma_n^Z} \{\Delta(z)\}$,

$$\begin{aligned}
&\left\| \max_{u \in G(\widehat{Z}_n)} \mathbb{1}_{\{u < t^*(Z_n)\}} \left| J(v_{n+1}, g)(Z_n, u) - \widehat{J}_{n+1}(\widehat{v}_{n+1}, g)(\widehat{Z}_n, u) \right| \right\|_p \\
&\quad \leq \big(3C_g E_1 + 2[v_{n+1}]_1 E_2 + E_3\big)\left\|Z_n - \widehat{Z}_n\right\|_p \\
&\quad\quad + 2C_g \Big\{ \frac{2}{\eta}\|S_{n+1} - \widehat{S}_{n+1}\|_p + C_\lambda \eta + \frac{3L_{t^*}\|Z_n - \widehat{Z}_n\|_p}{\eta} \Big\} \\
&\quad\quad + 2L_{v_{n+1}}\left\|Z_{n+1} - \widehat{Z}_{n+1}\right\|_p + \|V_{n+1} - \widehat{V}_{n+1}\|_p.
\end{aligned} \qquad [7.34]$$

PROOF.– We have clearly

$$\begin{aligned}&\Big\| \max_{u\in G(\widehat{Z}_n)} \mathbb{1}_{\{u<t^*(Z_n)\}}\Big|J(v_{n+1},g)(Z_n,u)-\widehat{J}_{n+1}(\widehat{v}_{n+1},g)(\widehat{Z}_n,u)\Big|\Big\|_p\\ &\leq \Big\| \max_{u\in G(\widehat{Z}_n)} \mathbb{1}_{\{u<t^*(Z_n)\}}\Big|J(v_{n+1},g)(Z_n,u)-J(v_{n+1},g)(\widehat{Z}_n,u)\Big|\Big\|_p\\ &+\Big\| \max_{u\in G(\widehat{Z}_n)} \mathbb{1}_{\{u<t^*(Z_n)\}}\Big|J(v_{n+1},g)(\widehat{Z}_n,u)-\widehat{J}_{n+1}(v_{n+1},g)(\widehat{Z}_n,u)\Big|\Big\|_p\\ &+\Big\| \max_{u\in G(\widehat{Z}_n)} \mathbb{1}_{\{u<t^*(Z_n)\}}\Big|\widehat{J}_{n+1}(v_{n+1},g)(\widehat{Z}_n,u)-\widehat{J}_{n+1}(\widehat{v}_{n+1},g)(\widehat{Z}_n,u)\Big|\Big\|_p.\end{aligned} \quad [7.35]$$

Regarding the first-term in the right hand side of equation [7.35],

$$\begin{aligned}&\Big\| \max_{u\in G(\widehat{Z}_n)} \mathbb{1}_{\{u<t^*(Z_n)\}}\Big|J(v_{n+1},g)(Z_n,u)-J(v_{n+1},g)(\widehat{Z}_n,u)\Big|\Big\|_p\\ &\leq\big(C_gE_1+[v_{n+1}]_1E_2+E_3\big)\big\|Z_n-\widehat{Z}_n\big\|_p,\end{aligned} \quad [7.36]$$

by using lemma 7.4. Now the second-term in the right hand side of [7.35] satisfies

$$\begin{aligned}&\Big\| \max_{u\in G(\widehat{Z}_n)} \mathbb{1}_{\{u<t^*(Z_n)\}}\Big|J(v_{n+1},g)(\widehat{Z}_n,u)-\widehat{J}_{n+1}(v_{n+1},g)(\widehat{Z}_n,u)\Big|\Big\|_p\\ &\leq \Big\| \max_{u\in G(\widehat{Z}_n)} \mathbb{1}_{\{u<t^*(Z_n)\}}\Big|\mathbb{E}_{x_0}\big[w_{n+1}(Z_{n+1})\mathbb{1}_{\{S_{n+1}<u\wedge t^*(Z_n)\}}\big|Z_n=\widehat{Z}_n\big]\\ &-\mathbb{E}_{x_0}\big[w_{n+1}(\widehat{Z}_{n+1})\mathbb{1}_{\{\widehat{S}_{n+1}<u\wedge t^*(\widehat{Z}_n)\}}\big|\widehat{Z}_n\big]\Big|\Big\|_p,\end{aligned}$$

where $w_{n+1}=v_{n+1}+C_g$ and by using the fact that $\sigma(\widehat{Z}_n)\subset\sigma(Z_n,S_n)$

$$\begin{aligned}&\Big\| \max_{u\in G(\widehat{Z}_n)} \mathbb{1}_{\{u<t^*(Z_n)\}}\Big|J(v_{n+1},g)(\widehat{Z}_n,u)-\widehat{J}_{n+1}(v_{n+1},g)(\widehat{Z}_n,u)\Big|\Big\|_p\\ &\leq\Big\| \max_{u\in G(\widehat{Z}_n)} \mathbb{1}_{\{u<t^*(Z_n)\}}\Big|\mathbb{E}_{x_0}\big[w_{n+1}(Z_{n+1})\mathbb{1}_{\{S_{n+1}<u\wedge t^*(Z_n)\}}\big|Z_n=\widehat{Z}_n\big]\\ &-\mathbb{E}_{x_0}\big[w_{n+1}(Z_{n+1})\mathbb{1}_{\{S_{n+1}<u\wedge t^*(Z_n)\}}\big|\widehat{Z}_n\big]\Big|\Big\|_p\\ &+\Big\| \max_{u\in G(\widehat{Z}_n)} \Big|\mathbb{E}_{x_0}\big[w_{n+1}(Z_{n+1})\mathbb{1}_{\{S_{n+1}<u\wedge t^*(Z_n)\}}\big|\widehat{Z}_n\big]\\ &-\mathbb{E}_{x_0}\big[w_{n+1}(\widehat{Z}_{n+1})\mathbb{1}_{\{\widehat{S}_{n+1}<u\wedge t^*(\widehat{Z}_n)\}}\big|\widehat{Z}_n\big]\Big|\Big\|_p.\end{aligned}$$

Recalling that $u < t^*(\widehat{Z}_n) - \eta$ since $u \in G(\widehat{Z}_n)$, we obtain

$$\Big\| \max_{u \in G(\widehat{Z}_n)} \mathbb{1}_{\{u<t^*(Z_n)\}} \Big| J(v_{n+1}, g)(\widehat{Z}_n, u) - \widehat{J}_{n+1}(v_{n+1}, g)(\widehat{Z}_n, u) \Big| \Big\|_p$$

$$\leq \sup_{u \in \mathbb{R}^+} \Big\| Iw_{n+1}(\widehat{Z}_n, u) - Iw_{n+1}(Z_n, u) \Big\|_p$$

$$+ L_{w_{n+1}} \big\| Z_{n+1} - \widehat{Z}_{n+1} \big\|_p$$

$$+ C_{w_{n+1}} \Big\| \max_{u \in G(\widehat{Z}_n)} \mathbb{E}_{x_0} \Big[\big| \mathbb{1}_{\{S_{n+1} < u \wedge t^*(Z_n)\}} - \mathbb{1}_{\{\widehat{S}_{n+1} < u\}} \big| \Big| \widehat{Z}_n \Big] \Big\|_p.$$

Therefore, observing that $C_{w_{n+1}} = 2C_g, [w_{n+1}]_1 = [v_{n+1}]_1, L_{w_{n+1}} = L_{v_{n+1}}$, we obtain from Lemma 7.4

$$\Big\| \max_{u \in G(\widehat{Z}_n)} \mathbb{1}_{\{u<t^*(Z_n)\}} \Big| J(v_{n+1}, g)(\widehat{Z}_n, u) - \widehat{J}_{n+1}(v_{n+1}, g)(\widehat{Z}_n, u) \Big| \Big\|_p$$

$$\leq \big(2C_g E_1 + [v_{n+1}]_1 E_2\big) \big\| Z_n - \widehat{Z}_n \big\|_p + L_{v_{n+1}} \big\| Z_{n+1} - \widehat{Z}_{n+1} \big\|_p$$

$$+ 2C_g \Big\| \max_{u \in G(\widehat{Z}_n)} \mathbb{E}_{x_0} \Big[\mathbb{1}_{\{u<t^*(Z_n)\}} \big| \mathbb{1}_{\{S_{n+1} < u\}} - \mathbb{1}_{\{\widehat{S}_{n+1} < u\}} \big| \Big| \widehat{Z}_n \Big] \Big\|_p$$

$$+ 2C_g \Big\| \mathbb{1}_{\{t^*(Z_n) \leq t^*(\widehat{Z}_n) - \eta\}} \Big\|_p$$

implying by lemmas 7.11 and 7.12

$$\Big\| \max_{u \in G(\widehat{Z}_n)} \mathbb{1}_{\{u<t^*(Z_n)\}} \Big| J(v_{n+1}, g)(\widehat{Z}_n, u) - \widehat{J}_{n+1}(v_{n+1}, g)(\widehat{Z}_n, u) \Big| \Big\|_p$$

$$\leq \big(2C_g E_1 + [v_{n+1}]_1 E_2\big) \big\| Z_n - \widehat{Z}_n \big\|_p + L_{v_{n+1}} \big\| Z_{n+1} - \widehat{Z}_{n+1} \big\|_p$$

$$+ 2C_g \Big\{ \frac{2}{\eta} \| S_{n+1} - \widehat{S}_{n+1} \|_p + C_\lambda \eta + \frac{3L_{t^*} \| Z_n - \widehat{Z}_n \|_p}{\eta} \Big\}. \quad [7.37]$$

The third and last term in the right hand side of [7.35] satisfies

$$\Big\| \max_{u \in G(\widehat{Z}_n)} \mathbb{1}_{\{u<t^*(Z_n)\}} \Big| \widehat{J}_{n+1}(v_{n+1}, g)(\widehat{Z}_n, u) - \widehat{J}_{n+1}(\widehat{v}_{n+1}, g)(\widehat{Z}_n, u) \Big| \Big\|_p$$

$$= \Big\| \max_{u \in G(\widehat{Z}_n)} \mathbb{1}_{\{u<t^*(Z_n)\}} \Big| \mathbb{E}_{x_0} \Big[v_{n+1}(\widehat{Z}_{n+1}) \mathbb{1}_{\{\widehat{S}_{n+1} < u\}} \Big| \widehat{Z}_n \Big]$$

$$- \mathbb{E}_{x_0} \Big[\widehat{v}_{n+1}(\widehat{Z}_{n+1}) \mathbb{1}_{\{\widehat{S}_{n+1} < u\}} \Big| \widehat{Z}_n \Big] \Big| \Big\|_p$$

$$\leq L_{v_{n+1}} \| Z_{n+1} - \widehat{Z}_{n+1} \|_p + \| V_{n+1} - \widehat{V}_{n+1} \|_p. \quad [7.38]$$

Now, combining equations [7.35]–[7.38] and lemma 7.12, we obtain the result. □

THEOREM 7.2.– Set $n \in \{0, \ldots, N-2\}$ and suppose the discretization parameters are chosen such that there exists $0 < a < 1$ satisfying

$$\frac{\beta}{a} = \left\{ \frac{1}{C_\lambda} \Big(\frac{(4-3a)}{1-a} L_{t^*} \|\widehat{Z}_n - Z_n\|_p + 2\|S_{n+1} - \widehat{S}_{n+1}\|_p \Big) \right\}^{1/2}$$
$$< \min_{z \in \Gamma_n^z} \{\Delta(z)\}$$

Then we have

$$\begin{aligned}\left\|\overline{V}_n - V_n\right\|_p \leq & \left\|V_n - \widehat{V}_n\right\|_p + 2\left\|V_{n+1} - \overline{V}_{n+1}\right\|_p \\ & + 4L_{v_{n+1}}\left\|Z_{n+1} - \widehat{Z}_{n+1}\right\|_p \\ & + 2\|V_{n+1} - \widehat{V}_{n+1}\|_p + a_n\left\|Z_n - \widehat{Z}_n\right\|_p \\ & + 4C_g\Big\{C_\lambda\Big(\frac{(4-3a)}{1-a} L_{t^*}\|\widehat{Z}_n - Z_n\|_p + 2\|S_{n+1} - \widehat{S}_{n+1}\|_p\Big)\Big\}^{1/2}.\end{aligned}$$

with $a_n = 2C_gE_4 + 4[v_{n+1}]_1E_2 + 2[v_{n+1}]_*[Q] + 3C_gE_1 + E_3$.

PROOF.– For notational convenience, set $R_n = r_{n+1,\beta}(Z_n, S_n)$. The strong Markov property of the process (X_t) yield

$$\mathbb{E}_{x_0}[g(X_{\tau_{N-n-1}})|\mathcal{F}_{T_{n+1}}] = \mathbb{E}_{x_0}[g(X_{\tau_{N-n-1}})|\sigma(Z_{n+1}, S_{n+1})],$$

since by definition $\tau_{N-n-1} \geq T_{n+1}$. Consequently,

$$\begin{aligned}\overline{V}_n =& \mathbb{E}_{x_0}\left[g(\Phi(Z_n, R_n))\mathbb{1}_{\{S_{n+1} > R_n\}}\middle|\sigma(Z_n, S_n)\right] \\ & + \mathbb{E}_{x_0}\left[g(X_{\tau_{N-n-1}})\mathbb{1}_{\{S_{n+1} \leq R_n\}}\middle|\sigma(Z_n, S_n)\right] \\ =& \mathbb{E}_{x_0}\left[g(\Phi(Z_n, R_n))\mathbb{1}_{\{S_{n+1} > R_n\}}\middle|\sigma(Z_n, S_n)\right] \\ & + \mathbb{E}_{x_0}\left[\mathbb{E}_{x_0}[g(X_{\tau_{N-n-1}})|\sigma(Z_{n+1}, S_{n+1})]\mathbb{1}_{\{S_{n+1} \leq R_n\}}\middle|\sigma(Z_n, S_n)\right]\end{aligned}$$

and so,

$$\begin{aligned}\overline{V}_n =& \mathbb{E}_{x_0}\left[g(\Phi(Z_n, R_n))\mathbb{1}_{\{S_{n+1} > R_n\}}\middle|\sigma(Z_n, S_n)\right] \\ & + \mathbb{E}_{x_0}\left[\overline{V}_{n+1}\mathbb{1}_{\{S_{n+1} \leq R_n\}}\middle|\sigma(Z_n, S_n)\right]\end{aligned}$$

Observe that the following equalities hold

$$\mathbb{1}_{\{R_n < t^*(Z_n)\}}\mathbb{1}_{\{S_{n+1} > R_n\}} = \mathbb{1}_{\{R_n < t^*(Z_n)\}}\mathbb{1}_{\{S_{n+1} \geq R_n\}},$$

and

$$\mathbb{1}_{\{R_n<t^*(Z_n)\}}\mathbb{1}_{\{S_{n+1}\leq R_n\}} = \mathbb{1}_{\{R_n<t^*(Z_n)\}}\mathbb{1}_{\{S_{n+1}<R_n\}},$$

$\mathbb{P}_{x_0} - a.s.$ Therefore, we have

$$\overline{V}_n = \mathbb{1}_{\{R_n\geq t^*(Z_n)\}}\mathbb{E}_{x_0}\left[\overline{V}_{n+1}\middle|\sigma(Z_n,S_n)\right] + \mathbb{1}_{\{R_n<t^*(Z_n)\}}\Big(Hg(Z_n,R_n)$$

$$+\ \mathbb{E}_{x_0}\left[\overline{V}_{n+1}\mathbb{1}_{\{S_{n+1}<R_n\wedge t^*(Z_n)\}}\middle|\sigma(Z_n,S_n)\right]\Big).$$

However, our definition of $r_{n,\beta}$ with the special use of parameter β implies

$$\{R_n \geq t^*(Z_n)\} = \Big\{\widehat{K}_{n+1}\widehat{v}_{n+1}(\widehat{Z}_n) > \max_{s\in G(\widehat{Z}_n)} \widehat{J}_{n+1}(\widehat{v}_{n+1},g)(\widehat{Z}_n,s)\Big\}.$$

Consequently, we obtain

$$\overline{V}_n = \widehat{K}_{n+1}\widehat{v}_{n+1}(\widehat{Z}_n) \vee \max_{s\in G(\widehat{Z}_n)} \widehat{J}_{n+1}(\widehat{v}_{n+1},g)(\widehat{Z}_n,s)$$

$$+\mathbb{1}_{\{R_n\geq t^*(Z_n)\}}\Big[\mathbb{E}_{x_0}\left[\overline{V}_{n+1}\middle|\sigma(Z_n,S_n)\right] - \widehat{K}_{n+1}\widehat{v}_{n+1}(\widehat{Z}_n)\Big]$$

$$+\mathbb{1}_{\{R_n<t^*(Z_n)\}}\Big[Hg(Z_n,R_n) + \mathbb{E}_{x_0}\left[\overline{V}_{n+1}\mathbb{1}_{\{S_{n+1}<R_n\wedge t^*(Z_n)\}}\middle|\sigma(Z_n,S_n)\right]$$

$$- \max_{s\in G(\widehat{Z}_n)} \widehat{J}_{n+1}(\widehat{v}_{n+1},g)(\widehat{Z}_n,s)\Big]. \qquad [7.39]$$

Let us study the term with operator $\widehat{K}_{n+1}$. First, we insert V_n and by using our work of the previous section, we have

$$\Big|\mathbb{E}_{x_0}\left[\overline{V}_{n+1}\middle|\sigma(Z_n,S_n)\right] - \widehat{K}_{n+1}\widehat{v}_{n+1}(\widehat{Z}_n)\Big|$$

$$\leq \mathbb{E}_{x_0}\left[|\overline{V}_{n+1} - V_{n+1}|\middle|\sigma(Z_n,S_n)\right]$$

$$+\ \Big|Kv_{n+1}(Z_n) - \widehat{K}_{n+1}\widehat{v}_{n+1}(\widehat{Z}_n)\Big|,$$

since $Kv_{n+1}(Z_n) = \mathbb{E}_{x_0}\left[V_{n+1}\middle|\sigma(Z_n,S_n)\right]$. By using lemma 7.17, we have

$$\Big\|\mathbb{E}_{x_0}\left[\overline{V}_{n+1}\middle|\sigma(Z_n,S_n)\right] - \widehat{K}_{n+1}\widehat{v}_{n+1}(\widehat{Z}_n)\Big\|_p \leq \big\|V_{n+1} - \overline{V}_{n+1}\big\|_p$$

$$+\ 2\big(C_gE_4 + [v_{n+1}]_1E_2 + [v_{n+1}]_*[Q]\big)\big\|Z_n - \widehat{Z}_n\big\|_p$$

$$+\ 2L_{v_{n+1}}\big\|Z_{n+1} - \widehat{Z}_{n+1}\big\|_p + \big\|V_{n+1} - \widehat{V}_{n+1}\big\|_p. \qquad [7.40]$$

Now we turn to the term related to the operator $\widehat{J}_{n+1}$. We first study the case when $R_n = s^*_{n+1}(Z_n, S_n) < t^*(Z_n)$. By definition, we have

$$\widehat{J}_{n+1}(\widehat{v}_{n+1}, g)(\widehat{Z}_n, R_n) = \max_{s \in G(\widehat{Z}_n)} \widehat{J}_{n+1}(\widehat{v}_{n+1}, g)(\widehat{Z}_n, s)$$

and so, by using the definition of $J(v_{n+1}, g)(Z_n, R_n)$ it follows that

$$\begin{aligned}
&\Big\|\Big[Hg(Z_n, R_n) + \mathbb{E}_{x_0}\big[\overline{V}_{n+1}\mathbb{1}_{\{S_{n+1}<R_n\wedge t^*(Z_n)\}}\big|\sigma(Z_n, S_n)\big] \\
&\qquad - \max_{s\in G(\widehat{Z}_n)} \widehat{J}_{n+1}(\widehat{v}_{n+1}, g)(\widehat{Z}_n, s)\Big]\mathbb{1}_{\{R_n=s^*_{n+1}(Z_n,S_n)<t^*(Z_n)\}}\Big\|_p \\
&\leq \Big\|\mathbb{E}_{x_0}\big[(\overline{V}_{n+1} - V_{n+1})\mathbb{1}_{\{S_{n+1}<R_n\wedge t^*(Z_n)\}}\big|\sigma(Z_n, S_n)\big]\Big\|_p \\
&\quad + \Big\|\Big[J(v_{n+1}, g)(Z_n, R_n) - \widehat{J}_{n+1}(\widehat{v}_{n+1}, g)(\widehat{Z}_n, R_n)\Big] \\
&\quad \times \mathbb{1}_{\{R_n=s^*_{n+1}(Z_n,S_n)<t^*(Z_n)\}}\Big\|_p \\
&\leq \big\|V_{n+1} - \overline{V}_{n+1}\big\|_p \\
&\quad + \Big\|\max_{u\in G(\widehat{Z}_n)} \mathbb{1}_{\{u<t^*(Z_n)\}}\Big\{J(v_{n+1}, g)(Z_n, u) - \widehat{J}_{n+1}(\widehat{v}_{n+1}, g)(\widehat{Z}_n, u)\Big\}\Big\|_p.
\end{aligned}$$

Now, lemma 7.18 yields

$$\begin{aligned}
&\Big\|\Big[Hg(Z_n, R_n) + \mathbb{E}_{x_0}\big[\overline{V}_{n+1}\mathbb{1}_{\{S_{n+1}<R_n\wedge t^*(Z_n)\}}\big|\sigma(Z_n, S_n)\big] \\
&\qquad - \max_{s\in G(\widehat{Z}_n)} \widehat{J}_{n+1}(\widehat{v}_{n+1}, g)(\widehat{Z}_n, s)\Big]\mathbb{1}_{\{R_n=s^*_{n+1}(Z_n,S_n)<t^*(Z_n)\}}\Big\|_p \\
&\leq \big(3C_gE_1 + 2[v_{n+1}]_1E_2 + E_3\big)\big\|Z_n - \widehat{Z}_n\big\|_p \\
&\quad + 2C_g\Big\{\frac{2}{\eta}\|S_{n+1} - \widehat{S}_{n+1}\|_p + C_\lambda\eta + \frac{3L_{t^*}\|Z_n - \widehat{Z}_n\|_p}{\eta}\Big\} \\
&\quad + 2L_{v_{n+1}}\big\|Z_{n+1} - \widehat{Z}_{n+1}\big\|_p + \|V_{n+1} - \widehat{V}_{n+1}\|_p \\
&\quad + \|V_{n+1} - \overline{V}_{n+1}\|_p
\end{aligned} \qquad [7.41]$$

for all $0 < \eta < \min_{z\in\Gamma^Z_n}\{\Delta(z)\}$.

Finally, on $\{s^*_{n+1}(Z_n, S_n) \geq t^*(Z_n) = R_n + \beta\}$, by construction of the grid $G(\widehat{Z}_n)$ (see the comment after definition 7.2), we have for all

$$0 < \eta < \min_{z \in \Gamma_n^Z} \{\Delta(z)\},$$

$$R_n = t^*(Z_n) - \beta < s_{n+1}^*(Z_n, S_n) < t^*(\widehat{Z}_n) - \eta.$$

Consequently, choosing $\beta < \eta$

$$\Big\| \Big[Hg(Z_n, R_n) + \mathbb{E}_{x_0}\big[\overline{V}_{n+1} \mathbb{1}_{\{S_{n+1} < R_n \wedge t^*(Z_n)\}} \big| \sigma(Z_n, S_n)\big]$$

$$- \max_{s \in G(\widehat{Z}_n)} \widehat{J}_{n+1}(\widehat{v}_{n+1}, g)(\widehat{Z}_n, s) \Big] \mathbb{1}_{\{s_{n+1}^*(Z_n, S_n) \geq t^*(Z_n) = R_n + \beta\}} \Big\|_p$$

$$\leq 2C_g \big\| \mathbb{1}_{\{t^*(Z_n) < t^*(\widehat{Z}_n) - (\eta - \beta)\}} \big\|_p,$$

implying, by lemma 7.11

$$\Big\| \Big[\ Hg(Z_n, R_n) + \mathbb{E}_{x_0}\big[\overline{V}_{n+1} \mathbb{1}_{\{S_{n+1} < R_n \wedge t^*(Z_n)\}} \big| \sigma(Z_n, S_n)\big]$$

$$- \max_{s \in G(\widehat{Z}_n)} \widehat{J}_{n+1}(\widehat{v}_{n+1}, g)(\widehat{Z}_n, s) \Big] \mathbb{1}_{\{s_{n+1}^*(Z_n, S_n) \geq t^*(Z_n) = R_n + \beta\}} \Big\|_p$$

$$\leq 2C_g \frac{L_{t^*} \| Z_n - \widehat{Z}_n \|_p}{\eta - \beta}, \qquad [7.42]$$

for any $\eta > \beta$.

Now combining equations [7.39]–7.42] yields, for all $\beta < \eta < \min_{z \in \Gamma_n^Z} \{\Delta(z)\}$

$$\begin{aligned}
\big\| \overline{V}_n - \widehat{V}_n \big\|_p &\leq 2 \big\| V_{n+1} - \overline{V}_{n+1} \big\|_p + 2L_{v_{n+1}} \big\| Z_{n+1} - \widehat{Z}_{n+1} \big\|_p \\
&+ 2\big(C_g E_4 + [v_{n+1}]_1 E_2 + [v_{n+1}]_*[Q]\big) \big\| Z_n - \widehat{Z}_n \big\|_p \\
&+ \big(3C_g E_1 + 2[v_{n+1}]_1 E_2 + E_3\big) \big\| Z_n - \widehat{Z}_n \big\|_p \\
&+ 2C_g \Big\{ \frac{2}{\eta} \| S_{n+1} - \widehat{S}_{n+1} \|_p + C_\lambda \eta + \frac{3L_{t^*} \| Z_n - \widehat{Z}_n \|_p}{\eta} \Big\} \\
&+ 2L_{v_{n+1}} \big\| Z_{n+1} - \widehat{Z}_{n+1} \big\|_p + 2 \| V_{n+1} - \widehat{V}_{n+1} \|_p \\
&+ 2C_g \frac{L_{t^*} \| Z_n - \widehat{Z}_n \|_p}{\eta - \beta}.
\end{aligned}$$

Now suppose there exists $0 < a < 1$ such that $\eta = a^{-1}\beta$. Then the optimal choice for η satisfies

$$C_\lambda \eta = \frac{1}{\eta} \Big(\frac{4 - 3a}{1 - a} L_{t^*} \| \widehat{Z}_n - Z_n \|_p + 2 \| S_{n+1} - \widehat{S}_{n+1} \|_p \Big),$$

providing it also satisfies the condition $0 < \eta < \min_{z \in \Gamma_n^z} \{\Delta(z)\}$, hence the result.
□

Theorem 7.2 gives a recursive error estimation. Here is the initializing step.

THEOREM 7.3.– Suppose the discretization parameters are chosen such that there exists $0 < a < 1$ satisfying

$$\begin{aligned} \frac{\beta}{a} &= \left\{ \frac{1}{C_\lambda} \Big(\frac{4-3a}{1-a} L_{t^*} \|\widehat{Z}_{N-1} - Z_{N-1}\|_p + 2\|S_N - \widehat{S}_N\|_p \Big) \right\}^{1/2} \\ &< \min_{z \in \Gamma_{N-1}^z} \{\Delta(z)\}. \end{aligned}$$

Then we have

$$\begin{aligned} \left\|\overline{V}_{N-1} - V_{N-1}\right\|_p &\le \left\|\widehat{V}_{N-1} - V_{N-1}\right\|_p + 6L_g\left\|Z_N - \widehat{Z}_N\right\|_p \\ &\quad + a_{N-1}\left\|Z_{N-1} - \widehat{Z}_{N-1}\right\|_p \\ &\quad + 4C_g\Big\{C_\lambda\Big(\frac{4-3a}{1-a}L_{t^*}\|\widehat{Z}_{N-1} - Z_{N-1}\|_p + 2\|S_N - \widehat{S}_N\|_p\Big)\Big\}^{1/2} \end{aligned}$$

with $a_{N-1} = 2C_g E_4 + 4[g]_1 E_2 + 2[g]_*[Q] + 3C_g E_1 + E_3$.

PROOF.– For notational convenience, set $R_{N-1} = r_{N,\beta}(Z_{N-1}, S_{N-1})$. By using similar arguments as in the proof of theorem 7.2, it follows

$$\begin{aligned} \overline{V}_{N-1} &= \mathbb{E}_{x_0}\big[g(\Phi(Z_{N-1}, R_{N-1}))\mathbb{1}_{\{S_N > R_{N-1}\}}\big|\sigma(Z_{N-1}, S_{N-1})\big] \\ &\quad + \mathbb{E}_{x_0}\big[g(Z_N)\mathbb{1}_{\{S_N \le R_{N-1}\}}\big|\sigma(Z_{N-1}, S_{N-1})\big], \end{aligned}$$

and so

$$\begin{aligned} \overline{V}_{N-1} &= \widehat{K}_N \widehat{v}_N(\widehat{Z}_{N-1}) \vee \max_{s \in G(\widehat{Z}_{N-1})} \widehat{J}_N(\widehat{v}_N, g)(\widehat{Z}_{N-1}, s) \\ &\quad + \mathbb{1}_{\{R_{N-1} \ge t^*(Z_{N-1})\}}\Big[Kg(Z_{N-1}) - \widehat{K}_N g(\widehat{Z}_{N-1})\Big] \\ &\quad + \mathbb{1}_{\{R_{N-1} < t^*(Z_{N-1})\}}\Big[J(g,g)(Z_{N-1}, R_{N-1}) \\ &\quad - \max_{s \in G(\widehat{Z}_{N-1})} \widehat{J}_N(g,g)(\widehat{Z}_{N-1}, s)\Big], \end{aligned}$$

by recalling that $\widehat{v}_N = g$. Consequently, proceeding as in the proof of the previous theorem we obtain

$$\begin{aligned}
&\left\|\overline{V}_{N-1} - \widehat{V}_{N-1}\right\|_p \\
&\quad \leq \left\|Kg(Z_{N-1}) - \widehat{K}_N g(\widehat{Z}_{N-1})\right\|_p \\
&\qquad + \left\| \max_{u \in G(\widehat{Z}_{N-1})} \mathbb{1}_{\{u<t^*(Z_{N-1})\}} \left| J(g,g)(Z_{N-1},u) - \widehat{J}_N(g,g)(\widehat{Z}_{N-1},u) \right| \right\|_p \\
&\quad + 2C_g \frac{L_{t^*} \|Z_{N-1} - \widehat{Z}_{N-1}\|_p}{\eta - \beta}
\end{aligned}$$

implying,

$$\begin{aligned}
&\left\|\overline{V}_{N-1} - \widehat{V}_{N-1}\right\|_p \\
&\qquad \leq 2\big(C_g E_4 + [g]_1 E_2 + [g]_*[Q]\big) \left\|Z_{N-1} - \widehat{Z}_{N-1}\right\|_p \\
&\qquad + 3L_g \left\|Z_N - \widehat{Z}_N\right\|_p \\
&\qquad + \big(3C_g E_1 + 2[g]_1 E_2 + E_3\big) \left\|Z_{N-1} - \widehat{Z}_{N-1}\right\|_p \\
&\qquad + 2C_g \left\{ \frac{2}{\eta} \|S_N - \widehat{S}_N\|_p + C_\lambda \eta + \frac{3L_{t^*} \|Z_{N-1} - \widehat{Z}_{N-1}\|_p}{\eta} \right\} \\
&\qquad + 3L_g \left\|Z_N - \widehat{Z}_N\right\|_p + 2C_g \frac{L_{t^*} \|Z_{N-1} - \widehat{Z}_{N-1}\|_p}{\eta - \beta}
\end{aligned}$$

by combining lemmas 7.17, 7.18 and 7.11. Rearranging the different terms, we obtain the result. □

As in section 7.5, it is now clear that an adequate choice of discretization parameters yields arbitrarily small errors if we use the stopping-time τ_N.

7.7. Example

Now we apply the procedures described in sections 7.3 and 7.6 and algorithms 7.2 and 7.3 on the TCP process introduced in section 1.8.2 and present numerical results.

We use the same parameters $x_0 = 0$, $v = \alpha = 1$ and $\beta = 3$ and $N = 10$ as for the trajectories presented in Figure 1.3. The reward function g is defined on E by $g(x) = x$. Basically, we want to stop the process when it is as high as possible. However, recall that the higher the value of the process, the higher the probability of jumping back to the interval $[0, 1/2]$. Thus, this simple problem is not trivial. Our assumptions are clearly satisfied, and we are even in the special case when the flow is Lipschitz-continuous (see remark 7.1). All the constants involved in theorems 7.1 and 7.2 can be computed explicitly.

Pt	QE	Δ	$\widehat{V}_0$	$\overline{V}_0$	B_1	B_2	B_3
10	0.0943	0.151	0.7760	0.8173	0.1705	74.64	897.0
50	0.0418	0.100	0.8298	0.8785	0.1093	43.36	511.5
100	0.0289	0.083	0.8242	0.8850	0.1028	34.15	400.3
500	0.0133	0.056	0.8432	0.8899	0.0989	21.03	243.1
900	0.0102	0.049	0.8514	0.8968	0.0910	17.98	206.9

Pt Number of points in each quantization grid
QE Quantization error: $QE = \max_{0\leq k\leq N} \|\Theta_k - \widehat{\Theta}_k\|_2$
Δ For all z, $\Delta(z) = \Delta$
B_1 Empirical bound $\mathbb{E}_{x_0}\big[\sup_{0\leq t\leq T_N} g(X_t)\big] - \overline{V}_0$
B_2 Theoretical bound given by theorem 7.1
B_3 Theoretical bound given by theorems 7.2 and 7.3

Table 7.1. *Simulation results for optimal stopping of the TCP process*

The real value function $V_0 = v_0(x_0)$ is unknown, but, as our stopping rule τ_N is a stopping time dominated by T_N, we clearly have

$$\overline{V}_0 = \mathbf{E}_{x_0}\big[g(X_{\tau_N})\big] \leq V_0 = \sup_{\tau\in\mathcal{M}_N} \mathbb{E}_{x_0}\big[g(X_\tau)\big]$$
$$\leq \mathbb{E}_{x_0}\big[\sup_{0\leq t\leq T_N} g(X_t)\big]. \qquad [7.43]$$

The first and last terms can be evaluated by Monte Carlo simulations, which provides another indicator of the sharpness of our numerical procedure. For 10^6 Monte Carlo simulations, we obtain $\mathbb{E}_{x_0}\big[\sup_{0\leq t\leq T_N} g(X_t)\big] = 0.9878$. Simulation results (for $d = 2$, $x_0 = 0$, $v = \alpha = 1$, $\beta = 3$, up to the 10-th jump and for 10^5 Monte Carlo simulations) are given in Table 7.1. Note that, as expected, the theoretical errors decrease as the quantization error decreases.

The theoretical errors bounds are not sharp at all. This was to be expected as they hold for a very general class of PDMPs. From equation [7.43], it follows:

$$V_0 - \overline{V}_0 \leq \mathbb{E}_{x_0}\big[\sup_{0 \leq t \leq T_N} g(X_t)\big] - \overline{V}_0.$$

This provides an empirical upper bound for the error.

Other examples of application of our optimal procedure are detailed in Chapter 9.

8

Partially Observed Optimal Stopping Problem

8.1. Introduction

To apply the optimal stopping procedure described in Chapter 7, we need a perfect observation both of the jump times and the post-jump locations of the process. Such observations are often not available in practice, but we can have access to noisy measurements of the post-jump locations. This more realistic problem is mathematically more involved and falls into the category of optimal stopping problems under partial observation.

The aim of this chapter is thus to investigate the optimal stopping problem under partial observation for piecewise-deterministic Markov processes (PDMPs), both from theoretical and numerical points of view. In this chapter, we consider an optimal stopping problem for a partially observed PDMP $(X_t)_{t\geq 0}$. Generally speaking, the observation process $(Y_t)_{t\geq 0}$ is a point process defined through the embedded discrete time Markov chain $(Z_n, S_n)_{n\in\mathbb{N}}$. The inter-arrival times of $(Y_t)_{t\geq 0}$ are given by $(S_n)_{n\in\mathbb{N}}$ and the marks by a noisy function of $(Z_n)_{n\in\mathbb{N}}$. For a given reward function g and a computation horizon $N \in \mathbb{N}$, we study the following optimal stopping problem:

$$\sup_{\sigma\leq T_N} \mathbb{E}\left[g(X_\sigma)\right],$$

where T_N is the Nth jump time of the PDMP $(X_t)_{t\geq 0}$, σ is a stopping time with respect to the natural filtration $\mathfrak{F}^Y = (\mathfrak{F}_t^Y)_{t\geq 0}$ generated by the observations $(Y_t)_{t\geq 0}$. This is the main difference with the optimal stopping problem studied in Chapter 7 where the stopping times were taken with respect to the natural filtration of the original process $(X_t)_{t\geq 0}$.

In some applications, it may be more appropriate to consider a fixed optimization horizon t_f rather than the random horizon T_N. This is a difficult problem with few references in the literature (see, for instance, [EDO 10]), where the underlying process is not piecewise deterministic. Regarding PDMPs, this problem could be addressed using the same ideas as in Chapter 4. It involves the time-augmented process (X_t, t). Although this process is still a PDMP, its local characteristics may not have the same good properties as those of the original process leading to several new technical difficulties.

A general methodology to solve such a problem is to split it into two subproblems. The first step consists of deriving the filter process given by the conditional expectation of X_t with respect to the observed information $\mathfrak{F}_t^Y$. Its main objective is to transform the initial problem into a completely observed optimal stopping problem where the new state variable is the filter process. The second step consists of solving this reformulated problem, the new difficulty being its infinite dimension. Indeed, the filter process takes values in a set of probability measures.

Our work is inspired by [PHA 05] which deals with an optimal stopping problem under partial observation for a Markov chain with finite state space. The authors study the optimal filtering and convert their original problem into a standard optimal stopping problem for a continuous state space Markov chain. Then, they propose a discretization method based on the quantization technique to approximate the value function. However, their method cannot be directly applied to our problem for the following main reasons related to the specificities of PDMPs.

First, PDMPs are continuous time processes. Although the dynamics can be described by the discrete-time embedded Markov chain $(Z_n, S_n)_{n\in\mathbb{N}}$, this optimization problem remains intrinsically a *continuous-time* optimization problem. Indeed, the performance criterion is maximized over the set of stopping times defined with respect to the *continuous-time* filtration $(\mathfrak{F}_t^Y)_{t\geq 0}$. Consequently, our problem cannot be converted into a fully discrete time problem.

Second, the distribution of a PDMP combines both absolutely continuous and singular components. This is due to the existence of forced jumps when the process hits the boundary of the state space. As a result, the derivation of the filter process is not straightforward. In particular, the absolute continuity hypothesis **(H)** of [PHA 05] does not hold.

Third, in our context, the reformulated optimization problem is not standard, unlike in [PHA 05]. As already explained, this reformulated optimization problem combines *continuous-time* and *discrete-time* features. Consequently, this problem does not correspond to the classical optimal stopping problem of a discrete-time Markov chain. Moreover, it is different from the optimal stopping problem of a PDMP under complete observation mainly because the new state variables given by the Markov chain $(\Pi_n, S_n)_{n\geq 0}$ are not the underlying Markov chain of some PDMP. Therefore, the results of Chapter 7 and those of the literature [GUG 86, PHA 05] cannot be used.

Finally, a natural way to proceed with the numerical approximation is then to follow the ideas developed in Chapter 7 and [PHA 05] namely to replace the filter Π_n and the inter-jump time S_n by some finite state space approximations in the dynamic programming equation. However, a noticeable difference from Chapter 7 lies in the fact that the dynamic programming operators therein were Lipschitz continuous, whereas our new operators are only locally Lipschitz continuous between some points of discontinuity. We overcome this drawback by splitting the operators into their restrictions onto their continuity sets. This way, we obtain not only an approximation of the value function of the optimal stopping problem but also an ϵ-optimal stopping time with respect to the filtration $(\mathfrak{F}_t^Y)_{t\geq 0}$ that can be computed in practice.

The main material of this chapter comes from [BRA 13]. This chapter is organized as follows. Section 8.2 defines the optimal stopping problem we are interested in, especially the observation process and presents the assumptions. The recursive formulation of the filter process is derived in section 8.3. In section 8.4, we reduce our partially observed problem for the PDMP $(X_t)_{t\geq 0}$ to a completely observed one involving the process $(\Pi_n, S_n)_{n\in\mathbb{N}}$ for which we provide the dynamic programming equation and construct a family of ϵ-optimal stopping times. Then, our numerical methods to compute the value function and an ϵ-optimal stopping time are presented in section 8.5 where we also prove the convergence of our algorithms. Finally, an academic example is discussed in section 8.6.

8.2. Problem formulation and assumptions

The objective of this section is to introduce the optimal stopping problem for a partially observed PDMP.

We suppose in this chapter that assumption 1.4 holds and that the transition kernel Q satisfies the following hypothesis.

ASSUMPTION 8.1.– There exists a finite set $E_0 = \{x_1, \dots, x_q\} \subset E$ such that $X_0 \in E_0$ and for all $x \in \overline{E}$, we have $Q(x, E_0) = 1$

In other words, for all $n \in \mathbb{N}$, Z_n may only take its values in the finite set E_0. This assumption ensures that the filter process, defined in the next section, has finite dimensions. This is required to derive a tractable numerical method in section 8.5. When this assumption does not hold, we may consider a preliminary discretization of the transition kernel to introduce it. For all $m \in \{1, \dots, q\}$, denote $t^*_m = t^*(x_m)$ and assume that $x_1, \dots, x_q$ are numbered such that $t^*_1 \leq t^*_2 \leq \dots \leq t^*_q$. Moreover, let $t^*_0 = 0$.

Denote by $\mathcal{M}(E_0)$ the set of finite signed measures on E_0 and by $\mathcal{P}(E_0)$ the subset of probability measures on E_0. For any $\pi \in \mathcal{M}(E_0)$, the norm $|\pi|_1$ is given by $|\pi|_1 = \sum_{i=1}^q |\pi^i|$ where π^i denotes $\pi(\{x_i\})$.

We consider from now on a PDMP $(X_t)_{t\geq 0}$ of which the initial state, $X_0 = Z_0$ is a fixed point $x_0 \in E_0$. We assume that this PDMP is observed through noise and we now turn to the description of our observation procedure. For all $n \in \mathbb{N}$, we assume that S_n is perfectly observed but that Z_n is not (except for the initial state Z_0). In some examples, it seems reasonable to consider that the jump times of the process are observed (for instance, if the jumps correspond to changes of environment as in the corrosion example presented in section 2.2) and that, when a jump occurs, the actual post-jump location is measured with noise. The *observation* process of Z_n, denoted by Y_n, is assumed to be of the following form: $Y_0 = x_0$ (deterministic) and for $n \geq 1$,

$$Y_n = \varphi(Z_n) + W_n, \qquad [8.1]$$

where $\varphi : E_0 \to \mathbb{R}^d$ and the *noise* $(W_n)_{n\geq 1}$ is a sequence of $\mathbb{R}^d$-valued, i.i.d. random variables with bounded density function f_W that are also independent of $(Z_n, S_n)_{n\in\mathbb{N}}$. In order to define real-valued stopping times adapted to the observation process, we need to consider a continuous time version of the observation process. We, therefore, define the piecewise-constant process $(Y_t)_{t\geq 0}$ with a slight misuse of notation[1] as

$$Y_t = \sum_{j=0}^{+\infty} \mathbb{1}_{[T_j, T_{j+1})}(t) Y_j.$$

1 The quantity Y_n represents the value of the process $(Y_t)_{t\geq 0}$ at time $t = T_n$ and must not be confused with the value of the process at time $t = n$.

Let $\mathfrak{F}^Y = (\mathfrak{F}_t^Y)_{t\geq 0}$ be the filtration generated by $(Y_t)_{t\geq 0}$ (the *observed* filtration) and $\mathfrak{F} = (\mathfrak{F}_t)_{t\geq 0}$ be the filtration generated by $(X_t, Y_t)_{t\geq 0}$ (the *total* filtration). Without changing the notation, we then complete these filtrations with all the $\mathbb{P}$-null sets. This leads us to the following definition.

DEFINITION 8.1.– *Denote by Σ^Y the set of $(\mathfrak{F}_t^Y)_{t\geq 0}$-stopping times that are a.s. finite and for $n \in \mathbb{N}$, define*

$$\Sigma_n^Y = \left\{\sigma \in \Sigma^Y \text{ such that } \sigma \leq T_n \text{ a.s.}\right\}.$$

For all $n \in \mathbb{N}$, we define the filter $\Pi_n \in \mathcal{P}(E_0)$. The quantity $\Pi_n(\{x_i\})$, denoted by Π_n^i, represents the probability of the event $\{Z_n = x_i\}$ given the information available until time T_n, i.e.

$$\forall i \in \{1, \ldots, q\}, \qquad \Pi_n^i = \mathbb{E}[\mathbb{1}_{\{Z_n = x_i\}} | \mathfrak{F}_{T_n}^Y]. \qquad [8.2]$$

Finally, let $N \in \mathbb{N}$ be the *horizon* and $g \in \mathbf{B}(\overline{E})$ be the *reward function*, we are interested in maximizing the following *performance criterion*

$$\mathbb{E}\left[g(X_\sigma) | \Pi_0 = \pi\right],$$

with respect to the stopping times $\sigma \in \Sigma_N^Y$. The *value function* associated with this partially observed optimal stopping problem is given by

$$v(\pi) = \sup_{\sigma \in \Sigma_N^Y} \mathbb{E}\left[g(X_\sigma) | \Pi_0 = \pi\right], \qquad [8.3]$$

where π is a probability measure in $\mathcal{P}(E_0)$. The solution of our problem is then obtained by setting $\pi = \delta_{x_0}$. For some applications, it would be interesting to consider a more general form for the reward function such as an integral term also possibly depending on the observation process (see, for instance, [LUD 12]). However, this new setup would lead to several technical difficulties. In particular, the dynamic programming would be more complex. Thus, the derivation of the error bounds for the numerical approximation would be possibly intractable.

We will also need the following assumption about the reward function g associated with the optimal stopping problem. The function g is in $\mathbf{B}(\overline{E})$, i.e. bounded by C_g and there exists $[g]_2 \in \mathbb{R}^+$ such that for all $i \in \{1, \ldots, q\}$ and $t, u \in [0, t_i^*]$, we have

$$|g(\Phi(x_i, t)) - g(\Phi(x_i, u))| \leq [g]_2 |t - u|.$$

The objectives of this chapter are first to explicit the filter process $(\Pi_n)_{n\in\mathbb{N}}$ (section 8.3); second to rewrite the partially observed optimal stopping problem [8.3] as a totally observed one for a suitable Markov chain on $\mathcal{P}(E_0) \times \mathbb{R}^+$ (section 8.4.2); third to derive a dynamic programming equation and construct a family of ϵ-optimal stopping times (section 8.4.3); and finally to propose a numerical method to compute an approximation of the value function and an ϵ-optimal stopping time (section 8.5). As a starting point, we will derive, in the next section, a recursive construction of the optimal filter that is the key point of our approach.

8.3. Optimal filtering

The aim of this section is to obtain a recursive formulation of the filter Π_n. As far as we know, there is no result concerning the filter process for generic PDMPs. We may, however, refer to [ARJ 92] for a recursive formulation of the filter for point processes, which can be seen as a subclass of PDMPs.

For all $n \in \mathbb{N}$, we denote $\mathcal{G}_n = (Y_0, S_0, \ldots, Y_n, S_n)$. The continuous-time observation process $(Y_t)_{t\geq 0}$ being a point process in the sense developed in [BRÉ 81], we have $\mathfrak{F}^Y_{T_n} = \sigma(\mathcal{G}_n)$ (see [BRÉ 81, page 58, Theorem T2]). Moreover, $\mathfrak{F}_{T_n} = \sigma(Z_0, \ldots, Z_n) \vee \mathfrak{F}^Y_{T_n}$ (where $\mathcal{F} \vee \mathcal{G}$ denotes the σ-algebra generated by $\mathcal{F} \cup \mathcal{G}$). Concerning the filter Π_n, first notice that, since it is an $\mathfrak{F}^Y_{T_n}$-measurable random variable, there exists for all $n \in \mathbb{N}$ a measurable function $\pi_n : (\mathbb{R}^d \times \mathbb{R}^+)^{n+1} \to \mathcal{P}(E_0)$ such that $\Pi_n = \pi_n(\mathcal{G}_n)$. As in the case of the Kalman–Bucy filter, the iteration leading from Π_{n-1} to Π_n can be split into two steps: prediction and correction. For all $n \geq 1$, let μ_n^- be the conditional distribution of (Z_n, S_n) given $\mathfrak{F}^Y_{T_{n-1}}$. Thus, μ_n^- is a transition kernel defined on $(\mathbb{R}^d \times \mathbb{R}^+)^n \times \mathfrak{B}(E_0 \times \mathbb{R}^+)$ for all $j \in \{1, \ldots, q\}$ and $\gamma_{n-1} \in (\mathbb{R}^d \times \mathbb{R}^+)^n$ by

$$\mu_n^-(\gamma_{n-1}, \{x_j\}, ds) = \mathbb{P}(Z_n = x_j, S_n \in ds | \mathcal{G}_{n-1} = \gamma_{n-1}). \qquad [8.4]$$

In the remainder of this section, $\mathbb{P}_{\mathcal{G}_n}$ denotes the distribution of $\mathcal{G}_n$.

LEMMA 8.1.– We have the following equality of probability measures on $E_0 \times \mathbb{R}^d \times \mathbb{R}^+$, for all $j \in \{1, \ldots, q\}$,

$$\begin{aligned}\mathbb{P}(Z_n = x_j, Y_n \in\ & dy, S_n \in ds | \mathcal{G}_{n-1} = \gamma_{n-1}) \\ &= \mu_n^-(\gamma_{n-1}, \{x_j\}, ds) f_W(y - \varphi(x_j)) dy,\end{aligned}$$

$\mathbb{P}_{\mathcal{G}_n}$-$a.s.$ on $(\mathbb{R}^d \times \mathbb{R}^+)^n$.

PROOF.– Set h in $\mathbf{B}(E_0 \times \mathbb{R}^d \times \mathbb{R}^+)$, using equation [8.1] that defines Y_n, we have

$$\begin{aligned}
&\mathbb{E}\big[h(Z_n, Y_n, S_n)\big|\mathcal{G}_{n-1} = \gamma_{n-1}\big] \\
&= \sum_{j=1}^{q} \int h(x_j, \varphi(x_j) + w, s) \\
&\times \mathbb{P}(Z_n = x_j, S_n \in ds, W_n \in dw|\mathcal{G}_{n-1} = \gamma_{n-1}).
\end{aligned}$$

Moreover, W_n is independent of $\sigma(Z_n, S_n) \vee \mathfrak{F}^Y_{T_{n-1}}$ and admits the density function f_W. Consequently, we easily obtain the result by using the change of variable $y = \varphi(x_j) + w$. □

Integrating w.r.t. to the first variable in the previous lemma (i.e. summing w.r.t. x_j) yields the following result.

COROLLARY 8.1.– We have the following equality of probability measures on $\mathbb{R}^d \times \mathbb{R}^+$,

$$\begin{aligned}
&\mathbb{P}(Y_n \in dy, S_n \in ds|\mathcal{G}_{n-1} = \gamma_{n-1}) \\
&= \sum_{j=1}^{q} \mu_n^-(\gamma_{n-1}, \{x_j\}, ds) f_W(y - \varphi(x_j))dy,
\end{aligned}$$

$\mathbb{P}_{\mathcal{G}_n}$-$a.s.$ on $(\mathbb{R}^d \times \mathbb{R}^+)^n$.

LEMMA 8.2.– For any $n \geq 1$ and $j \in \{1, \ldots, q\}$, the distribution μ_n^-, defined in [8.4], satisfies

$$\begin{aligned}
&\mu_n^-(\gamma_{n-1}, \{x_j\}, ds) \\
&= \sum_{m=0}^{q-1} \mathbb{1}_{\{s \in (t^*_m; t^*_{m+1})\}} \\
&\times \sum_{i=m+1}^{q} \pi^i_{n-1}(\gamma_{n-1}) \lambda(\Phi(x_i, s)) e^{-\Lambda(x_i, s)} Q(\Phi(x_i, s), x_j) ds \\
&+ \sum_{m=1}^{q} \pi^m_{n-1}(\gamma_{n-1}) e^{-\Lambda(x_m, t^*_m)} Q(\Phi(x_m, t^*_m), x_j) \delta_{t^*_m}(ds),
\end{aligned}$$

$\mathbb{P}_{\mathcal{G}_n}$-$a.s.$ on $(\mathbb{R}^d \times \mathbb{R}^+)^n$.

PROOF.– Let h be a function of $\mathbf{B}(E_0 \times \mathbb{R}^+)$. Since $\sigma(\mathcal{G}_{n-1}) = \mathfrak{F}^Y_{T_{n-1}} \subset \mathfrak{F}_{T_{n-1}}$, the law of iterated conditional expectations yields

$$\mathbb{E}\left[h(Z_n, S_n)\middle|\mathcal{G}_{n-1} = \gamma_{n-1}\right] = \mathbb{E}\left[\mathbb{E}\left[h(Z_n, S_n)\middle|\mathfrak{F}_{T_{n-1}}\right]\middle|\mathcal{G}_{n-1} = \gamma_{n-1}\right].$$

Besides, $\mathfrak{F}_{T_{n-1}} \subset \sigma(Z_0, S_0, W_0, \ldots, Z_{n-1}, S_{n-1}, W_{n-1})$ so that

$$\mathbb{E}\left[h(Z_n, S_n)\middle|\mathfrak{F}_{T_{n-1}}\right] = \mathbb{E}\left[h(Z_n, S_n)\middle|Z_0, S_0, \ldots, Z_{n-1}, S_{n-1}\right],$$

by independence of the sequences $(W_n)_{n\in\mathbb{N}}$ and $(Z_n, S_n)_{n\in\mathbb{N}}$. Now, we apply the Markov property of $(Z_n, S_n)_{n\in\mathbb{N}}$ and a well-known special feature of the transition kernel of the underlying Markov chain of a PDMP (see section 1.6) to obtain

$$\mathbb{E}\left[h(Z_n, S_n)\middle|\mathfrak{F}_{T_{n-1}}\right] = \mathbb{E}\left[h(Z_n, S_n)\middle|Z_{n-1}\right].$$

Moreover, the transition kernel can be explicitly expressed in terms of the local characteristics of the PDMP (see lemma 1.4), and this yields the following equations

$$\begin{aligned}
&\mathbb{E}[h(Z_n, S_n)|\mathcal{G}_{n-1} = \gamma_{n-1}] \\
&\quad = \mathbb{E}\Big[\sum_{i=1}^{q} \mathbb{1}_{\{Z_{n-1}=x_i\}}\mathbb{E}[h(Z_n, S_n)|Z_{n-1} = x_i]\Big|\mathcal{G}_{n-1} = \gamma_{n-1}\Big] \\
&\quad = \mathbb{E}\Big[\sum_{i=1}^{q} \mathbb{1}_{\{Z_{n-1}=x_i\}} \\
&\qquad \times \sum_{j=1}^{q}\Big\{\int_{\mathbb{R}^+} h(x_j, s)\lambda(\Phi(x_i, s))e^{-\Lambda(x_i,s)}\mathbb{1}_{\{s<t^*_i\}} \times Q(\Phi(x_i, s), x_j)ds \\
&\qquad \times + h(x_j, t^*_i)e^{-\Lambda(x_i,t^*_i)}Q(\Phi(x_i, t^*_i), x_j)\Big\}\Big|\mathcal{G}_{n-1} = \gamma_{n-1}\Big] \\
&\quad = \sum_{j=1}^{q}\Big(\int_{\mathbb{R}^+} h(x_j, s) \\
&\qquad \times \sum_{i=1}^{q} \pi^i_{n-1}(\gamma_{n-1})\lambda(\Phi(x_i, s))e^{-\Lambda(x_i,s)}\mathbb{1}_{\{s<t^*_i\}}Q(\Phi(x_i, s), x_j)ds \\
&\qquad + \sum_{i=1}^{q} h(x_j, t^*_i)\pi^i_{n-1}(\gamma_{n-1})e^{-\Lambda(x_i,t^*_i)}Q(\Phi(x_i, t^*_i), x_j)\Big).
\end{aligned}$$

This can be written equivalently as:

$$\mathbb{E}\Big[h(Z_n, S_n)\big|\mathcal{G}_{n-1} = \gamma_{n-1}\Big]$$

$$= \sum_{j=1}^{q}\Big\{\sum_{m=0}^{q-1}\Big(\int_{t_m^*}^{t_{m+1}^*} h(x_j, s)$$

$$\times \sum_{i=m+1}^{q} \pi_{n-1}^i(\gamma_{n-1})\lambda(\Phi(x_i, s))e^{-\Lambda(x_i,s)}Q(\Phi(x_i, s), x_j)\Big)ds$$

$$+ \sum_{i=1}^{q} h(x_j, t_i^*)\pi_{n-1}^i(\gamma_{n-1})e^{-\Lambda(x_i,t_i^*)}Q(\Phi(x_i, t_i^*), x_j)\Big\}.$$

Hence, the result. □

We now state the main result of this section, namely the recursive formulation of the filter sequence $(\Pi_n)_{n\in\mathbb{N}}$.

PROPOSITION 8.1.– Let $\Psi = (\Psi^1, \ldots, \Psi^q) : \mathcal{P}(E_0) \times \mathbb{R}^d \times \mathbb{R}^+ \to \mathcal{P}(E_0)$ be defined as follows: for all $j \in \{1, \ldots, q\}$,

$$\Psi^j(\pi, y, s) = \sum_{m=0}^{q-1} \mathbb{1}_{\{s\in(t_m^*;t_{m+1}^*)\}} \frac{\Psi_m^j(\pi, y, s)}{\overline{\Psi}_m(\pi, y, s)} + \sum_{m=1}^{q} \mathbb{1}_{\{s=t_m^*\}} \frac{\Psi_m^{*j}(y)}{\overline{\Psi}_m^*(y)},$$

where

$$\Psi_m^j(\pi, y, s)$$

$$= \sum_{i=m+1}^{q} \pi^i\lambda(\Phi(x_i, s))e^{-\Lambda(x_i,s)}Q(\Phi(x_i, s), x_j)f_W(y - \varphi(x_j)),$$

$$\overline{\Psi}_m(\pi, y, s) = \sum_{k=1}^{q} \Psi_m^k(\pi, y, s),$$

$$\Psi_m^{*j}(y) = Q(\Phi(x_m, t_m^*), x_j)f_W(y - \varphi(x_j)),$$

$$\overline{\Psi}_m^*(y) = \sum_{k=1}^{q} \Psi_m^{*k}(y).$$

Then, the filter, defined in equation [8.2], satisfies $\Pi_0^j = \mathbb{P}(Z_0 = x_j)$ and the following recursion: for all $n \geq 1$,

$$\mathbb{P}\text{-a.s.}, \Pi_n = \Psi(\Pi_{n-1}, Y_n, S_n).$$

PROOF.– Fix γ_{n-1} in $(\mathbb{R}^d \times \mathbb{R}^+)^n$. Bayes formula yields for all $j \in \{1, \ldots, q\}$,

$$\mathbb{P}(Z_n = x_j, Y_n \in dy, S_n \in ds \big| \mathcal{G}_{n-1} = \gamma_{n-1})$$
$$= \mathbb{P}\big(Z_n = x_j \big| \mathcal{G}_n = (\gamma_{n-1}, y, s)\big)\mathbb{P}(Y_n \in dy, S_n \in ds \big| \mathcal{G}_{n-1} = \gamma_{n-1}).$$

Lemma 8.1 and corollary 8.1 yield

$$\mu_n^-(\gamma_{n-1}, \{x_j\}, ds) f_W(y - \varphi(x_j))dy$$
$$= \mathbb{P}\left(Z_n = x_j \big| \mathcal{G}_n = (\gamma_{n-1}, y, s)\right)$$
$$\times \left[\sum_{k=1}^{q} \mu_n^-(\gamma_{n-1}, \{x_k\}, ds) f_W(y - \varphi(x_k))\right] dy.$$

From lemma 8.2, it follows that almost surely on $\mathbb{R}^d \times \left(\mathbb{R}^+ \backslash \{t_1^*, \ldots, t_q^*\}\right)$ with respect to the Lebesgue measure

$$\mathbb{P}\big(Z_n = x_j \big| \mathcal{G}_n = (\gamma_{n-1}, y, s)\big)$$
$$= \sum_{m=1}^{q} \mathbb{1}_{(t_{m-1}^*, t_m^*)}(s) \frac{f_m(\gamma_{n-1}, x_j, s) f_W(y - \varphi(x_j))}{\sum_{k=1}^{q} f_m(\gamma_{n-1}, x_k, s) f_W(y - \varphi(x_k))}$$

with

$$f_m(\gamma_{n-1}, x_j, s) = \sum_{i=m+1}^{q} \pi_{n-1}^i(\gamma_{n-1}) \lambda(\Phi(x_i, s)) e^{-\Lambda(x_i, s)} Q(\Phi(x_i, s), x_j).$$

Again by using lemma 8.2,

$$\mathbb{P}\big(Z_n = x_j \big| \mathcal{G}_n = (\gamma_{n-1}, y, t_m^*)\big)$$
$$= \frac{\pi_{n-1}^m(\gamma_{n-1}) e^{-\Lambda(x_m, t_m^*)} Q(\Phi(x_m, t_m^*), x_j) f_W(y - \varphi(x_j))}{\sum_{k=1}^{q} \pi_{n-1}^m(\gamma_{n-1}) e^{-\Lambda(x_m, t_m^*)} Q(\Phi(x_m, t_m^*), x_k) f_W(y - \varphi(x_k))}$$
$$= \frac{Q(\Phi(x_m, t_m^*), x_j) f_W(y - \varphi(x_j))}{\sum_{k=1}^{q} Q(\Phi(x_m, t_m^*), x_k) f_W(y - \varphi(x_k))}$$

almost surely on $\mathbb{R}^d$ with respect to the Lebesgue measure. Therefore,

$$\pi_n(\gamma_{n-1}, y, s) = \Psi(\pi_{n-1}(\gamma_{n-1}), y, s). \qquad [8.5]$$

almost surely on $\mathbb{R}^d \times \left(\mathbb{R}^+ - \{t_1^*, \ldots, t_q^*\}\right)$ with respect to the Lebesgue measure, and for any $m \in \{1, \ldots, q\}$

$$\pi_n(\gamma_{n-1}, y, t_m^*) = \Psi(\pi_{n-1}(\gamma_{n-1}), y, t_m^*). \qquad [8.6]$$

almost surely on $\mathbb{R}^d \times \left(\mathbb{R}^+ - \{t_1^*, \ldots, t_q^*\}\right)$ with respect to the Lebesgue measure. However, the measure $\mathbb{P}\left(Y_n \in dy, S_n \in ds \middle| \mathcal{G}_{n-1} = \gamma_{n-1}\right)$ is absolutely continuous with respect to the Lebesgue measure on $\mathbb{R}^d \times \left(\mathbb{R}^+ - \{t_1^*, \ldots, t_q^*\}\right)$ and for any $m \in \{1, \ldots, q\}$, the measure $\mathbb{P}\left(Y_n \in dy, S_n = t_m^* \middle| \mathcal{G}_{n-1} = \gamma_{n-1}\right)$ is absolutely continuous with respect to the Lebesgue measure on $\mathbb{R}^d$. Finally, we can conclude from equations [8.5] and [8.6] that

$$\pi_n(\mathcal{G}_{n-1}, Y_n, S_n) = \Psi(\pi_{n-1}(\mathcal{G}_{n-1}), Y_n, S_n).$$

The result follows since $\pi_n(\mathcal{G}_{n-1}, Y_n, S_n) = \Pi_n$ and $\pi_{n-1}(\mathcal{G}_{n-1}) = \Pi_{n-1}$. □

This proposition will play a crucial part in the next section. On the one hand, this result will enable us to prove the Markov property of the sequence $(\Pi_n, S_n)_{n \geq 0}$ w.r.t. the observed filtration. On the other hand, the recursive formulation allows for simulation of the process $(\Pi_n)_{n \geq 0}$ which is crucial to obtain numerical approximations. Finally, notice that the specific structure of the PDMP appears in the recursive formulation of the filter which contains both an absolutely continuous part and some weighted points.

8.4. Dynamic programming

The main objective of this section is to derive the dynamic programming equation for the value function of the partially observed optimal stopping problem [8.3]. Generally speaking, the proof of this result can be decomposed into two steps. The first point consists of converting the partially observed optimal stopping problem into an optimal stopping problem under complete observation where the state variables are described by the *discrete-time* Markov chain $(\Pi_n, S_n)_{n \geq 0}$ (see section 8.4.2). It is important to note that under this new formulation, the optimization problem remains intrinsically a *continuous-time* optimization problem because the performance criterion is maximized over the set of stopping times with respect to the *continuous-time* filtration $(\mathfrak{F}_t^Y)_{t \geq 0}$. We show in the second step (see section 8.4.3) that the value function associated with the optimal stopping problem [8.3] can be calculated by iterating a functional operator, labeled L (see definition 8.2). As a by-product, we also provide a family of ϵ-optimal stopping times.

We would like to emphasize that the results obtained in this section are not straightforward to obtain due to the specific structure of this optimization problem. Indeed, as already explained, it combines *continuous-time* and *discrete-time* features. Consequently, this problem does not correspond to the classical optimal stopping problem of a discrete-time Markov chain. Moreover, it is different from the optimal stopping problem of a PDMP under complete observation mainly because the new state variables given by the Markov chain $(\Pi_n, S_n)_{n\geq 0}$ are not the underlying Markov chain of some PDMP. Therefore, the results presented in Chapter 7 and [GUG 86] cannot be used.

These derivations first require some technical results about the structure of the stopping times in Σ_N^Y, the Markov property of the filtering process and the conditional expectation of the state process.

8.4.1. *Preliminary results*

In this section, we study the special structure of $(\mathfrak{F}_t^Y)_{t\geq 0}$-stopping times. Our aim is to obtain similar results as in section 1.7.

LEMMA 8.3.– For all $n \in \mathbb{N}$, T_n is an $(\mathfrak{F}_t^Y)_{t\geq 0}$-stopping time.

PROOF.– Notice that for all $n \in \mathbb{N}$, $\mathbb{P}(Y_n = Y_{n+1}) = 0$. This arises from the independence and the absolute continuity of the distribution of the random variables $(W_n)_{n\in\mathbb{N}}$ since

$$\{Y_n = Y_{n+1}\} \subset \bigcup_{1\leq i,j\leq q} \{W_n - W_{n+1} = \varphi(x_i) - \varphi(x_j)\}.$$

Hence, for all $n \in \mathbb{N}$ and $t \in \mathbb{R}^+$, we have $\mathbb{P}$ a.s. $\{T_n \leq t\} = \{N_t \geq n\}$ where we denote $N_t = \sum_{0\leq s\leq t} \mathbb{1}_{\{Y_s \neq Y_{s-}\}}$. The process $(N_t)_{t\geq 0}$ is $\mathfrak{F}^Y$-adapted, thus $\{N_t \geq n\} \in \mathfrak{F}_t^Y$ and since the filtration $\mathfrak{F}^Y$ contains the $\mathbb{P}$-null sets, we have $\{T_n \leq t\} \in \mathfrak{F}_t^Y$. For all $n \in \mathbb{N}$, T_n is, therefore, an $(\mathfrak{F}_t^Y)_{t\geq 0}$-stopping time. □

We now state some results concerning the structure of $(\mathfrak{F}_t^Y)_{t\geq 0}$-stopping times. These results can be easily obtain by using similar arguments as for those of section 1.7.

THEOREM 8.1.– σ is an $(\mathfrak{F}_t^Y)_{t\geq 0}$-stopping time if and only if there exists a sequence of non-negative $(\mathfrak{F}_{T_n}^Y)_{n\in\mathbb{N}}$-adapted random variables $(R_n)_{n\in\mathbb{N}}$ satisfying $R_{n+1} = 0$ on $\{S_{n+1} > R_n\}$ and

$$\sigma = \sum_{n=1}^{\infty} R_{n-1} \wedge S_n.$$

We outline the following result, which is a direct consequence of the above theorem, because it will be used several times in our derivation.

LEMMA 8.4.– Let σ be an $(\mathfrak{F}_t^Y)_{t\geq 0}$-stopping time and $(R_n)_{n\in\mathbb{N}}$ be the sequence of random variables associated with σ as introduced in theorem 8.1. For all $n \in \mathbb{N}$, $\{T_n \leq \sigma < T_{n+1}\} = \{T_n \leq \sigma\} \cap \{S_{n+1} > R_n\}$.

PROOF.– Theorem 8.1 states that on the event $\{T_n \leq \sigma\}$, we have $\sigma \wedge T_{n+1} = T_n + (R_n \wedge S_{n+1})$ so that, still on the event $\{T_n \leq \sigma\}$, we have $\{\sigma < T_{n+1}\} \Leftrightarrow \{R_n < S_{n+1}\}$. We deduce the result from this observation. □

We now investigate the effect of the translation operator of the Markov chain $(\Pi_n, Y_n, S_n)_{n\in\mathbb{N}}$ on the $(\mathfrak{F}_t^Y)_{t\geq 0}$-stopping times. Let us consider the canonical space $\Omega = (\mathcal{P}(E_0) \times \mathbb{R}^d \times \mathbb{R}^+)^{\mathbb{N}}$ of the $(\mathfrak{F}_{T_n}^Y)_{n\in\mathbb{N}}$-Markov chain $(\Pi_n, Y_n, S_n)_{n\in\mathbb{N}}$. Thus, for $\omega = (\omega_0, \omega_1, \ldots) \in \Omega$, we have $(\Pi_n, Y_n, S_n)(\omega) = \omega_n$. Besides, we define the *translation operator*

$$\theta : \begin{cases} \Omega & \rightarrow \quad \Omega \\ (\omega_0, \omega_1, \ldots) & \rightarrow (\omega_1, \omega_2, \ldots) \end{cases}$$

We then define $\theta^0 = Id_\Omega$ and recursively for $l \geq 2$, $\theta^l = \theta \circ \theta^{l-1}$. Thus, for all $n \in \mathbb{N}^*$ and $l \in \mathbb{N}$, we have $(\Pi_n, Y_n, S_n) \circ \theta^l = (\Pi_{n+l}, Y_{n+l}, S_{n+l})$. As $T_0 = 0$, we have

$$T_n \circ \theta^l = \sum_{k=1}^{n} S_k \circ \theta^l = \sum_{k=1}^{n} S_{k+l} = T_{n+l} - T_l,$$

for $n \in \mathbb{N}^*$.

The following result will be important for the derivation of the dynamic programming equation.

PROPOSITION 8.2.– Consider σ an $(\mathfrak{F}_t^Y)_{t\geq 0}$-stopping time bounded by T_N. Then, there exists a mapping $\widetilde{\sigma} : (\mathbb{R}^d \times \mathbb{R}^+)^2 \times \Omega \rightarrow \mathbb{R}^+$ such that $\widetilde{\sigma}(\gamma, \cdot)$ is a $(\mathfrak{F}_t^Y)_{t\geq 0}$-stopping time for any $\gamma \in (\mathbb{R}^d \times \mathbb{R}^+)^2$ and

$$\sigma = T_1 + \widetilde{\sigma}(Y_0, S_0, Y_1, S_1, \theta),$$

on $\{\sigma \geq T_1\}$. Moreover, there exists a family $(R_n)_{n\in\{0,\ldots,N-1\}}$ of non-negative random variable $\mathfrak{F}^Y_{T_n}$-measurable and a family $(\widetilde{R}_n)_{n\in\{0,\ldots,N-2\}}$ of measurable mappings defined on $\left(\mathbb{R}^d \times \mathbb{R}^+\right)^2 \times \Omega$ with values on $\mathbb{R}^+$ such that $\widetilde{R}_n(\gamma,\cdot)$ is $\mathfrak{F}^Y_{T_n}$-measurable for any $\gamma \in \left(\mathbb{R}^d \times \mathbb{R}^+\right)^2$ and

$$\sigma = \sum_{n=1}^{N} R_{n-1} \wedge S_n,$$

$$\widetilde{\sigma}(\gamma,\cdot) = \sum_{n=1}^{N-1} \widetilde{R}_{n-1}(\gamma,\cdot) \wedge S_n,$$

and for any $n \in \{1,\ldots,N\}$

$$R_n = \widetilde{R}_{n-1}(Y_0, S_0, Y_1, S_1, \theta).$$

We now present a result required in the sequel, investigating the Markov property of the filter process.

PROPOSITION 8.3.– The sequences $(\Pi_n, Y_n, S_n)_{n\in\mathbb{N}}$, $(\Pi_n, S_n)_{n\in\mathbb{N}}$ and $(\Pi_n)_{n\in\mathbb{N}}$ are $(\mathfrak{F}^Y_{T_n})_{n\in\mathbb{N}}$-Markov chains. Moreover, for any $n \in \mathbb{N}^*$

$$\mathbb{E}[h(\Pi_n, Y_n, S_n)|\mathfrak{F}^Y_{T_{n-1}}] = \mathbb{E}\big[h(\Pi_n, Y_n, S_n)\big|\Pi_{n-1}\big].$$

PROOF.– Let $h \in \mathbf{B}(\mathcal{P}(E_0) \times \mathbb{R}^d \times \mathbb{R}^+)$. The law of iterated conditional expectations yields

$$\mathbb{E}[h(\Pi_n, Y_n, S_n)|\mathfrak{F}^Y_{T_{n-1}}] = \mathbb{E}\big[\mathbb{E}[h(\Pi_n, Y_n, S_n)|\mathfrak{F}_{T_{n-1}}]\big|\mathfrak{F}^Y_{T_{n-1}}\big].$$

From proposition 8.1 and equation [8.1], which defines Y_n, we obtain

$$\begin{aligned}
&\mathbb{E}[h(\Pi_n, Y_n, S_n)|\mathfrak{F}_{T_{n-1}}] \\
&\quad= \mathbb{E}\big[h\big(\Psi(\Pi_{n-1}, \varphi(Z_n) + W_n, S_n), \varphi(Z_n) + W_n, S_n\big)\big|\mathfrak{F}_{T_{n-1}}\big] \\
&\quad= \sum_{j=1}^{q} \int_{\mathbb{R}^d} h\big(\Psi(\Pi_{n-1}, \varphi(x_j) + w, s), \varphi(x_j) + w, s\big) \\
&\qquad \times \mathbb{P}(Z_n = x_j, W_n \in dw, S_n \in ds|\mathfrak{F}_{T_{n-1}}).
\end{aligned}$$

Yet, W_n is independent of $\sigma(Z_n, S_n) \vee \mathfrak{F}_{T_{n-1}}$ and admits the density function f_W. As in the proof of lemma 8.1, we thus obtain

$$\begin{aligned}&\mathbb{E}[h(\Pi_n, Y_n, S_n)|\mathfrak{F}_{T_{n-1}}]\\&= \sum_{j=1}^{q} \int_{\mathbb{R}^d} h(\Psi(\Pi_{n-1}, y, s), y, s)\mathbb{P}(Z_n = x_j, S_n \in ds|\mathfrak{F}_{T_{n-1}})\\&\times f_W(y - \varphi(x_j))dy.\end{aligned}$$

Besides, we have $\mathbb{P}(Z_n = x_j, S_n \in ds|\mathfrak{F}_{T_{n-1}}) = \mathbb{P}(Z_n = x_j, S_n \in ds|Z_{n-1})$ as in the proof of lemma 8.2, so that we have

$$\begin{aligned}&\mathbb{E}[h(\Pi_n, Y_n, S_n)|\mathfrak{F}_{T_{n-1}}]\\&= \sum_{i=1}^{q} \mathbb{1}_{\{Z_{n-1}=x_i\}} \sum_{j=1}^{q} \int_{\mathbb{R}^d} \Big(\int_0^{t_i^*} h(\Psi(\Pi_{n-1}, y, s), y, s)\\&\times \lambda\big(\Phi(x_i, s)\big)e^{-\Lambda(x_i,s)}Q\big(\Phi(x_i, s), x_j\big)ds\\&+ h\big(\Psi(\Pi_{n-1}, y, t_i^*), y, t_i^*\big)e^{-\Lambda(x_i,t_i^*)}Q\big(\Phi(x_i, t_i^*), x_j\big)\Big) f_W(y - \varphi(x_j))dy.\end{aligned}$$

Take now the conditional expectation w.r.t. $\mathfrak{F}^Y_{T_{n-1}}$, to obtain

$$\begin{aligned}&\mathbb{E}[h(\Pi_n, Y_n, S_n)|\mathfrak{F}^Y_{T_{n-1}}]\\&= \sum_{i=1}^{q} \Pi_{n-1}^i \sum_{j=1}^{q} \int_{\mathbb{R}^d} \Big(\int_0^{t_i^*} h(\Psi(\Pi_{n-1}, y, s), y, s)\\&\times \lambda\big(\Phi(x_i, s)\big)e^{-\Lambda(x_i,s)}Q\big(\Phi(x_i, s), x_j\big)ds\\&+ h\big(\Psi(\Pi_{n-1}, y, t_i^*), y, t_i^*\big)e^{-\Lambda(x_i,t_i^*)}Q\big(\Phi(x_i, t_i^*), x_j\big)\Big) f_W(y - \varphi(x_j))dy.\end{aligned}$$

Hence, $\mathbb{E}[h(\Pi_n, Y_n, S_n)|\mathfrak{F}^Y_{T_{n-1}}]$ is merely a function of Π_{n-1} yielding the result for the three processes. □

Finally, we derive a property of the conditional expectation of the PDMP with respect to the filtration generated by the state and observation processes.

LEMMA 8.5.– Let $(s_k)_{k\in\mathbb{N}}$ be a sequence of $(\mathfrak{F}_{T_k})_{k\in\mathbb{N}}$-adapted random variables. Then, we have

$$\mathbb{E}[\mathbb{1}_{\{S_{k+1}>s_k\}}|\mathfrak{F}_{T_k}] = \mathbb{1}_{\{s_k<t^*(Z_k)\}}e^{-\Lambda(Z_k,s_k)}.$$

PROOF.– First, recall some results concerning the random variables $(S_k)_{k\in\mathbb{N}}$, whose details may be found in [DAV 93]. After a jump of the process to the point $z \in E$, the survival function of the time until the next jump is

$$\psi(t,z) = \begin{cases} 1 & \text{if } t \leq 0, \\ e^{-\Lambda(z,t)} & \text{if } 0 \leq t < t^*(z), \\ 0 & \text{if } t \geq t^*(z). \end{cases}$$

Define $\overline{\psi}(u,z) = \inf\{t \geq 0 \text{ such that } \psi(t,z) \leq u\}$. Then, for all $k \in \mathbb{N}$, we have $S_{k+1} = \overline{\psi}(\Upsilon_k, Z_k)$, where Υ_k are i.i.d. random variables with uniform distribution on $[0;1]$ independent of $\mathfrak{F}_{T_k}$. Thus, $\mathbb{E}[\mathbb{1}_{\{S_{k+1}>s_k\}}|\mathfrak{F}_{T_k}] = \mathbb{E}[f(\Upsilon_k, Z_k, s_k)|\mathfrak{F}_{T_k}]$ where $f(u,z,r) = \mathbb{1}_{\{\overline{\psi}(u,z)>r\}}$. As (Z_k, s_k) is $\mathfrak{F}_{T_k}$-measurable, Υ_k is independent of $\mathfrak{F}_{T_k}$ and $\mathbb{E}[\mathbb{1}_{\{\overline{\psi}(\Upsilon_k,z)>r\}}] = \mathbb{1}_{\{r<t^*(z)\}}e^{-\Lambda(z,r)}$, [OUV 04, Proposition 11.2] yields the result. □

8.4.2. *Optimal stopping problem under complete observation*

In this section, we show how our optimal stopping problem under partial observation for the process $(X_t)_{t\geq 0}$ can be converted into an optimal stopping problem under complete observation involving the Markov chain $(\Pi_n, S_n)_{0\leq n\leq N}$. More precisely, for a fixed stopping time $\sigma \in \Sigma_N^Y$, we show in proposition 8.4 that the performance criterion $\mathbb{E}[g(X_\sigma)|\Pi_0 = \pi]$ can be expressed in terms of the discrete-time Markov chain $(\Pi_n, S_n)_{0\leq n\leq N}$. We would like to emphasize the following important fact. Although the performance criterion can be written in terms of *discrete-time* process, the optimization problem remains intrinsically a *continuous-time* optimization problem. Indeed, the performance criterion is maximized over the set of stopping times with respect to the *continuous-time* filtration $(\mathfrak{F}_t^Y)_{t\geq 0}$.

PROPOSITION 8.4.– Let $\sigma \in \Sigma^Y$ and $n \geq 1$. For all $\pi \in \mathcal{P}(E_0)$, we have

$$\begin{aligned} &\mathbb{E}[g(X_{\sigma\wedge T_n})|\Pi_0 = \pi] \\ &= \sum_{k=0}^{n-1}\sum_{i=1}^{q} \mathbb{E}[\mathbb{1}_{\{T_k\leq\sigma\}}\mathbb{1}_{\{R_k<t_i^*\}} g\circ\Phi(x_i,R_k)e^{-\Lambda(x_i,R_k)}\Pi_k^i|\Pi_0 = \pi] \\ &\quad + \sum_{i=1}^{q} \mathbb{E}[\mathbb{1}_{\{T_n\leq\sigma\}} g(x_i)\Pi_n^i|\Pi_0 = \pi], \end{aligned}$$

where $(R_k)_{k\in\mathbb{N}}$ is the sequence of non-negative random variables associated with σ as introduced in theorem 8.1.

PROOF.– We split $\mathbb{E}[g(X_{\sigma \wedge T_n})|\Pi_0 = \pi]$ into several terms depending on the position of σ w.r.t. the jump times T_k

$$
\begin{aligned}
&\mathbb{E}[g(X_{\sigma\wedge T_n})\big|\Pi_0 = \pi] \\
&= \sum_{k=0}^{n-1}\sum_{i=1}^{q} \mathbb{E}[\mathbb{1}_{\{T_k \le \sigma < T_{k+1}\}}\mathbb{1}_{\{Z_k = x_i\}} g\circ\Phi(x_i, R_k)|\Pi_0 = \pi] \\
&+ \sum_{i=1}^{q} \mathbb{E}[\mathbb{1}_{\{T_n \le \sigma\}}\mathbb{1}_{\{Z_n = x_i\}} g(x_i)|\Pi_0 = \pi].
\end{aligned}
$$

For notational convenience, consider

$$
\begin{cases} A_{k,i} &= \mathbb{1}_{\{T_k \le \sigma < T_{k+1}\}}\mathbb{1}_{\{Z_k = x_i\}} g\circ\Phi(x_i, R_k), \\ B_i &= \mathbb{1}_{\{T_n \le \sigma\}}\mathbb{1}_{\{Z_n = x_i\}} g(x_i). \end{cases}
$$

On the one hand, we have $\mathbb{E}[B_i|\mathfrak{F}^Y_{T_n}] = g(x_i)\mathbb{1}_{\{T_n\le\sigma\}}\Pi^i_n$ since $\{T_n \le \sigma\} \in \mathfrak{F}^Y_{T_n}$ (see, for instance, [BRÉ 81, p. 298, Theorem T7]). On the other hand, to compute $\mathbb{E}[A_{k,i}|\mathfrak{F}^Y_{T_k}]$, we use lemma 8.4 to obtain

$$
\begin{aligned}
\mathbb{E}[A_{k,i}|\mathfrak{F}^Y_{T_k}] &= \mathbb{1}_{\{T_k\le\sigma\}} g\circ\Phi(x_i, R_k)\mathbb{E}[\mathbb{1}_{\{S_{k+1}>R_k\}}\mathbb{1}_{\{Z_k=x_i\}}|\mathfrak{F}^Y_{T_k}] \\
&= \mathbb{1}_{\{T_k\le\sigma\}} g\circ\Phi(x_i, R_k)\mathbb{E}\big[\mathbb{1}_{\{Z_k=x_i\}}\mathbb{E}[\mathbb{1}_{\{S_{k+1}>R_k\}}|\mathfrak{F}_{T_k}]\big|\mathfrak{F}^Y_{T_k}\big] \\
&= \mathbb{1}_{\{T_k\le\sigma\}} g\circ\Phi(x_i, R_k)\mathbb{E}[\mathbb{1}_{\{Z_k=x_i\}}\mathbb{1}_{\{R_k<t^*(Z_k)\}} \\
&\quad \times e^{-\Lambda(Z_k,R_k)}|\mathfrak{F}^Y_{T_k}] \\
&= \mathbb{1}_{\{T_k\le\sigma\}} g\circ\Phi(x_i, R_k)\mathbb{1}_{\{R_k<t^*_i\}}e^{-\Lambda(x_i,R_k)}\Pi^i_k.
\end{aligned}
$$

where we used lemma 8.5 to get the third equality. The result confirmed. □

8.4.3. *Dynamic programming equation*

Based on the new formulation, the main objective of this section is to derive the backward dynamic programming equation. It involves some operators introduced in definition 8.2. By iterating the operator, labeled L, we define a sequence of real-valued functions $(v_n)_{0\le n\le N}$ in definition 8.3. Theorem 8.2 establishes that v_n is the value function of our partially observed optimal stopping problem with horizon T_{N-n} and in particular that v_0 is the value function of problem defined in equation [8.3].

Another important result of this section is given by theorem 8.3 which constructs a sequence of ϵ-optimal stopping times.

DEFINITION 8.2.– *The operators* $G : \mathbf{B}(\mathcal{P}(E_0)) \to \mathbf{B}(\mathcal{P}(E_0) \times \mathbb{R}^+)$, $H : \mathbf{B}(E) \to \mathbf{B}(\mathcal{P}(E_0) \times \mathbb{R}^+)$, $J : \mathbf{B}(\mathcal{P}(E_0)) \times \mathbf{B}(E) \to \mathbf{B}(\mathcal{P}(E_0) \times \mathbb{R}^+)$, *and* $L : \mathbf{B}(\mathcal{P}(E_0)) \times \mathbf{B}(E) \to \mathbf{B}(\mathcal{P}(E_0))$ *are defined for all* $(v, h) \in \mathbf{B}(\mathcal{P}(E_0)) \times \mathbf{B}(E)$ *and* $(\pi, u) \in \mathcal{P}(E_0) \times \mathbb{R}^+$ *by*

$$\begin{aligned} Gv(\pi, u) &= \mathbb{E}[v(\Pi_1)\mathbb{1}_{\{S_1 \leq u\}}|\Pi_0 = \pi], \\ Hh(\pi, u) &= \sum_{i=1}^{q} h \circ \Phi(x_i, u)\mathbb{1}_{\{u < t_i^*\}}e^{-\Lambda(x_i, u)}\pi^i, \\ J(v, h)(\pi, u) &= Hh(\pi, u) + Gv(\pi, u), \\ L(v, h)(\pi) &= \sup_{u \geq 0} J(v, h)(\pi, u). \end{aligned}$$

DEFINITION 8.3.– *The sequence* $(v_n)_{0 \leq n \leq N}$ *of real-valued functions is defined on* $\mathcal{P}(E_0)$ *by*

$$\begin{cases} v_N(\pi) = \sum_{i=1}^{q} g(x_i)\pi^i, \\ v_{n-1}(\pi) = L(v_n, g)(\pi), \quad 1 \leq n \leq N. \end{cases}$$

The following theorem is the main result of this section showing that the operator L is the dynamic programming operator associated with the initial optimization problem.

THEOREM 8.2.– For all $1 \leq n \leq N$ and $\pi \in \mathcal{P}(E_0)$, we have

$$\sup_{\sigma \in \Sigma_n^Y} \mathbb{E}[g(X_\sigma)|\Pi_0 = \pi] = v_{N-n}(\pi).$$

PROOF.– The proof of this result is based on proposition 8.5 and theorem 8.3. Proposition 8.5 proves that v_{N-n} is an upper bound for the value function of the problem with horizon T_n. The reverse inequality is derived in theorem 8.3 by constructing a sequence of ϵ-optimal stopping times. □

PROPOSITION 8.5.– For all $1 \leq n \leq N$ and $\pi \in \mathcal{P}(E_0)$, we have

$$\sup_{\sigma \in \Sigma_n^Y} \mathbb{E}[g(X_\sigma)|\Pi_0 = \pi] \leq v_{N-n}(\pi).$$

PROOF.– Let $\sigma \in \Sigma_N^Y$. Consider $(R_k)_{k \in \mathbb{N}}$ the sequence associated with σ as introduced in theorem 8.1. We prove the theorem by induction on n. For $n = 1$, proposition 8.4 yields

$$\mathbb{E}[g(X_{\sigma\wedge T_1})|\Pi_0=\pi] = \sum_{i=1}^{q}\mathbb{E}[\mathbb{1}_{\{R_0<t_i^*\}}g\circ\Phi(x_i,R_0)e^{-\Lambda(x_i,R_0)}\Pi_0^i|\Pi_0=\pi]$$
$$+\sum_{i=1}^{q}\mathbb{E}[\mathbb{1}_{\{T_1\le\sigma\}}g(x_i)\Pi_1^i|\Pi_0=\pi]. \qquad [8.7]$$

Since R_0 is deterministic and by using lemma 8.7, we recognize that the first term of the right-hand side of equation [8.7] is $Hg(\pi,R_0)$. We now turn to the second term of the right-hand side of equation [8.7] which is given by

$$\mathbb{E}[\mathbb{1}_{\{S_1\le R_0\}}\sum_{i=1}^{q}g(x_i)\Pi_1^i|\Pi_0=\pi] = \mathbb{E}[v_N(\Pi_1)\mathbb{1}_{\{S_1\le R_0\}}|\Pi_0=\pi]$$
$$= Gv_N(\pi,R_0),$$

from lemma 8.4 and the definition of G. Recall that from definition 8.2 we have $J(v_N,g)=Hg+Gv_N$ thus, we obtain

$$\mathbb{E}[g(X_{\sigma\wedge T_1})|\Pi_0=\pi] = J(v_N,g)(\pi,R_0)\le\sup_{u\ge 0}J(v_N,g)(\pi,u)$$
$$= L(v_N,g)(\pi)=v_{N-1}(\pi).$$

Set now $2\le n\le N$ and assume that $\mathbb{E}[g(X_\tau)|\Pi_0=\pi]\le v_{N-(n-1)}(\pi)$, for all $\tau\in\Sigma_{n-1}^Y$. Proposition 8.4 yields

$$\mathbb{E}[g(X_{\sigma\wedge T_n})|\Pi_0=\pi]$$
$$=\sum_{k=0}^{n-1}\sum_{i=1}^{q}\mathbb{E}[\mathbb{1}_{\{T_k\le\sigma\}}\mathbb{1}_{\{R_k<t_i^*\}}g\circ\Phi(x_i,R_k)e^{-\Lambda(x_i,R_k)}\Pi_k^i|\Pi_0=\pi]$$
$$+\sum_{i=1}^{q}\mathbb{E}[\mathbb{1}_{\{T_n\le\sigma\}}g(x_i)\Pi_n^i|\Pi_0=\pi].$$

As in the case $n=1$, the term for $k=0$ equals $Hg(\pi,R_0)$. Notice that for $k\ge 1$, $\mathbb{1}_{\{T_k\le\sigma\}}=\mathbb{1}_{\{T_k\le\sigma\}}\mathbb{1}_{\{T_1\le\sigma\}}$ and that $\mathbb{1}_{\{T_1\le\sigma\}}=\mathbb{1}_{\{S_1\le R_0\}}$ is $\mathfrak{F}_{T_1}^Y$-measurable. By taking the conditional expectation w.r.t. $\mathfrak{F}_{T_1}^Y$, it follows that

$$\mathbb{E}[g(X_{\sigma\wedge T_n})|\Pi_0=\pi]=Hg(\pi,R_0)+\mathbb{E}[\Xi\mathbb{1}_{\{S_1\le R_0\}}|\Pi_0=\pi] \qquad [8.8]$$

with

$$\Xi = \mathbb{E}\Big[\sum_{k=1}^{n-1}\sum_{i=1}^{q} \mathbb{1}_{\{T_k \le \sigma\}} \mathbb{1}_{\{R_k < t_i^*\}} g \circ \Phi(x_i, R_k) e^{-\Lambda(x_i, R_k)} \Pi_k^i$$
$$+ \sum_{i=1}^{q} \mathbb{1}_{\{T_n \le \sigma\}} g(x_i) \Pi_n^i \big| \mathfrak{F}_{T_1}^Y \Big].$$

We now use the Markov property of the chain $(\Pi_n, Y_n, S_n)_{n\in\mathbb{N}}$. Indeed, for $k \ge 1$, we have $\Pi_k = \Pi_{k-1} \circ \theta$, where θ is the translation operator of the $(\mathfrak{F}_{T_n}^Y)_{n\in\mathbb{N}}$-Markov chain $(\Pi_n, Y_n, S_n)_{n\in\mathbb{N}}$. Moreover, from proposition 8.2, we have the existence of $\widetilde{\sigma}$ and $(\widetilde{R}_k)_{k\in\{0,\cdots,N-2\}}$ satisfying $\sigma = T_1 + \widetilde{\sigma}(Y_0, S_0, Y_1, S_1, \theta)$ and $R_k = \widetilde{R}_{k-1}(Y_0, S_0, Y_1, S_1, \theta)$ on $\{T_1 \le \sigma\}$. Since for $k \ge 1$, $T_k = T_1 + T_{k-1} \circ \theta$, we have $\mathbb{1}_{\{T_k \le \sigma\}} = \mathbb{1}_{\{T_{k-1} \le \widetilde{\sigma}(Y_0, S_0, Y_1, S_1, \theta)\}}$. However, by using a property of the conditional expectation, we have $\Xi = \widetilde{\Xi}(Y_0, S_0, Y_1, S_1)$ where

$$\widetilde{\Xi}(\gamma) = \mathbb{E}\Big[\sum_{k=1}^{n-1}\sum_{i=1}^{q} \mathbb{1}_{\{T_{k-1}\circ\theta \le \widetilde{\sigma}(\gamma,\theta)\}} \mathbb{1}_{\{\widetilde{R}_{k-1}(\gamma,\theta) < t_i^*\}} g \circ \Phi(x_i, \widetilde{R}_{k-1}(\gamma, \theta))$$
$$\times e^{-\Lambda(x_i, \widetilde{R}_{k-1}(\gamma,\theta))} \Pi_{k-1}^i \circ \theta \big| \mathfrak{F}_{T_1}^Y \Big]$$
$$+ \mathbb{E}\Big[\sum_{i=1}^{q} \mathbb{1}_{\{T_{n-1}\circ\theta \le \widetilde{\sigma}(\gamma,\theta)\}} g(x_i) \Pi_{n-1}^i \circ \theta \big| \mathfrak{F}_{T_1}^Y \Big],$$

for $\gamma \in (\mathbb{R}^d \times \mathbb{R}^+)^2$. Now, by using the Markov property of the chain $(\Pi_n, Y_n, S_n)_{n\in\mathbb{N}}$ (see proposition 8.3), we obtain that

$$\widetilde{\Xi}(\gamma) = w(\gamma, \Pi^1)$$

where

$$w(\gamma, \pi) = \mathbb{E}\Big[\sum_{k=1}^{n-1}\sum_{i=1}^{q} \mathbb{1}_{\{T_{k-1} \le \widetilde{\sigma}(\gamma,\cdot)\}} \mathbb{1}_{\{\widetilde{R}_{k-1}(\gamma,\cdot) < t_i^*\}}$$
$$\times g \circ \Phi(x_i, \widetilde{R}_{k-1}(\gamma, \cdot)) e^{-\Lambda(x_i, \widetilde{R}_{k-1}(\gamma,\cdot))} \Pi_{k-1}^i \big| \Pi_0 = \pi \Big]$$
$$+ \mathbb{E}\Big[\sum_{i=1}^{q} \mathbb{1}_{\{T_{n-1} \le \widetilde{\sigma}(\gamma,\cdot)\}} g(x_i) \Pi_{n-1}^i \big| \Pi_0 = \pi \Big],$$

and from proposition 8.4, it follows

$$w(\gamma, \pi) = \mathbb{E}[g(X_{\widetilde{\sigma}(\gamma,\cdot)\wedge T_{n-1}}) | \Pi_0 = \pi].$$

Therefore,

$$\Xi = w(Y_0, S_0, Y_1, S_1, \Pi_1)$$

and so,

$$\Xi \leq v_{N-(n-1)}(\Pi_1) \qquad [8.9]$$

from the induction hypothesis. Finally, combining equations [8.8] and [8.9], we have

$$\begin{aligned}&\mathbb{E}[g(X_{\sigma \wedge T_n})|\Pi_0 = \pi] \\ &\leq Hg(\pi, R_0) + \mathbb{E}[v_{N-(n-1)}(\Pi_1)\mathbb{1}_{\{S_1 \leq R_0\}}|\Pi_0 = \pi].\end{aligned}$$

In the second term, we recognize the operator G implying

$$\begin{aligned}\mathbb{E}[g(X_{\sigma \wedge T_n})|\Pi_0 = \pi] &\leq Hg(\pi, R_0) + Gv_{N-(n-1)}(\pi, R_0) \\ &= J(v_{N-(n-1)}, g)(\pi, R_0) \\ &\leq \sup_{u \geq 0} J(v_{N-(n-1)}, g)(\pi, u) \\ &= L(v_{N-(n-1)}, g)(\pi) \;=\; v_{N-n}(\pi),\end{aligned}$$

that proves the induction. □

We now prove the reverse inequality by constructing a sequence of ϵ-optimal stopping times.

DEFINITION 8.4.– *For $\epsilon > 0$, $0 \leq n \leq N-1$ and for $\pi \in \mathcal{P}(E_0)$, we define*

$$r_n^\epsilon(\pi) = \inf\{u > 0 : J(v_{N-n}, g)(\pi, u) > v_{N-n-1}(\pi) - \epsilon\}.$$

Consider $R_{1,0}^\epsilon = r_0^\epsilon(\Pi_0)$ and for $2 \leq n \leq N$,

$$\begin{cases} R_{n,0}^\epsilon &= r_{n-1}^{\epsilon/2}(\Pi_0), \\ R_{n,k}^\epsilon &= r_{n-1-k}^{\epsilon/(2^{k+1})}(\Pi_k)\mathbb{1}_{\{R_{n,k-1}^\epsilon \geq S_k\}} \text{ for } 1 \leq k \leq n-2, \\ R_{n,n-1}^\epsilon &= r_0^{\epsilon/(2^{n-1})}(\Pi_{n-1})\mathbb{1}_{\{R_{n,n-2}^\epsilon \geq S_{n-1}\}}, \end{cases}$$

and finally set

$$U_n^\epsilon = \sum_{k=1}^{n} R_{n,k-1}^\epsilon \wedge S_k.$$

The following lemma describes the effect of the translation operator θ on the sequence $(R_{n,k}^\epsilon)_{1 \leq n \leq N, 0 \leq k \leq n-1}$.

LEMMA 8.6.– For $\epsilon > 0, n \geq 2$ and $1 \leq k \leq n-1$, on the set $\{T_1 \leq U_n^{2\epsilon}\}$, we have

$$R_{n,k}^{2\epsilon} = R_{n-1,k-1}^{\epsilon} \circ \theta.$$

PROOF.– For $n = 2$, we just have to prove that on the event $\{T_1 \leq U_2^{2\epsilon}\}$, we have $R_{1,0}^{\epsilon} \circ \theta = R_{2,1}^{2\epsilon}$. Yet, from the definition of the sequence $(R_{n,k}^{\epsilon})_{1 \leq n \leq N, 0 \leq k \leq n-1}$, we have $R_{1,0}^{\epsilon} \circ \theta = r_0^{\epsilon}(\Pi_1)$ and $R_{2,1}^{2\epsilon} = r_0^{\frac{2\epsilon}{2}}(\Pi_1)\mathbb{1}_{\{R_{2,0}^{2\epsilon} \geq S_1\}}$. The result follows since we are on the event $\{T_1 \leq U_2^{2\epsilon}\} = \{R_{2,0}^{2\epsilon} \geq S_1\}$. For a fixed $n \geq 3$, we prove the lemma by induction on $1 \leq k \leq n-1$. Set $k = 1$. We have from the definition on the sequence $(R_{n,k}^{\epsilon})_{1 \leq n \leq N, 0 \leq k \leq n-1}$, $R_{n-1,0}^{\epsilon} \circ \theta = r_{n-2}^{\frac{\epsilon}{2}}(\Pi_1)$ and $R_{n,1}^{2\epsilon} = r_{n-2}^{\frac{2\epsilon}{4}}(\Pi_1)\mathbb{1}_{\{R_{n,0}^{2\epsilon} \geq S_1\}}$. We obtain $R_{n-1,0}^{\epsilon} \circ \theta = R_{n,1}^{2\epsilon}$ because we have assumed that we are on the event $\{T_1 \leq U_n^{2\epsilon}\} = \{R_{n,0}^{2\epsilon} \geq S_1\}$. The propagation of the induction is similar to the case $k = 1$. □

Equipped with this preliminary result, we may now prove that $(U_n^{\epsilon})_{1 \leq n \leq N}$ is a sequence of ϵ-optimal stopping times with respect to the filtration generated by the observations.

THEOREM 8.3.– For all $1 \leq n \leq N$ and $\epsilon > 0$, we have $U_n^{\epsilon} \in \Sigma_n^Y$ and

$$\mathbb{E}[g(X_{U_n^{\epsilon}})|\Pi_0 = \pi] \geq v_{N-n}(\pi) - \epsilon.$$

PROOF.– Let $n \in \{1, \ldots, N\}$. First, notice that, as a direct consequence of theorem 8.1, U_n^{ϵ} is an $(\mathfrak{F}_t^Y)_{t \geq 0}$-stopping time since, by construction, the $R_{n,k}^{\epsilon}$ are $\mathfrak{F}_{T_k}^Y$-measurable and satisfy the condition $R_{n,k}^{\epsilon} = 0$ on the event $\{S_k > R_{n,k-1}^{\epsilon}\}$. It is also clear that $U_n^{\epsilon} \leq \sum_{k=1}^n S_k = T_n$. Thus, we have $U_n^{\epsilon} \in \Sigma_n^Y$. Let us now prove the second assessment by induction. Set $n = 1$. Let $\pi \in \mathcal{P}(E_0)$, we denote $r_0^{\epsilon} = r_0^{\epsilon}(\pi)$. Since $R_{1,0}^{\epsilon} = r_0^{\epsilon}$ is deterministic, we have clearly $R_{1,0}^{\epsilon} \in \Sigma^Y$. Consequently, by using the same arguments as in the proof of proposition 8.4, we obtain

$$\mathbb{E}[g(X_{R_{1,0}^{\epsilon} \wedge S_1})|\Pi_0 = \pi] = Hg(\pi, r_0^{\epsilon}) + Gv_N(\pi, r_0^{\epsilon}) = J(v_N, g)(\pi, r_0^{\epsilon}).$$

Finally, the definition of r_0^{ϵ} yields $J(v_N, g)(\pi, r_0^{\epsilon}) \geq v_{N-1}(\pi) - \epsilon$ thus, we have

$$\mathbb{E}[g(X_{R_{1,0}^{\epsilon} \wedge S_1})|\Pi_0 = \pi] \geq v_{N-1}(\pi) - \epsilon.$$

Now, set $2 \le n \le N$ and assume that $\mathbb{E}[g(X_{U^\epsilon_{n-1}})|\Pi_0 = \pi] \ge v_{N-(n-1)}(\pi) - \epsilon$, for all $\epsilon > 0$. Proposition 8.4 yields

$$
\begin{aligned}
&\mathbb{E}[g(X_{U^{2\epsilon}_n})|\Pi_0 = \pi] \\
&= \sum_{k=0}^{n-1}\sum_{i=1}^{q} \mathbb{E}\left[\mathbb{1}_{\{T_k \le U^{2\epsilon}_n\}}\mathbb{1}_{\{R^{2\epsilon}_{n,k}<t^*_i\}} g\circ\Phi(x_i, R^{2\epsilon}_{n,k})e^{-\Lambda(x_i,R^{2\epsilon}_{n,k})}\Pi^i_k \big|\Pi_0 = \pi\right] \\
&\quad + \sum_{i=1}^{q} \mathbb{E}[\mathbb{1}_{\{T_n \le U^{2\epsilon}_n\}} g(x_i)\Pi^i_n|\Pi_0 = \pi].
\end{aligned}
$$

Denote $r^\epsilon_{n-1} = r^\epsilon_{n-1}(\pi)$. Similarly to the case $n = 1$, the term for $k = 0$ equals $Hg(\pi, r^\epsilon_{n-1})$ since $R^{2\epsilon}_{n,0} = r^\epsilon_{n-1}(\Pi_0)$. Take the conditional expectation w.r.t. $\mathfrak{F}^Y_{T_1}$ in the other terms. We have then,

$$
\mathbb{E}[g(X_{U^{2\epsilon}_n})|\Pi_0 = \pi] = Hg(\pi, r^\epsilon_{n-1}) + \mathbb{E}[\Xi'\mathbb{1}_{\{T_1 \le U^{2\epsilon}_n\}}|\Pi_0 = \pi], \quad [8.10]
$$

with

$$
\begin{aligned}
\Xi' = \mathbb{E}\Big[&\sum_{k=1}^{n-1}\sum_{i=1}^{q} \mathbb{1}_{\{T_k \le U^{2\epsilon}_n\}}\mathbb{1}_{\{R^{2\epsilon}_{n,k}<t^*_i\}} g\circ\Phi(x_i, R^{2\epsilon}_{n,k})e^{-\Lambda(x_i,R^{2\epsilon}_{n,k})}\Pi^i_k \\
&+ \sum_{i=1}^{q} \mathbb{1}_{\{T_n \le U^{2\epsilon}_n\}} g(x_i)\Pi^i_n \big|\mathfrak{F}^Y_{T_1}\Big].
\end{aligned}
$$

Our objective is to apply the Markov property of $(\Pi_k)_{k\in\mathbb{N}}$ in the term Ξ'. Recall that, from lemma 8.6, we have $R^\epsilon_{n-1,k-1}\circ\theta = R^{2\epsilon}_{n,k}$ for $n \ge 2$ and $1 \le k \le n-1$ on the event $\{T_1 \le U^{2\epsilon}_n\} = \{S_1 \le R^{2\epsilon}_{n,0}\}$ (the equality of these events arises from lemma 8.4). Thus, on this set, we have

$$
\begin{aligned}
U^{2\epsilon}_n &= S_1 + \sum_{k=2}^{n} R^{2\epsilon}_{n,k-1}\wedge S_k = T_1 + \sum_{k=2}^{n}(R^\epsilon_{n-1,k-2}\circ\theta)\wedge(S_{k-1}\circ\theta) \\
&= T_1 + U^\epsilon_{n-1}\circ\theta.
\end{aligned}
$$

Recall that $T_k = T_1 + T_{k-1}\circ\theta$, for $k \ge 1$. Consequently, on the set $\{T_1 \le U^{2\epsilon}_n\}$, we have $\mathbb{1}_{\{T_k \le U^{2\epsilon}_n\}} = \mathbb{1}_{\{T_{k-1} \le U^\epsilon_{n-1}\}}\circ\theta$ and thus, combining the Markov property of the chain $(\Pi_k)_{k\ge 0}$ and proposition 8.4, we have

$$
\Xi'(\Pi_1) = w'(\Pi_1),
$$

with $w'(\pi) = \mathbb{E}\left[g(X_{U^{\epsilon}_{n-1}})\middle|\Pi_0 = \pi\right]$. Moreover, due to the induction assumption, we have $w'(\pi) \geq v_{N-(n-1)}(\pi) - \epsilon$ so that we obtain

$$\Xi' \geq v_{N-(n-1)}(\Pi_1) - \epsilon. \qquad [8.11]$$

Finally, combining equations [8.10] and [8.11] and noticing that, by lemma 8.4, $\{T_1 \leq U^{2\epsilon}_n\} = \{S_1 \leq r^{\epsilon}_{n-1}\}$, we obtain

$$\begin{aligned}
\mathbb{E}[g(X_{U^{2\epsilon}_n}) \ |\Pi_0 = \pi] & \\
&\geq Hg(\pi, r^{\epsilon}_{n-1}) + \mathbb{E}[v_{N-(n-1)}(\Pi_1)\mathbb{1}_{\{S_1 \leq r^{\epsilon}_{n-1}\}}|\Pi_0 = \pi] - \epsilon \\
&= J(v_{N-(n-1)}, g)(\pi, r^{\epsilon}_{n-1}) - \epsilon \\
&\geq v_{N-n}(\pi) - 2\epsilon,
\end{aligned}$$

from the definition of r^{ϵ}_{n-1}, showing the result. □

8.5. Numerical approximation by quantization

In this section, we are interested in the computational issue for our optimal stopping problem under partial observation. Indeed, we want to compute a numerical approximation of the value function [8.3] and propose a computable ϵ-optimal stopping time.

As we have seen in the previous section, the value function v can be obtained by iterating the dynamic programming operator L. However, the operator L involves conditional expectations that are in essence difficult to compute and iterate numerically. We manage to overcome this difficulty by combining two special properties of our problem. On the one hand, the underlying process (Π_n, S_n) in the expression of the operator L is a Markov chain. Therefore, it can be discretized using a quantization technique which is a powerful method suitable for numerical computation and iteration of conditional expectations. On the other hand, the recursion on the functions $(v_n)_{0\leq n\leq N}$ involving the operator L can be transformed into a recursion on suitably defined random variables as in section 7.2. Thus, they are easier to iterate numerically as we do not need to compute an approximation of each v_n on the whole state space.

This section is organized as follows. First, we explain how the recursion on the functions $(v_n)_{0\leq n\leq N}$ can be transformed into a recurrence on random variables involving only the Markov chain (Π_n, S_n). Afterward, we construct a quantized version of the main operators in definition 8.9 that is used to build an approximation of the value function in definition 8.10, and a computable ϵ-optimal stopping time in definition 8.11. The main results of this section are theorems 8.4 and 8.6 that prove the convergence of our approximation scheme and provide a rate of convergence for the value function and the optimal stopping rule, respectively.

We first explain how the dynamic programming equation, as introduced in definition 8.3, yields a recursion on the random variables $\big(v_n(\Pi_n)\big)_{0\leq n\leq N}$. The sequence $(V_n)_{0\leq n\leq N}$ of random variables defined by

$$V_n = v_n(\Pi_n).$$

In other words, we have $V_N = \sum_{i=1}^{q} g(x_i)\Pi_N^i$, and

$$\begin{aligned} V_n &= \sup_{u\geq 0}\sum_{i=1}^{q} g\circ\Phi(x_i,u)\mathbb{1}_{\{u<t_i^*\}}e^{-\Lambda(x_i,u)}\Pi_n^i + \mathbb{E}\big[V_{n+1}\mathbb{1}_{\{S_{n+1}\leq u\}}|\Pi_n\big] \\ &= Hg(\Pi_n,u) + \mathbb{E}\big[V_{n+1}\mathbb{1}_{\{S_{n+1}\leq u\}}|\Pi_n\big], \end{aligned} \quad [8.12]$$

for $0\leq n\leq N-1$. Notice that V_N is known and the expression of V_n involves only V_{n+1} and the Markov chain (Π_n, S_n). Thus, the sequence $(V_n)_{0\leq n\leq N}$ is completely characterized by the system [8.12]. In addition, $V_0 = v_0(\Pi_0) = v(\Pi_0)$. Thus, to approximate the value function v at the initial point of our process, it is sufficient to provide an approximation of the sequence of random variables $(V_N)_{0\leq n\leq N}$.

8.5.1. *Lipschitz properties*

In this section, we derive the Lipschitz properties of our operators in order to obtain them for the value functions $(v_n)_{0\leq n\leq N}$. Similarly to the proof of proposition 8.3, we first derive the integral form of operators G and H.

LEMMA 8.7.– For all $h\in\mathbf{B}(E)$, $v\in\mathbf{B}(\mathcal{P}(E_0))$ and $(\pi,u)\in\mathcal{P}(E_0)\times\mathbb{R}^+$, we have

$$\begin{aligned} Gv(\pi,u) &= Iv(\pi,u) + \sum_{i=1}^{q}\pi^i\mathbb{1}_{\{t_i^*\leq u\}}e^{-\Lambda(x_i,t_i^*)} \\ &\quad\times\int_{\mathbb{R}^d} v(\Psi(\pi,y',t_i^*))\sum_{j=1}^{q} Q\big(\Phi(x_i,t_i^*),x_j\big)f_W(y'-\varphi(x_j))dy', \end{aligned}$$

where

$$Iv(\pi, u) = \sum_{i=1}^{q} \pi^i \int_0^{u \wedge t_i^*} \Big(\lambda \circ \Phi(x_i, s') e^{-\Lambda(x_i, s')}$$

$$\times \int_{\mathbb{R}^d} v(\Psi(\pi, y', s')) \sum_{j=1}^{q} Q(\Phi(x_i, s'), x_j) f_W(y' - \varphi(x_j)) dy' \Big) ds'.$$

Now, note that the functions $Hh(\pi, \cdot)$ and $Gv(\pi, \cdot)$ are not continuous. However, they are càdlàg with a finite number of jumps. Therefore, they can be rewritten as sums of continuous functions as follows.

DEFINITION 8.5.– *For all $m \in \{0, \ldots, q-1\}$, we define the operators G^m: $\mathbf{B}(\mathcal{P}(E_0)) \to \mathbf{B}(\mathcal{P}(E_0) \times \mathbb{R}^+)$ and H^m: $\mathbf{B}(E) \to \mathbf{B}(\mathcal{P}(E_0) \times \mathbb{R}^+)$ as follows*

– *if $u < t_m^*$, $G^m v(\pi, u) = Gv(\pi, t_m^*)$ and $H^m h(\pi, u) = Hh(\pi, t_m^*)$,*

– *if $u \geq t_m^*$,*

$$G^m v(\pi, u) = Iv(\pi, u \wedge t_{m+1}^*) + \sum_{i=1}^{m} \pi^i e^{-\Lambda(x_i, t_i^*)}$$

$$\times \int_{\mathbb{R}^d} v(\Psi(\pi, y', t_i^*)) \sum_{j=1}^{q} Q(\Phi(x_i, t_i^*), x_j) f_W(y' - \varphi(x_j)) dy',$$

$$H^m h(\pi, u) = \sum_{i=m+1}^{q} \pi^i e^{-\Lambda(x_i, u \wedge t_{m+1}^*)} h \circ \Phi(x_i, u \wedge t_{m+1}^*).$$

We also define $J^m(v, h)(\pi, u) = H^m h(\pi, u) + G^m v(\pi, u)$.

REMARK 8.1.– For all $m \in \{0, \ldots, q-1\}$ and for all $h \in \mathbf{B}(E)$, $v \in \mathbf{B}(\mathcal{P}(E_0))$ and $(\pi, u) \in \mathcal{P}(E_0) \times \mathbb{R}^+$, the functions $u \to G^m v(\pi, u)$, $u \to H^m h(\pi, u)$ and $u \to J^m(v, h)(\pi, u)$ are continuous. Moreover, they are constant on $[0; t^*_m]$ and on $[t^*_{m+1}; +\infty)$ and we have

$$Gv(\pi, u) = \sum_{m=0}^{q-1} \mathbb{1}_{[t^*_m, t^*_{m+1})}(u) G^m v(\pi, u),$$

$$Hh(\pi, u) = \sum_{m=0}^{q-1} \mathbb{1}_{[t^*_m, t^*_{m+1})}(u) H^m h(\pi, u),$$

$$J(v, h)(\pi, u) = \sum_{m=0}^{q-1} \mathbb{1}_{[t^*_m, t^*_{m+1})}(u) J^m(v, h)(\pi, u).$$

We now investigate the Lipschitz properties of our operators.

PROPOSITION 8.6.– For $m \in M$, $((\pi, u), (\tilde{\pi}, \tilde{u})) \in (\mathcal{P}(E_0) \times \mathbb{R}^+)^2$, we have

$$|H^m g(\pi, u) - H^m g(\tilde{\pi}, \tilde{u})| \leq \sqrt{q} C_g |\pi - \tilde{\pi}| + ([g]_2 + C_g C_\Lambda)|u - \tilde{u}|.$$

PROOF.– Since the function $u \to H^m h(\pi, u)$ is constant on the intervals $[0; t^*_m]$ and $[t^*_{m+1}; +\infty)$, we may assume that $u, \tilde{u} \in [t^*_m; t^*_{m+1}]$ so that we have $H^m g(\pi, u) = \sum_{i=m+1}^{q} \pi^i e^{-\Lambda(x_i, u)} g \circ \Phi(x_i, u)$, and similarly for $H^m g(\tilde{\pi}, \tilde{u})$. Then, on the one hand, we have

$$\begin{aligned} |H^m g(\pi, u) - H^m g(\tilde{\pi}, u)| &= \Big| \sum_{i=m+1}^{q} (\pi^i - \tilde{\pi}^i) e^{-\Lambda(x_i, u)} g \circ \Phi(x_i, u) \Big| \\ &\leq C_g \sum_{i=m+1}^{q} |\pi^i - \tilde{\pi}^i|. \end{aligned}$$

and on the other hand, lemma 7.2 in Chapter 7 yields

$$|e^{-\Lambda(x_i, u)} g \circ \Phi(x_i, u) - e^{-\Lambda(x_i, \tilde{u})} g \circ \Phi(x_i, \tilde{u})| \leq ([g]_2 + C_g C_\Lambda)|u - \tilde{u}|,$$

showing the result. □

The following technical lemma will be useful to derive the Lipschitz properties of the operator I. The first part of its proof is adapted from [PHA 05].

LEMMA 8.8.– For all $\pi, \tilde{\pi} \in \mathcal{P}(E_0)$ and $m \in M$, we have

$$\sum_{m=0}^{q-1} \int_{t^*_m}^{t^*_{m+1}} \int_{\mathbb{R}^d} |\Psi(\pi, y', s') - \Psi(\tilde{\pi}, y', s')|_1 \overline{\Psi}_m(\pi, y', s') dy' ds' \leq 2|\pi - \tilde{\pi}|_1.$$

PROOF.– Let $s' \in (t^*_m; t^*_{m+1})$ and $y' \in \mathbb{R}^d$. In the following computation, we denote $\tau = (\pi, y', s')$ and $\tilde{\tau} = (\tilde{\pi}, y', s')$, we have

$$|\Psi(\tau) - \Psi(\tilde{\tau})|_1 \overline{\Psi}_m(\tau) = \sum_{j=1}^{q} \left| \frac{\Psi^j_m(\tau)}{\overline{\Psi}_m(\tau)} - \frac{\Psi^j_m(\tilde{\tau})}{\overline{\Psi}_m(\tilde{\tau})} \right| \overline{\Psi}_m(\tau)$$

$$= \sum_{j=1}^{q} \left| \frac{\Psi^j_m(\tau)\overline{\Psi}_m(\tilde{\tau}) - \Psi^j_m(\tilde{\tau})\overline{\Psi}_m(\tau)}{\overline{\Psi}_m(\tilde{\tau})} \right|$$

$$\leq \sum_{j=1}^{q} \left| \Psi^j_m(\tau) - \Psi^j_m(\tilde{\tau}) \right| + \sum_{j=1}^{q} \frac{\Psi^j_m(\tilde{\tau})}{\overline{\Psi}_m(\tilde{\tau})} \left| \overline{\Psi}_m(\tau) - \overline{\Psi}_m(\tilde{\tau}) \right|.$$

Note that $\sum_{j=1}^{q} \Psi^j_m(\tilde{\tau}) = \overline{\Psi}_m(\tilde{\tau})$ so that the second sum above reduces to $|\overline{\Psi}_m(\tau) - \overline{\Psi}_m(\tilde{\tau})|_1 = \sum_{j=1}^{q} \left| \Psi^j_m(\tau) - \Psi^j_m(\tilde{\tau}) \right|$. Finally, we have

$$|\Psi(\tau) - \Psi(\tilde{\tau})|_1 \overline{\Psi}_m(\tau) \leq 2 \sum_{j=1}^{q} |\Psi^j_m(\tau) - \Psi^j_m(\tilde{\tau})|.$$

As $\int_{\mathbb{R}^d} f_W\big(y' - \varphi(x_j)\big) dy' = 1$ and $\sum_{j=1}^{q} Q\big(\Phi(x_i, s'), x_j\big) = 1$, we obtain

$$\sum_{m=0}^{q-1} \int_{t^*_m}^{t^*_{m+1}} \int_{\mathbb{R}^d} \left|\Psi(\pi, y', s') - \Psi(\tilde{\pi}, y', s')\right|_1 \overline{\Psi}_m(\pi, y', s') dy' ds'$$

$$\leq 2 \sum_{m=0}^{q-1} \int_{t^*_m}^{t^*_{m+1}} \sum_{j=1}^{q} \int_{\mathbb{R}^d} \left|\Psi^j_m(\pi, y', s') - \Psi^j_m(\tilde{\pi}, y', s')\right| dy' ds'$$

$$\leq 2 \sum_{m=0}^{q-1} \sum_{i=m+1}^{q} \int_{t^*_m}^{t^*_{m+1}} \sum_{j=1}^{q} \int_{\mathbb{R}^d} |\pi^i - \tilde{\pi}^i| \lambda(\Phi(x_i, s')) e^{-\Lambda(x_i, s')}$$
$$\times\, Q(\Phi(x_i, s'), x_j) f_W(y' - \varphi(x_j)) dy' ds'$$

$$\leq 2 \sum_{m=0}^{q-1} \sum_{i=m+1}^{q} |\pi^i - \tilde{\pi}^i| \int_{t^*_m}^{t^*_{m+1}} \lambda(\Phi(x_i, s')) e^{-\Lambda(x_i, s')} ds'$$

$$\leq 2 \sum_{i=1}^{q} |\pi^i - \tilde{\pi}^i| \int_0^{t^*_i} \lambda(\Phi(x_i, s')) e^{-\Lambda(x_i, s')} ds'.$$

We obtain the result as $\int_0^{t^*_i} \lambda(\Phi(x_i, s')) e^{-\Lambda(x_i, s')} ds' = 1 - e^{-\Lambda(x_i, t^*_i)} \leq 1$. □

PROPOSITION 8.7.– For $v \in \mathbf{L}(\mathcal{P}(E_0))$ and $((\pi, u), (\tilde{\pi}, \tilde{u})) \in (\mathcal{P}(E_0) \times \mathbb{R}^+)^2$, we have

$$|Iv(\pi, u) - Iv(\tilde{\pi}, \tilde{u})| \leq \sqrt{q}(C_v + 2L_v)|\pi - \tilde{\pi}| + C_v C_\lambda |u - \tilde{u}|.$$

PROOF.– On the one hand, we clearly have

$$|Iv(\pi, u) - Iv(\pi, \tilde{u})| \leq \sum_{i=1}^{q} \pi^i \big|u \wedge t_i^* - \tilde{u} \wedge t_i^*\big| C_v C_\lambda \leq C_v C_\lambda |u - \tilde{u}|.$$

On the other hand, we have

$$\begin{aligned} &|Iv(\pi, u) - Iv(\tilde{\pi}, u)| \\ &\leq C_v |\pi - \tilde{\pi}|_1 + \sum_{i=1}^{q} \pi^i \int_0^{t_i^*} \int_{\mathbb{R}^d} \big|v\big(\Psi(\pi, y', s')\big) - v\big(\Psi(\tilde{\pi}, y', s')\big)\big| \\ &\quad \times \sum_{j=1}^{q} Q\big(\Phi(x_i, s'), x_j\big) f_W(y' - \varphi(x_j)) \lambda \circ \Phi(x_i, s') e^{-\Lambda(x_i, s')} dy' ds'. \end{aligned}$$

Besides, we have assumed that v is Lipschitz continuous so that we have

$$\big|v\big(\Psi(\pi, y', s')\big) - v\big(\Psi(\tilde{\pi}, y', s')\big)\big| \leq L_v \big|\Psi(\pi, y', s') - \Psi(\tilde{\pi}, y', s')\big|_1.$$

Thus, we have

$$\begin{aligned} &\big|v(\pi, y, s, u) - Iv(\tilde{\pi}, y, s, u)\big| \\ &\leq C_v |\pi - \tilde{\pi}|_1 + L_v \sum_{i=1}^{q} \pi^i \int_0^{t_i^*} \int_{\mathbb{R}^d} \Big|\Psi(\pi, y', s') - \Psi(\tilde{\pi}, y', s')\Big|_1 \\ &\quad \times \sum_{j=1}^{q} Q\big(\Phi(x_i, s'), x_j\big) f_W(y' - \varphi(x_j)) \lambda \circ \Phi(x_i, s') e^{-\Lambda(x_i, s')} dy' ds' \\ &\leq C_v |\pi - \tilde{\pi}|_1 + L_v \sum_{m=0}^{q-1} \sum_{i=m+1}^{q} \pi^i \int_{t_m^*}^{t_{m+1}^*} \int_{\mathbb{R}^d} \Big|\Psi(\pi, y', s') - \Psi(\tilde{\pi}, y', s')\Big|_1 \\ &\quad \times \sum_{j=1}^{q} Q\big(\Phi(x_i, s'), x_j\big) f_W(y' - \varphi(x_j)) \lambda \circ \Phi(x_i, s') e^{-\Lambda(x_i, s')} dy' ds' \end{aligned}$$

$$\leq L_v \sum_{m=0}^{q-1} \int_{t_m^*}^{t_{m+1}^*} \int_{\mathbb{R}^d} \left|\Psi(\pi, y', s') - \Psi(\tilde{\pi}, y', s')\right|_1 \overline{\Psi}_m(\pi, y', s') dy' ds'$$
$$+ C_v |\pi - \tilde{\pi}|_1$$

The previous lemma provides the result. □

PROPOSITION 8.8.– For $m \in M$, $v \in \mathbf{L}(\mathcal{P}(E_0))$ and $((\pi, u), (\tilde{\pi}, \tilde{u})) \in (\mathcal{P}(E_0) \times \mathbb{R}^+)^2$, we have

$$|G^m v(\pi, u) - G^m v(\tilde{\pi}, \tilde{u})| \leq 2\sqrt{q}(C_v + L_v)|\pi - \tilde{\pi}| + C_v C_\lambda |u - \tilde{u}|.$$

PROOF.– As in the proof of proposition 8.6, we may assume without loss of generality that $u, \tilde{u} \in [t_m^*; t_{m+1}^*]$ so that we have

$$G^m v(\pi, u) = Iv(\pi, u) + \sum_{i=1}^{m} \pi^i e^{-\Lambda(x_i, t_i^*)}$$
$$\times \sum_{j=1}^{q} \int_{\mathbb{R}^d} v(\Psi(\pi, y', t_i^*)) Q(\Phi(x_i, t_i^*), x_j) f_W(y' - \varphi(x_j)) dy',$$

and similarly for $G^m v(\tilde{\pi}, \tilde{u})$. The second term does not depend on u, thus

$$|G^m v(\pi, u) - G^m v(\pi, \tilde{u})| \leq |Iv(\pi, u) - Iv(\pi, \tilde{u})| \leq C_v C_\lambda |u - \tilde{u}|,$$
$$|G^m v(\pi, u) - G^m v(\tilde{\pi}, u)| \leq |Iv(\pi, u) - Iv(\tilde{\pi}, u)| + C_v |\pi - \tilde{\pi}|_1,$$

as $\Psi(\pi, y', t_i^*) = \Psi(\tilde{\pi}, y', t_i^*)$ by proposition 8.1. This yields the result. □

PROPOSITION 8.9.– For all $v \in \mathbf{L}(\mathcal{P}(E_0))$ and $(\pi, \tilde{\pi}) \in \mathcal{P}(E_0)^2$, we have

$$|Kv(\pi) - Kv(\tilde{\pi})| \leq 2\sqrt{q}(C_v +)|\pi - \tilde{\pi}|.$$

PROOF.– As $Kv(\pi) = Gv(\pi, t_q^*)$, this is a consequence of proposition 8.8. □

PROPOSITION 8.10.– For $v \in \mathbf{L}(\mathcal{P}(E_0))$ and $(\pi, \tilde{\pi}) \in \mathcal{P}(E_0)^2$, we have

$$|L(v, g)(\pi) - L(v, g)(\tilde{\pi})| \leq (C_g + 2C_v + 2L_v) |\pi - \tilde{\pi}|.$$

PROOF.– We have

$$|L(v, g)(\pi) - L(v, g)(\tilde{\pi})|$$
$$\leq \max_{m \in M} \Big\{ \sup_{u \in [t_m^*; t_{m+1}^*[} |J^m(v, g)(\pi, u) - J^m(v, g)(\tilde{\pi}, u)| \Big\}$$
$$\vee |Kv(\pi) - Kv(\tilde{\pi})|$$
$$\leq \sqrt{q} (C_g + 2C_v + 2L_v) |\pi - \tilde{\pi}|,$$

using propositions 8.6, 8.8 and 8.9 since $J^m(v,g) = H^m g + G^m v$. □

PROPOSITION 8.11.– For all $n \in \{0,\dots,N\}$, we have $v_n \in \mathbf{L}(\mathcal{P}(E_0))$ with $C_{v_n} \leq C_g$ and $L_{v_n} \leq 2(2\sqrt{q})^{N-n} C_g \sqrt{q}$.

PROOF.– We proved that v_n is the value function of the optimal stopping problem with horizon T_{N-n} thus we have $v_n(\pi) = \sup_{\sigma \in \Sigma^Y_{N-n}} \mathbb{E}[g(X_\sigma) | \Pi_0 = \pi] \leq C_g$. Therefore, v_n is bounded and $C_{v_n} \leq C_g$. The second assessment is proved by backward induction. Let π, $\tilde{\pi} \in \mathcal{P}(E_0)$. We have

$$|v_N(\pi) - v_N(\tilde{\pi})| \leq \sum_{j=1}^{N} g(x_j)|\pi^j - \tilde{\pi}^j| \leq \sqrt{q} C_g |\pi - \tilde{\pi}|.$$

Therefore, we have the result for $n = N$ with $L_{v_N} \leq \sqrt{q} C_g$. Moreover, since $v_n = L(v_{n+1}, g)$ for $0 \leq n \leq N-1$, proposition 8.10 yields $L_{v_n} \leq 3\sqrt{q} C_g + 2\sqrt{q} L_{v_{n+1}}$ which proves the propagation of the induction. □

8.5.2. *Discretization scheme*

Our approximation method is based on the optimal quantization of the Markov chain $(\Pi_k, S_k)_{k \leq N}$, see Chapter 3. Thus, from now on, we will denote, for $0 \leq k \leq N$, $\Theta_k = (\Pi_k, S_k)$. The quantized process $(\widehat{\Theta}_k)_{k \leq N} = (\widehat{\Pi}_k, \widehat{S}_k)_{k \leq N}$ with value for each k in the finite grid Γ_k of $\mathcal{P}(E_0) \times \mathbb{R}^+$ is then defined by

$$(\widehat{\Pi}_k, \widehat{S}_k) = proj_{\Gamma_k}(\Pi_k, S_k).$$

We will also denote by Γ_k^Π, the projection of Γ_k on $\mathcal{P}(E_0)$, and by Γ_k^S, the projection of Γ_k on $\mathbb{R}^+$.

In practice, we begin with the computation of the quantization grids, which merely requires to be able to simulate the process (see algorithm 3.3). Notice that in our case, what is actually simulated is the sequence of observation $(Y_k, S_k)_{0 \leq k \leq N}$. We are then able to compute the filter $(\Pi_k)_{0 \leq k \leq N}$ due to the recursive equation provided by proposition 8.1. The grids are only computed once and for all and may be stored offline. Our schemes are then based on the following simple idea: we replace the process by its quantized approximation within the different recursions as in the previous chapters. The computation is thus carried out in a very simple way since the quantized process has finite state space.

Our approximation scheme of the sequence $(V_n)_{0\leq n\leq N}$ follows the same lines as in Chapter 7, but once more, the results therein cannot be applied directly as the Markov chain $(\Theta_k)_{k\in\mathbb{N}}$ is not the underlying Markov chain of some PDMPs. Our approach decomposes into two steps. The first step will be to discretize the time-continuous maximization of the operator L to obtain a maximization over a finite set. The second step consists of replacing the Markov chain $(\Theta_n)_{n\in\mathbb{N}} = (\Pi_n, S_n),_{n\in\mathbb{N}}$ by its quantized approximation $(\widehat{\Theta}_n)_{n\in\mathbb{N}} = (\widehat{\Pi}_n, \widehat{S}_n)_{n\in\mathbb{N}}$ within the dynamic programming equation. Thus, the conditional expectations will become easily tractable finite sums.

Let us first build a finite time grid to discretize the continuous-time maximization in the expression of the operator L. The maximum is originally taken over the set $[0,\infty)$. However, it can be seen from definition 8.2 that $J(v,h)(\pi,u) = J(v,h)(\pi,t_q^*)$ for all $u \geq t_q^*$. Indeed, the random variable S_1 is bounded by the greatest deterministic exit time t_q^* that is finite. Therefore, the maximization set can be reduced to the compact set $[0,t_q^*]$. Instead of directly discretizing the set $[0,t_q^*]$, we will actually discretize the subsets (t_m^*, t_{m+1}^*) following the same idea as for definition 7.2. The reason why we want to exclude the points t_m^* from our grid is technical and will be explained with lemma 8.12. Now, it seems natural to distinguish whether $t_m^* = t_{m+1}^*$ or $t_m^* < t_{m+1}^*$.

DEFINITION 8.6.– *Let $M \subset \{0,\ldots,q-1\}$ be the set of indices m such that $t_m^* < t_{m+1}^*$.*

Notice that M is not empty because it contains at least the index 0 since we assumed that $t_1^* > 0 = t_0^*$. We can now build our approximation grid.

DEFINITION 8.7.– *Let $\Delta > 0$ be such that*

$$\Delta < \frac{1}{2}\min\left\{|t_i^* - t_j^*| \text{ with } 0 \leq i,j, \leq q \text{ such that } t_i^* \neq t_j^*\right\}. \qquad [8.13]$$

For all $m \in M$, let $Gr_m(\Delta)$ be the finite grid on $(t_m^*; t_{m+1}^*)$ defined as follows

$$Gr_m(\Delta) = \{t_m^* + i\Delta, 1 \leq i \leq i_m\} \cup \{t_{m+1}^* - \Delta\},$$

where $i_m = \max\{i \in \mathbb{N}$ such that $t_m^* + i\Delta \leq t_{m+1}^* - \Delta\}$. We also denote $Gr(\Delta) = \cup_{m\in M} Gr_m(\Delta)$.

REMARK 8.2.– Let $m \in M$. Notice that, due to equation [8.13], $Gr_m(\Delta)$ is not empty. Moreover, it satisfies two properties that will be crucial in the following:

a) for all $t \in [t_m^*; t_{m+1}^*]$, there exists $u \in Gr_m(\Delta)$ such that $|u-t| \leq \Delta$,

b) for all $u \in Gr_m(\Delta)$ and $0 < \eta < \Delta$, we have $[u-\eta; u+\eta] \subset (t^*_m; t^*_{m+1})$.

A discretized maximization operator L^d is then defined as follows.

DEFINITION 8.8.– *Let* L^d: $\mathbf{B}(\mathcal{P}(E_0)) \times \mathbf{B}(E) \to \mathbf{B}(\mathcal{P}(E_0))$ *be defined for all* $\pi \in \mathcal{P}(E_0)$ *by*

$$L^d(v,h)(\pi) = \max_{m \in M} \{ \max_{u \in Gr_m(\Delta)} \{J(v,h)(\pi,u)\}\} \vee Kv(\pi),$$

with $Kv(\pi) = J(v,h)(\pi, t^*_q) = Gv(\pi, t^*_q) = \mathbb{E}[v(\Pi_1)|\Pi_0 = \pi]$.

We now proceed to our second step: replacing $(\Theta_n)_{n\in\mathbb{N}} = (\Pi_n, S_n)_{,n\in\mathbb{N}}$ by its quantized approximation $(\widehat{\Theta}_n)_{n\in\mathbb{N}} = (\widehat{\Pi}_n, \widehat{S}_n)_{n\in\mathbb{N}}$ within the operators involved in the construction of the value function.

DEFINITION 8.9.– *We define the* quantized operators $\widehat{G}_n$, $\widehat{J}_n$, $\widehat{K}_n$ *and* $\widehat{L}^d_n$ *for* $n \in \{1,\ldots,N\}$, $v \in \mathbf{B}(\Gamma_n)$, $h \in \mathbf{B}(E)$, $\pi \in \Gamma^{\Pi}_{n-1}$ *and* $u \geq 0$ *as follows*

$$\begin{aligned}
\widehat{G}_n v(\pi,u) &= \mathbb{E}[v(\widehat{\Pi}_n)\mathbb{1}_{\{\widehat{S}_n \leq u\}}|\widehat{\Pi}_{n-1} = \pi],\\
\widehat{J}_n(v,h)(\pi,u) &= Hh(\pi,u) + \widehat{G}_n v(\pi,u),\\
\widehat{K}_n v(\pi) &= \widehat{J}_n(v,h)(\pi,t^*_q) = \mathbb{E}[v(\widehat{\Pi}_n)|\widehat{\Pi}_{n-1} = \pi],\\
\widehat{L}^d_n(v,h)(\pi) &= \max_{m \in M} \{ \max_{u \in Gr_m(\Delta)} \{\widehat{J}_n(v,h)(\pi,u)\}\} \vee \widehat{K}_n v(\pi).
\end{aligned}$$

The quantized approximation of the value functions naturally follows.

DEFINITION 8.10.– *For* $0 \leq n \leq N$, *define the functions* $\widehat{v}_n$ *on* Γ^{Π}_n *as follows*

$$\begin{cases} \widehat{v}_N(\pi) = \sum_{i=1}^q g(x_i)\pi^i \text{ for all } \pi \in \Gamma^{\Pi}_N,\\ \widehat{v}_{n-1}(\pi) = \widehat{L}^d_n(\widehat{v}_n, g)(\pi) \text{ for all } \pi \in \Gamma^{\Pi}_{n-1} \text{and } 1 \leq n \leq N. \end{cases}$$

For $0 \leq n \leq N$, *let* $\widehat{V}_n = \widehat{v}_n(\widehat{\Pi}_n)$.

Our approximation procedure is described in details in algorithm 8.1. We may now state our main result for the numerical approximation.

THEOREM 8.4.– Suppose that for all $0 \leq n \leq N-2$,

$$\Delta > (2C_\lambda)^{-1/2}\|S_{n+1} - \widehat{S}_{n+1}\|_p^{1/2}, \qquad [8.14]$$

then, we have the following bound for the approximation error

$$\|V_n - \widehat{V}_n\|_p \leq \|V_{n+1} - \widehat{V}_{n+1}\|_p + a\Delta + b\|S_{n+1} - \widehat{S}_{n+1}\|_p^{\frac{1}{2}}$$
$$+c_n\|\Pi_n - \widehat{\Pi}_n\|_p + 2L_{v_{n+1}}\|\Pi_{n+1} - \widehat{\Pi}_{n+1}\|_p,$$

where $a = [g]_2 + 2C_g C_\lambda$, $b = 2C_g(2C_\lambda)^{\frac{1}{2}}$ and $c_n = L_{v_n} + 2\sqrt{q}(2C_g + L_{v_{n+1}})$.

Algorithm 8.1: Approximation of the value function $v(\delta_{x_0})$

---

input : Quantization grids $(\Gamma_n^\Theta)_{0\leq n\leq N}$, Transition matrices $(\widehat{P}_n^\Theta)_{1\leq n\leq N}$, cost function g, time discretization step Δ

output : $\widehat{V}_0$

begin
- **for** $\pi \in \Gamma_N^\Pi$ **do**
  - $\widehat{v}_N(\pi) \leftarrow \sum_{i=1}^{q} g(x_i)\pi^i$
- **end**
- **for** $n \leftarrow N-1$ **to** 0 **do**
  - **for** $\pi \in \Gamma_n^\Pi$ **do**
    - $\widehat{K}\widehat{v}_{n+1}(\pi) \leftarrow \sum_{(\pi',s')\in\Gamma_{n+1}^\Theta} \widehat{P}_{n+1}(\pi,(\pi',s'))v_{n+1}(\pi')$
    - **for** $u \in \cup_{m\in M}(\{t_m^* + i\Delta, 1 \leq i \leq i_m\} \cup \{t_{m+1}^* - \Delta\})$ **do**
      - $\widehat{J}\widehat{v}_{n+1}(\pi,u) \leftarrow \sum_{(\pi',s')\in\Gamma_{n+1}^\Theta} \widehat{P}_{n+1}(\pi,(\pi',s'))(v_{n+1}(\pi'))$
        $+\sum_{i=1}^{q} g\circ\Phi(x_i,u)\mathbb{1}_{\{u<t^*i\}}e^{-\Lambda(x_i,u)}\pi^i$
    - **end**
    - $\widehat{v}_n(\pi) \leftarrow \max_u\{\widehat{J}\widehat{v}_{n+1}(\pi,u)\} \vee \widehat{K}\widehat{v}_{n+1}(\pi)$
  - **end**
- **end**
- **return** : $\widehat{v}_0$

end

Theorem 8.4 establishes the convergence of our approximation scheme and provides a bound for the rate of convergence. More precisely, it gives a rate for the L_p convergence of $\widehat{V}_0$ toward V_0. Indeed, we have $\|V_N - \widehat{V}_N\|_p = \|\sum_{i=1}^q g(x_i)(\Pi_N^i - \widehat{\Pi}_N^i)\|_p \leq C_g\|\Pi_N - \widehat{\Pi}_N\|_p$, so by virtue of theorem 8.4

$\left|V_0 - \widehat{V}_0\right|$ can be made arbitrarily small when the quantization errors $(\|\Theta_n - \widehat{\Theta}_n\|_p)_{0\leq n\leq N}$ go to zero, i.e. when the number of points in the quantization grids goes to infinity. In order to prove theorem 8.4, we proceed similarly to Chapter 7 and split the approximation error into four terms $\|V_n - \widehat{V}_n\|_p \leq \Xi_1 + \Xi_2 + \Xi_3 + \Xi_4$, with

$$\begin{aligned}
\Xi_1 &= \|v_n(\Pi_n) - v_n(\widehat{\Pi}_n)\|_p, \\
\Xi_2 &= \|L(v_{n+1}, g)(\widehat{\Pi}_n) - L^d(v_{n+1}, g)(\widehat{\Pi}_n)\|_p, \\
\Xi_3 &= \|L^d(v_{n+1}, g)(\widehat{\Pi}_n) - \widehat{L}^d_{n+1}(v_{n+1}, g)(\widehat{\Pi}_n)\|_p, \\
\Xi_4 &= \|\widehat{L}^d_{n+1}(v_{n+1}, g)(\widehat{\Pi}_n) - \widehat{L}^d_{n+1}(\widehat{v}_{n+1}, g)(\widehat{\Pi}_n)\|_p.
\end{aligned}$$

Proposition 8.11 proves that the value functions v_n are Lipschitz continuous, yielding a bound for the first term.

LEMMA 8.9.– The first term Ξ_1 is bounded as follows

$$\|v_n(\Pi_n) - v_n(\widehat{\Pi}_n)\|_p \leq L_{v_n} \|\Pi_n - \widehat{\Pi}_n\|_p.$$

The other error terms are studied separately in the following lemmas. For the second error term, we investigate the consequences of replacing the continuous maximization in operator L by a discrete one on $Gr(\Delta)$.

LEMMA 8.10.– For all $m \in M, v \in \mathbf{B}(\mathcal{P}(E_0))$ and $\pi \in \mathcal{P}(E_0)$ we have

$$\begin{aligned}
&\Big| \sup_{u\in[t^*_m;t^*_{m+1}[} J(v,g)(\pi,u) - \max_{u\in Gr_m(\Delta)} J(v,g)(\pi,u)\Big| \\
&\qquad\qquad \leq ([g]_2 + C_g C_\lambda + C_v C_\lambda)\,\Delta.
\end{aligned}$$

PROOF.– We use definition 8.5 to split operator J into a sum of continuous operators J^m. Thus, we have

$$\sup_{u\in[t^*_m;t^*_{m+1}[} J(v,g)(\pi,u) = \sup_{u\in[t^*_m;t^*_{m+1}]} J^m(v,g)(\pi,u).$$

The function $u \to J^m(v,h)(\pi,u)$ being continuous, there exists $\bar{t} \in [t^*_m; t^*_{m+1}]$ such that $\sup_{u\in[t^*_m;t^*_{m+1}]} J^m(v,h)(\pi,u) = J^m(v,h)(\pi,\bar{t})$. Moreover, from remark 8.2(a), we may chose $\overline{u} \in Gr_m(\Delta)$ so that $|\overline{u} - \bar{t}| \leq \Delta$. Propositions 8.6 and 8.8 stating the Lipschitz continuity of J^m

then yield

$$
\begin{aligned}
0 \leq & \sup_{u \in [t_m^*; t_{m+1}^*]} J^m(v,h)(\pi,u) - \max_{u \in Gr_m(\Delta)} J^m(v,h)(\pi,u) \\
& \leq J^m(v,h)(\pi,\bar{t}) - J^m(v,h)(\pi,\bar{u}) \\
& \leq ([g]_2 + C_g C_\lambda + C_v C_\lambda) |\bar{t} - \bar{u}| \leq ([g]_2 + C_g C_\lambda + C_v C_\lambda) \Delta,
\end{aligned}
$$

showing the result. □

LEMMA 8.11.– The second term Ξ_2 is bounded as follows

$$
\|L(v_{n+1}, g)(\widehat{\Pi}_n) - L^d(v_{n+1}, g)(\widehat{\Pi}_n)\|_p \leq ([g]_2 + 2C_g C_\lambda) \Delta.
$$

PROOF.– This is a straightforward consequence of the previous lemma once it has been noticed that for all $a, b, c, d \in \mathbb{R}$, we have $|a \vee b - c \vee d| \leq |a - c| \vee |b - d|$. Notice also that proposition 8.11 provides $C_{v_{n+1}} \leq C_g$. □

To investigate the third error term, we use the properties of quantization to bound the error made by replacing an operator by its quantized approximation. As in Chapter 7, we must first deal with non-continuous indicator functions. The fact that the t_m^* and a small neighborhood around them do not belong to the discretization grid $Gr(\Delta)$ is crucial to obtain the following lemma, which is an analogous to lemma 7.12.

LEMMA 8.12.– For all $0 \leq n \leq N-1$, $m \in M$ and $0 < \eta < \Delta$, we have

$$
\begin{aligned}
\Big\| \max_{u \in Gr_m(\Delta)} & \mathbb{E}[|\mathbb{1}_{\{S_{n+1} \leq u\}} - \mathbb{1}_{\{\widehat{S}_{n+1} \leq u\}} | \widehat{\Pi}_n] \Big\|_p \\
& \leq \eta^{-1} \|S_{n+1} - \widehat{S}_{n+1}\|_p + 2\eta C_\lambda.
\end{aligned}
$$

PROOF.– Let $0 < \eta < \Delta$. The difference of the indicator functions equals 1 if and only if S_{n+1} and $\widehat{S}_{n+1}$ are on different sides of u. Therefore, if the difference of the indicator functions equals 1, either $|S_{n+1} - u| \leq \eta$, or $|S_{n+1} - u| > \eta$ and in the latter case $|S_{n+1} - \widehat{S}_{n+1}| > \eta$ too since $|S_{n+1} - \widehat{S}_{n+1}| > |S_{n+1} - u|$. We have $|\mathbb{1}_{\{S_{n+1} \leq u\}} - \mathbb{1}_{\{\widehat{S}_{n+1} \leq u\}}| \leq \mathbb{1}_{\{|S_{n+1} - \widehat{S}_{n+1}| > \eta\}} + \mathbb{1}_{\{|S_{n+1} - u| \leq \eta\}}$, leading to

$$
\begin{aligned}
\Big\| \max_{u \in Gr_m(\Delta)} & \mathbb{E}\big[|\mathbb{1}_{\{S_{n+1} \leq u\}} - \mathbb{1}_{\{\widehat{S}_{n+1} \leq u\}}| |\widehat{\Pi}_n\big] \Big\|_p \\
& \leq \|\mathbb{1}_{\{|S_{n+1} - \widehat{S}_{n+1}| > \eta\}}\|_p + \Big\| \max_{u \in Gr_m(\Delta)} \mathbb{E}[\mathbb{1}_{\{|S_{n+1} - u| \leq \eta\}} | \widehat{\Pi}_n] \Big\|_p.
\end{aligned}
$$

On the one hand, Markov inequality yields

$$\|\mathbb{1}_{\{|S_{n+1}-\widehat{S}_{n+1}|>\eta\}}\|_p = \mathbb{P}(|S_{n+1} - \widehat{S}_{n+1}| > \eta)^{\frac{1}{p}} \leq \|S_{n+1} - \widehat{S}_{n+1}\|_p \eta^{-1}.$$

On the other hand, since $u \in Gr_m(\Delta)$, we have $[u - \eta; u + \eta] \subset (t_m^*; t_{m+1}^*)$ from remark 8.2(b), thus S_{n+1} has an absolutely continuous distribution on the interval $[u - \eta; u + \eta]$ since it does not contain any of the t_i^*. Recall that $\widehat{\Theta}_n = proj_{\Gamma_n}(\Theta_n)$, hence, the following inclusions of σ-fields $\sigma(\widehat{\Pi}_n) \subset \sigma(\widehat{\Theta}_n) \subset \sigma(\Theta_n)$. We also have $\sigma(\Theta_n) \subset \mathfrak{F}_{T_n}^Y \subset \mathfrak{F}_{T_n}$, the law of iterated conditional expectations provides

$$\begin{aligned}\mathbb{E}[\mathbb{1}_{\{|S_{n+1}-u|\leq\eta\}}|\widehat{\Pi}_n] &= \mathbb{E}\big[\mathbb{E}\big[\mathbb{E}[\mathbb{1}_{\{|S_{n+1}-u|\leq\eta\}}|\mathfrak{F}_{T_n}]\big|\mathfrak{F}_{T_n}^Y\big]\big|\widehat{\Pi}_n\big] \\ &\leq \mathbb{E}\Big[\mathbb{E}\Big[\int_{u-\eta}^{u+\eta} \lambda\big(\Phi(Z_n, s)\big)ds\Big|\mathfrak{F}_{T_n}^Y\Big]\Big|\widehat{\Pi}_n\Big] \\ &= \mathbb{E}\Big[\sum_{i=1}^{q} \Pi_n^i \int_{u-\eta}^{u+\eta} \lambda\big(\Phi(x_i, s)\big)ds\Big|\widehat{\Pi}_n\Big].\end{aligned}$$

Finally, we obtain $\mathbb{E}[\mathbb{1}_{\{|S_{n+1}-u|\leq\eta\}}|\widehat{\Pi}_n] \leq 2\eta C_\lambda$, showing the result. □

LEMMA 8.13.– For all $0 \leq n \leq N - 1$, we have

$$\begin{aligned}&|Kv_{n+1}(\widehat{\Pi}_n) - \widehat{K}_{n+1}v_{n+1}(\widehat{\Pi}_n)| \\ &\leq L_{v_{n+1}}\mathbb{E}\big[|\Pi_{n+1}-\widehat{\Pi}_{n+1}|\big|\widehat{\Pi}_n\big]+2\sqrt{q}(C_g+L_{v_{n+1}})\mathbb{E}\big[|\Pi_n-\widehat{\Pi}_n|\big|\widehat{\Pi}_n\big].\end{aligned}$$

PROOF.– By the definitions of operators K and $\widehat{K}_{n+1}$, we have

$$\begin{aligned}&|Kv_{n+1}(\widehat{\Pi}_n) - \widehat{K}_{n+1}v_{n+1}(\widehat{\Pi}_n)| \\ &= |\mathbb{E}[v_{n+1}(\Pi_{n+1})|\Pi_n = \widehat{\Pi}_n] - \mathbb{E}[v_{n+1}(\widehat{\Pi}_{n+1})|\widehat{\Pi}_n]| \\ &\leq |\mathbb{E}[v_{n+1}(\Pi_{n+1})|\Pi_n = \widehat{\Pi}_n] - \mathbb{E}[v_{n+1}(\Pi_{n+1})|\widehat{\Pi}_n]| \\ &\quad +|\mathbb{E}[v_{n+1}(\Pi_{n+1}) - v_{n+1}(\widehat{\Pi}_{n+1})|\widehat{\Pi}_n]|.\end{aligned} \qquad [8.15]$$

The second term on the right-hand side of equation [8.15] is readily bounded by using proposition 8.11 stating that v_{n+1} is Lipschitz continuous

$$|\mathbb{E}[v_{n+1}(\Pi_{n+1}) - v_{n+1}(\widehat{\Pi}_{n+1})|\widehat{\Pi}_n]| \leq L_{v_{n+1}}\mathbb{E}\big[|\Pi_{n+1} - \widehat{\Pi}_{n+1}|\big|\widehat{\Pi}_n\big].$$

To deal with the first term on the right-hand side of equation [8.15], we need to use the special properties of quantization. Indeed, we have $(\widehat{\Pi}_n, \widehat{S}_n) = proj_{\Gamma_n}(\Pi_n, S_n)$ so that we have $\sigma(\widehat{\Pi}_n) \subset \sigma(\Pi_n, S_n)$. The law of iterated conditional expectations gives

$$\mathbb{E}[v_{n+1}(\Pi_{n+1})|\widehat{\Pi}_n] = \mathbb{E}\big[\mathbb{E}[v_{n+1}(\Pi_{n+1})|(\Pi_n, S_n)]|\widehat{\Pi}_n\big].$$

Moreover, we have $\mathbb{E}[v_{n+1}(\Pi_{n+1})|(\Pi_n, S_n)] = \mathbb{E}[v_{n+1}(\Pi_{n+1})|\Pi_n]$ from proposition 8.3, as the conditional distribution of Π_{n+1} w.r.t. (Π_n, S_n) merely depends on Π_n. In addition, $|\mathbb{E}[v_{n+1}(\Pi_{n+1})|\Pi_n = \widehat{\Pi}_n]$ is $\sigma(\widehat{\Pi}_n)$-measurable. We have then

$$\begin{aligned}
&|\mathbb{E}[v_{n+1}(\Pi_{n+1})|\Pi_n = \widehat{\Pi}_n] - \mathbb{E}[v_{n+1}(\Pi_{n+1})|\widehat{\Pi}_n]| \\
&= \Big|\mathbb{E}\big[\mathbb{E}[v_{n+1}(\Pi_{n+1})|\Pi_n = \widehat{\Pi}_n] - \mathbb{E}[v_{n+1}(\Pi_{n+1})|\Pi_n]|\widehat{\Pi}_n\big]\Big| \\
&= |\mathbb{E}[Kv_{n+1}(\widehat{\Pi}_n) - Kv_{n+1}(\Pi_n)|\widehat{\Pi}_n]|,
\end{aligned}$$

by definition of K. Finally, we have

$$\begin{aligned}
&|\mathbb{E}[v_{n+1}(\Pi_{n+1})|\Pi_n = \widehat{\Pi}_n] - \mathbb{E}[v_{n+1}(\Pi_{n+1})|\widehat{\Pi}_n]| \\
&\leq 2\sqrt{q}(C_g + L_{v_{n+1}})\mathbb{E}\big[|\Pi_n - \widehat{\Pi}_n||\widehat{\Pi}_n\big],
\end{aligned}$$

due to propositions 8.9 and 8.11 stating the Lipschitz continuity of operator K and function v_{n+1}. □

LEMMA 8.14.– If Δ satisfies condition [8.14], an upper bound for the third term Ξ_3 is

$$\begin{aligned}
&\|L^d(v_{n+1}, g)(\widehat{\Pi}_n) - \widehat{L}^d_{n+1}(v_{n+1}, g)(\widehat{\Pi}_n)\|_p \\
&\leq L_{v_{n+1}}\|\Pi_{n+1} - \widehat{\Pi}_{n+1}\|_p + 2\sqrt{q}(2C_g + L_{v_{n+1}})\|\Pi_n - \widehat{\Pi}_n\|_p \\
&\quad +2C_g\Big\{2C_\lambda\|S_{n+1} - \widehat{S}_{n+1}\|_p\Big\}^{1/2}.
\end{aligned}$$

PROOF.– We have

$$
\begin{aligned}
&|L^d(v_{n+1},g)(\widehat{\Pi}_n) - \widehat{L}^d_{n+1}(v_{n+1},g)(\widehat{\Pi}_n)| \\
&\leq \max_{m\in M}\{\max_{u\in Gr_m(\Delta)} |J(v_{n+1},g)(\widehat{\Pi}_n,u) - \widehat{J}_{n+1}(v_{n+1},g)(\widehat{\Pi}_n,u)|\} \\
&\vee|Kv_{n+1}(\widehat{\Pi}_n) - \widehat{K}_{n+1}v_{n+1}(\widehat{\Pi}_n)|.
\end{aligned}
$$

The term involving operator K was studied in the previous lemma. Let us now study the term involving operator J. Set m in M, u in $Gr_m(\Delta)$ and define $\alpha(\pi,\pi',s') = \sum_{i=1}^q \pi^i g(\Phi(x_i,u))e^{-\Lambda(x_i,u)} + v_{n+1}(\pi')\mathbb{1}_{\{s'\leq u\}}$. We have then

$$
\begin{aligned}
&|J(v_{n+1},g)(\widehat{\Pi}_n,u) - \widehat{J}_{n+1}(v_{n+1},g)(\widehat{\Pi}_n,u)| \\
&\quad = |J^m(v_{n+1},g)(\widehat{\Pi}_n,u) - \widehat{J}_{n+1}(v_{n+1},g)(\widehat{\Pi}_n,u)| \\
&\quad = \left|\mathbb{E}[\alpha(\Pi_n,\Pi_{n+1},S_{n+1})|\Pi_n = \widehat{\Pi}_n] - \mathbb{E}[\alpha(\widehat{\Pi}_n,\widehat{\Pi}_{n+1},\widehat{S}_{n+1})|\widehat{\Pi}_n]\right| \\
&\quad \leq A + B,
\end{aligned}
$$

where

$$
\begin{aligned}
A &= \left|\mathbb{E}[\alpha(\Pi_n,\Pi_{n+1},S_{n+1})|\Pi_n = \widehat{\Pi}_n] - \mathbb{E}[\alpha(\Pi_n,\Pi_{n+1},S_{n+1})|\widehat{\Pi}_n]\right|, \\
B &= |\mathbb{E}[\alpha(\Pi_n,\Pi_{n+1},S_{n+1}) - \alpha(\widehat{\Pi}_n,\widehat{\Pi}_{n+1},\widehat{S}_{n+1})|\widehat{\Pi}_n]|.
\end{aligned}
$$

Using the boundedness of g and v_{n+1} as well as the Lipschitz continuity of v_{n+1} given in proposition 8.11, we get an upper bound for the second term

$$
\begin{aligned}
B \leq\ & \sqrt{q}C_g\mathbb{E}\big[|\Pi_n - \widehat{\Pi}_n|\big|\widehat{\Pi}_n\big] + \mathbb{E}\big[|\Pi_{n+1} - \widehat{\Pi}_{n+1}|\big|\widehat{\Pi}_n\big] \\
& + C_g\mathbb{E}\left[|\mathbb{1}_{\{S_{n+1}\leq u\}} - \mathbb{1}_{\{\widehat{S}_{n+1}\leq u\}}|\big|\widehat{\Pi}_n\right].
\end{aligned}
\quad [8.16]
$$

For the first term, we use the properties of quantization as in the previous proof to obtain

$$
A = \left|\mathbb{E}\big[\mathbb{E}[\alpha(\Pi_n,\Pi_{n+1},S_{n+1})|\Pi_n = \widehat{\Pi}_n] - \mathbb{E}[\alpha(\Pi_n,\Pi_{n+1},S_{n+1})|\Pi_n]\big|\widehat{\Pi}_n\big]\right|.
$$

We now recognize operator J^m, and from propositions 8.6 and 8.8, we have

$$
\begin{aligned}
A &= \left|\mathbb{E}[J^m(v_{n+1},g)(\widehat{\Pi}_n,u) - J^m(v_{n+1},g)(\Pi_n,u)|\widehat{\Pi}_n]\right| \\
&\leq \sqrt{q}(3C_g + 2L_{v_{n+1}})\mathbb{E}\big[|\widehat{\Pi}_n - \Pi_n|\widehat{\Pi}_n\big].
\end{aligned}
\quad [8.17]
$$

We gather the bounds provided by equations [8.16] and [8.17] to obtain

$$\begin{aligned}&|J(v_{n+1},g)(\widehat{\Pi}_n,u)-\widehat{J}_{n+1}(v_{n+1},g)(\widehat{\Pi}_n,u)|\\&\quad\leq 2\sqrt{q}(2C_g+L_{v_{n+1}})\mathbb{E}\big[|\Pi_n-\widehat{\Pi}_n||\widehat{\Pi}_n\big]+L_{v_{n+1}}\mathbb{E}\big[|\Pi_{n+1}-\widehat{\Pi}_{n+1}||\widehat{\Pi}_n\big]\\&\quad+C_g\mathbb{E}\left[|\mathbb{1}_{\{S_{n+1}\leq u\}}-\mathbb{1}_{\{\widehat{S}_{n+1}\leq u\}}||\widehat{\Pi}_n\right].\end{aligned}\qquad [8.18]$$

Finally, by combining the result for operators J and lemma 8.13, we obtain

$$\begin{aligned}&|L^d(v_{n+1},g)(\widehat{\Pi}_n)-\widehat{L}^d_{n+1}(v_{n+1},g)(\widehat{\Pi}_n)|\\&\quad\leq L_{v_{n+1}}\mathbb{E}\left[|\Pi_{n+1}-\widehat{\Pi}_{n+1}||\widehat{\Pi}_n\right]+2\sqrt{q}(2C_g+L_{v_{n+1}})\mathbb{E}\left[|\Pi_n-\widehat{\Pi}_n||\widehat{\Pi}_n\right]\\&\quad+C_g\max_{u\in Gr(\Delta)}\mathbb{E}\left[|\mathbb{1}_{\{S_{n+1}\leq u\}}-\mathbb{1}_{\{\widehat{S}_{n+1}\leq u\}}||\widehat{\Pi}_n\right].\end{aligned}$$

We conclude by taking the L^p norm in the above equation and by using lemma 8.12 to bound the last term

$$\begin{aligned}&\|L^d(v_{n+1},g)(\widehat{\Pi}_n)-\widehat{L}^d_{n+1}(v_{n+1},g)(\widehat{\Pi}_n)\|_p\\&\quad\leq L_{v_{n+1}}\|\Pi_{n+1}-\widehat{\Pi}_{n+1}\|_p+2\sqrt{q}(2C_g+L_{v_{n+1}})\|\Pi_n-\widehat{\Pi}_n\|_p\\&\quad+C_g(\eta^{-1}\|S_{n+1}-\widehat{S}_{n+1}\|_p+2\eta C_\lambda),\end{aligned}$$

for some $0<\eta<\Delta$. The best choice for η minimizing the error is when η satisfies

$$\eta^{-1}\|S_{n+1}-\widehat{S}_{n+1}\|_p=2\eta C_\lambda,$$

which yields $\eta=(2C_\lambda)^{-1/2}(\|S_{n+1}-\widehat{S}_{n+1}\|_p{}^{1/2}$. If Δ satisfies condition [8.14], we have $\eta<\Delta$ as required for this optimal choice. □

Finally, the fourth error term is bounded using Lipschitz properties.

LEMMA 8.15.– The fourth term Ξ_4 is bounded as follows

$$\begin{aligned}&\|\widehat{L}^d_{n+1}(v_{n+1},g)(\widehat{\Pi}_n)-\widehat{L}^d_{n+1}(\widehat{v}_{n+1},g)(\widehat{\Pi}_n)\|_p\\&\quad\leq L_{v_{n+1}}\|\Pi_{n+1}-\widehat{\Pi}_{n+1}\|_p+\|V_{n+1}-\widehat{V}_{n+1}\|_p.\end{aligned}$$

PROOF.– We have

$$\|\widehat{L}^d_{n+1}(v_{n+1},g)(\widehat{\Pi}_n) - \widehat{L}^d_{n+1}(\widehat{v}_{n+1},g)(\widehat{\Pi}_n)\|_p$$

$$= \Big\| \max_{m\in M} \max_{u\in Gr_m(\Delta)} \{Hg(\widehat{\Pi}_n,u) + \widehat{G}_{n+1}v_{n+1}(\widehat{\Pi}_n,u)\} \vee \widehat{K}_{n+1}v_{n+1}(\widehat{\Pi}_n)$$

$$- \max_{m\in M} \max_{u\in Gr_m(\Delta)} \{Hg(\widehat{\Pi}_n,u) + \widehat{G}_{n+1}\widehat{v}_{n+1}(\widehat{\Pi}_n,u)\} \vee \widehat{K}_{n+1}\widehat{v}_{n+1}(\widehat{\Pi}_n)\Big\|_p$$

$$\leq \Big\| \max_{m\in M} \max_{u\in Gr_m(\Delta)} \mathbb{E}\left[\left(v_{n+1}(\widehat{\Pi}_{n+1}) - \widehat{v}_{n+1}(\widehat{\Pi}_{n+1})\right)\mathbb{1}_{\{\widehat{S}_{n+1}\leq u\}}\middle|\widehat{\Pi}_n\right]$$

$$\vee\, \mathbb{E}[v_{n+1}(\widehat{\Pi}_{n+1}) - \widehat{v}_{n+1}(\widehat{\Pi}_{n+1})|\widehat{\Pi}_n]\Big\|_p$$

$$\leq \|v_{n+1}(\widehat{\Pi}_{n+1}) - \widehat{v}_{n+1}(\widehat{\Pi}_{n+1})\|_p. \qquad [8.19]$$

We now introduce $v_{n+1}(\Pi_{n+1})$ to split this term into two differences. The Lipschitz continuity of v_{n+1} stated by proposition 8.11 allows us to bound the first term, while we recognize V_{n+1} and $\widehat{V}_{n+1}$ in the second one.

$$\|\widehat{L}^d_{n+1}(v_{n+1},g)(\widehat{\Pi}_n) - \widehat{L}^d_{n+1}(\widehat{v}_{n+1},g)(\widehat{\Pi}_n)\|_p$$

$$\leq \|v_{n+1}(\widehat{\Pi}_{n+1}) - v_{n+1}(\Pi_{n+1})\|_p + \|v_{n+1}(\Pi_{n+1}) - \widehat{v}_{n+1}(\widehat{\Pi}_{n+1})\|_p$$

$$\leq L_{v_{n+1}}\left\|\Pi_{n+1} - \widehat{\Pi}_{n+1}\right\|_p + \|V_{n+1} - \widehat{V}_{n+1}\|_p.$$

Hence, the result. □

8.5.3. *Numerical construction of an ϵ-optimal stopping time*

As in the previous section, we follow the idea of Chapter 7 and we use both the Markov chain $(\Theta_n)_{0\leq n\leq N}$ and its quantized approximation $(\widehat{\Theta}_n)_{0\leq n\leq N}$ to approximate the expression of the ϵ-optimal stopping time introduced in definition 8.4. We check that we thus obtain actual stopping times for the observed filtration $(\mathfrak{F}^Y_t)_{t\geq 0}$ and that the expected reward when stopping according to this rule is a good approximation of the value function V_0. For all $(\pi,s) \in \mathcal{P}(E_0)\times\mathbb{R}^+$ and $0 \leq n \leq N$, we denote $(\widehat{\pi}_n,\widehat{s}_n) = proj_{\Gamma_n}(\pi,s)$. For $1\leq n\leq N$ and $\pi\in\mathcal{P}(E_0)$, we define

$$\widehat{s}^*_n(\pi,s) = \min\{t\in Gr(\Delta) : \widehat{J}_n(\widehat{v}_n,g)(\widehat{\pi}_{n-1},t) = \max_{u\in Gr(\Delta)} \widehat{J}_n(\widehat{v}_n,g)(\widehat{\pi}_{n-1},u)\}.$$

and

$$\widehat{r}_n(\pi,s) = \begin{cases} t^*_q & \text{if } \widehat{K}_n\widehat{v}_n(\widehat{\pi}_{n-1}) > \max_{u\in Gr(\Delta)}\widehat{J}_n(\widehat{v}_n,g)(\widehat{\pi}_{n-1},u), \\ \widehat{s}^*_n(\pi,s) & \text{otherwise.}\end{cases}$$

DEFINITION 8.11.– *For $n \in \{0, \cdots, N-1\}$, the sequence of random variables $\widehat{U}_{N-n}$ is given by*

$$\widehat{U}_{N-n} = \begin{cases} T_n + \widehat{r}_{n+1}(\Pi_n, S_n) & \text{if } \widehat{r}_{n+1}(\Pi_n, S_n) < S_{n+1}, \\ T_{n+1} + \widehat{r}_{n+2}(\Pi_{n+1}, S_{n+1}) & \text{if } \widehat{r}_{n+2}(\Pi_{n+1}, S_{n+1}) < S_{n+2}, \\ \vdots \\ T_{N-1} + \widehat{r}_N(\Pi_{N-1}, S_{N-1}) & \text{if } \widehat{r}_N(\Pi_{N-1}, S_{N-1}) < S_N, \\ T_N & \text{otherwise.} \end{cases}$$

So, we have

$$\begin{aligned} \widehat{U}_{N-n} &= \big(T_n + r_n(\Pi_n, S_n)\big)\mathbf{1}_{\{r_n(\Pi_n,S_n)<S_{n+1}\}} \\ &\quad + \widehat{U}_{N-n-1}\mathbf{1}_{\{r_n(\Pi_n,S_n)\geq S_{n+1}\}}. \end{aligned}$$

or equivalently

$$\widehat{U}_{N-n} = \sum_{k=1}^{n} C_{t^*} \wedge S_k + \sum_{k=n+1}^{N} \widehat{R}_k \wedge S_k$$

with $\widehat{R}_{n+1} = r_{n+1}(\Pi_n, S_n)$ and $\widehat{R}_k = r_k(\Pi_{k+1}, S_{k+1})\mathbf{1}_{\{U_{k-1}\geq S_{k-1}\}}$ for $n+2 \leq k \leq N-1$.

Algorithm 8.1 can be modified along the same lines as algorithm 7.2 to store the results of the comparisons between operators $\widehat{K}_n$ and $\widehat{J}_n$. With similar notation, the construction of an ϵ-optimal stopping time is given in algorithm 8.2.

The following result is a direct consequence of proposition 8.1. It is a very strong result as it states that the numerically computable random variables $\widehat{U}_n$ are actual $(\mathfrak{F}_t^Y)_{t\geq 0}$-stopping times.

THEOREM 8.5.– For $0 \leq n \leq N$, $\widehat{U}_n$ is an $(\mathfrak{F}_t^Y)_{t\geq 0}$-stopping time.

We now intend to prove that stopping at time $\widehat{U}_N$ provides a good approximation of the value function V_0. For any $\pi \in \mathcal{P}(E_0)$, $s \in \mathbb{R}^+$ and $0 \leq n \leq N-1$ we, therefore, introduce the performance when abiding by the stopping rule $(\widehat{U}_{N-n})_{0\leq n\leq N-1}$ and the corresponding random variables

$$\overline{v}_n(\pi, s) = \mathbb{E}[g(X_{\widehat{U}_{N-n}})|\Pi_n = \pi, S_n = s], \qquad \overline{V}_n = \overline{v}_n(\Pi_n, S_n).$$

Algorithm 8.2: Computation of an ϵ-optimal stopping time

input : Quantization grids $(\Gamma_n^\Theta)_{0\leq n\leq N}$, Outputs of (extended) algorithm 8.1: $s_n^*, i_n, 1 \leq n \leq N$, simulator of trajectories of the filtered PDMP, filter function Ψ, projection operators p_n onto the quantization grids

output: Stopping date τ

```
begin
    n ← 0
    τ ← NaN
    Π_0 = δ_{x_0}
    while n < N do
        simulate (Y_n, S_n, S_{n+1})
        compute filter Π_n = Ψ(Π_{n-1}, Y_n, S_n) (if n > 0)
        (π, s) ← p_n(Π_n, S_n)
        if i_{n+1}(π) = 1 then
            r ← t_q^*
        else
            r ← s_{n+1}(π)
        end
        if r < S_{n+1} then
            τ ← T_n + r
            n ← N
        else
            n ← n + 1
        end
        if n = N and τ = NaN then
            τ ← T_N
        end
    end
    return: τ
end
```

THEOREM 8.6.– Suppose that for all $0 \leq n \leq N-2$,

$$\Delta > (2C_\lambda)^{-1/2} \|S_{n+1} - \widehat{S}_{n+1}\|_p^{1/2},$$

we have then the following bound for the error between the expected reward $\overline{V}_n$ when stopping at time $\widehat{U}_{N-n}$ and the value function V_n

$$\|V_n - \overline{V}_n\|_p \leq \left\|V_{n+1} - \overline{V}_{n+1}\right\|_p + \left\|V_n - \widehat{V}_n\right\|_p + \left\|V_{n+1} - \widehat{V}_{n+1}\right\|_p$$
$$+ d_n\|\Pi_n - \widehat{\Pi}_n\|_p + 2L_{v_{n+1}}\|\Pi_{n+1} - \widehat{\Pi}_{n+1}\|_p$$
$$+ b\|S_{n+1} - \widehat{S}_{n+1}\|_p^{1/2},$$

where $b = 2C_g\left(2C_\lambda\right)^{1/2}$ and $d_n = \sqrt{q}\left(7C_g + 4L_{v_{n+1}}\right)$.

It is important to note that $\overline{v}_N(\pi) = \sum_{i=1}^q g(x_i)\pi^i = v_N(\pi)$ and thus $\overline{V}_N = V_N$. Therefore, the previous theorem proves that $|V_0 - \overline{V}_0|$ goes to zero when the quantization errors $(\|\Theta_n - \widehat{\Theta}_n\|_p)_{0\leq n\leq N}$ go to zero. In other words, the expected reward $\overline{V}_0$ when stopping at the random time $\widehat{U}_N$ can be made arbitrarily close to the value function V_0 of the partially observed optimal stopping problem [8.3] and hence $\widehat{U}_N$ is an ϵ-optimal stopping time.

PROOF.– The first step consists of finding a recursion satisfied by $(\overline{V}_n)_{0\leq n\leq N}$ in order to compare it with the dynamic programming equation giving $(\widehat{V}_n)_{0\leq n\leq N}$. Let $0 \leq n \leq N-2$. First, by definition of $\widehat{U}_{N-n}$

$$\begin{aligned}\overline{V}_n &= \mathbb{E}[g(X_{\widehat{U}_{N-n}})|\Pi_n, S_n] \\ &= \mathbb{E}[g\circ\Phi(Z_n, \widehat{r}_{n+1}(\Pi_n, S_n))\mathbb{1}_{\{\widehat{r}_{n+1}(\Pi_n,S_n)<S_{n+1}\}}|\Pi_n, S_n] \\ &\quad + \mathbb{E}[g(X_{\widehat{U}_{N-n-1}})\mathbb{1}_{\{\widehat{r}_{n+1}(\Pi_n,S_n)\geq S_{n+1}\}}|\Pi_n, S_n].\end{aligned}$$

However, we have by using lemma 8.5

$$\begin{aligned}&\mathbb{E}[g\circ\Phi(Z_n, \widehat{r}_{n+1}(\Pi_n, S_n))\mathbb{1}_{\{\widehat{r}_{n+1}(\Pi_n,S_n)<S_{n+1}\}}|\Pi_n, S_n] \\ &= \sum_{i=1}^q g\circ\Phi(x_i, \widehat{r}_{n+1}(\Pi_n, S_n))\mathbb{E}[\mathbb{1}_{\{Z_n=x_i\}}\mathbb{1}_{\{\widehat{r}_{n+1}(\Pi_n,S_n)<S_{n+1}\}}|\Pi_n, S_n] \\ &= \sum_{i=1}^q g\circ\Phi(x_i, \widehat{r}_{n+1}(\Pi_n, S_n))\mathbb{1}_{\{\widehat{r}_{n+1}(\Pi_n,S_n)<t_i^*\}}e^{-\Lambda(x_i,\widehat{r}_{n+1}(\Pi_n,S_n))}\Pi_n^i \\ &= Hg(\Pi_n, \widehat{r}_{n+1}(\Pi_n, S_n)).\end{aligned}$$

Moreover, by using the Markov property of the Markov chain $(\Pi_n, S_n)_{n\in\mathbb{N}}$, it can be shown easily following the same arguments as in the proof of proposition 8.4 that

$$\mathbb{E}[g(X_{\widehat{U}_{N-n-1}})|\mathfrak{F}^Y_{T_{n+1}}] = \mathbb{E}[g(X_{\widehat{U}_{N-n-1}})|\Pi_{n+1}, S_{n+1}] = \overline{V}_{n+1}.$$

Therefore,

$$\overline{V}_n = Hg(\Pi_n, \widehat{r}_{n+1}(\Pi_n, S_n)) + \mathbb{E}[\overline{V}_{n+1}\mathbb{1}_{\{\widehat{r}_{n+1}(\Pi_n,S_n)\geq S_{n+1}\}}|\Pi_n, S_n].$$

We are now able to study the error between $\overline{V}_n$ and $\widehat{V}_n$. Let us recall that, by definition of $\widehat{V}_n$ and $\widehat{r}_{n+1}(\Pi_n, S_n)$, we have

$$\widehat{V}_n = \mathbb{1}_{\{\widehat{r}_{n+1}(\Pi_n,S_n)=t_q^*\}}\widehat{K}_{n+1}\widehat{v}_{n+1}(\widehat{\Pi}_n) + \mathbb{1}_{\{\widehat{r}_{n+1}(\Pi_n,S_n)=\widehat{s}^*_{n+1}(\Pi_n,S_n)\}}\widehat{J}_{n+1}(\widehat{v}_{n+1}, g)(\widehat{\Pi}_n, \widehat{s}^*_{n+1}(\Pi_n, S_n))$$

Moreover,

$$\overline{V}_n = \mathbb{1}_{\{\widehat{r}_{n+1}(\Pi_n,S_n)=t_q^*\}}\mathbb{E}[\overline{V}_{n+1}|\Pi_n, S_n] + \mathbb{1}_{\{\widehat{r}_{n+1}(\Pi_n,S_n)=\widehat{s}^*_{n+1}(\Pi_n,S_n)\}}\Big\{Hg(\Pi_n, \widehat{s}^*_{n+1}(\Pi_n, S_n)) + \mathbb{E}[\overline{V}_{n+1}\mathbb{1}_{\{\widehat{s}^*_{n+1}(\Pi_n,S_n)\geq S_{n+1}\}}|\Pi_n, S_n]\Big\}.$$

Eventually, we have

$$|\overline{V}_n - \widehat{V}_n| \leq \mathbb{1}_{\{\widehat{r}_{N-n-1}(\Pi_n,S_n)=t_q^*\}}A \quad [8.20]$$

$$+ \mathbb{1}_{\{\widehat{r}_{N-n-1}(\Pi_n,S_n)=\widehat{s}^*_{N-n-1}(\Pi_n,S_n)\}}B \quad [8.21]$$

with

$$A = \big|\mathbb{E}[\overline{V}_{n+1}|\Pi_n, S_n] - \widehat{K}_{n+1}\widehat{v}_{n+1}(\widehat{\Pi}_n)\hat{A}\big|,$$

$$B = \Big|Hg(\Pi_n, \widehat{s}^*_{n+1}(\Pi_n, S_n)) + \mathbb{E}[\overline{V}_{n+1}\mathbb{1}_{\{\widehat{s}^*_{n+1}(\Pi_n,S_n)\geq S_{n+1}\}}|\Pi_n, S_n] - \widehat{J}_{n+1}(\widehat{v}_{n+1}, g)(\widehat{\Pi}_n, \widehat{s}^*_{n+1}(\Pi_n, S_n))\Big|.$$

To bound the first term A, we introduce the function v_{n+1}. Moreover, observe that $\mathbb{E}[V_{n+1}|\Pi_n, S_n] = Kv_{n+1}(\Pi_n)$ and so,

$$\begin{aligned}A &\leq |\mathbb{E}[\overline{V}_{n+1}|\Pi_n, S_n] - \mathbb{E}[V_{n+1}|\Pi_n, S_n]| + |Kv_{n+1}(\Pi_n) - Kv_{n+1}(\widehat{\Pi}_n)| \\ &\quad + |Kv_{n+1}(\widehat{\Pi}_n) - \widehat{K}_{n+1}v_{n+1}(\widehat{\Pi}_n)| + |\widehat{K}_{n+1}v_{n+1}(\widehat{\Pi}_n) - \widehat{K}_{n+1}\widehat{v}_{n+1}(\widehat{\Pi}_n)| \\ &\leq (a) + (b) + (c) + (d).\end{aligned}$$

The first term (a) is bounded by $\mathbb{E}\left[|\overline{V}_{n+1} - V_{n+1}||\Pi_n, S_n\right]$. For the second term (b), we use proposition 8.9 stating the Lipschitz continuity of the operator K. The third term (c) is bounded by lemma 8.13 and an upper bound of the fourth term (d) is given by equation [8.19]. Thus, we obtain

$$\|A\|_p \leq \left\|V_{n+1} - \overline{V}_{n+1}\right\|_p + \left\|V_{n+1} - \widehat{V}_{n+1}\right\|_p$$
$$+ 4\sqrt{q}(C_g + L_{v_{n+1}})\|\Pi_n - \widehat{\Pi}_n\|_p + 2L_{v_{n+1}}\|\Pi_{n+1} - \widehat{\Pi}_{n+1}\|_p.$$

We now turn to the second term B.

$$B \leq \mathbb{E}\Big[\left|\overline{V}_{n+1} - V_{n+1}\right| \mathbb{1}_{\{\widehat{s}^*_{n+1}(\Pi_n, S_n) \geq S_{n+1}\}} |\Pi_n, S_n\Big]$$
$$+ \left|J(v_{n+1}, g)(\Pi_n, \widehat{s}^*_{n+1}(\Pi_n, S_n)) - \widehat{J}_{n+1}(\widehat{v}_{n+1}, g)(\widehat{\Pi}_n, \widehat{s}^*_{n+1}(\Pi_n, S_n))\right|.$$

For the last term in the previous inequality, we use propositions 8.6 and 8.8 for the Lipschitz continuity of J and equations [8.18] and [8.19] to obtain

$$\|B\|_p \leq \left\|V_{n+1} - \overline{V}_{n+1}\right\|_p + \left\|V_{n+1} - \widehat{V}_{n+1}\right\|_p$$
$$+ \sqrt{q}(7C_g + 4L_{v_{n+1}})\|\Pi_n - \widehat{\Pi}_n\|_p + 2L_{v_{n+1}}\|\Pi_{n+1} - \widehat{\Pi}_{n+1}\|_p$$
$$+ 2C_g\left\{2C_\lambda\|S_{n+1} - \widehat{S}_{n+1}\|_p\right\}^{1/2},$$

after optimizing η, the result follows by taking the maximum between A and B. □

Finally, a similar calculation gives at the initializing step $N-1$ the error between the expected reward $\overline{V}_{N-1}$ when stopping at time $\widehat{U}_1$ and the value function V_{N-1}

THEOREM 8.7.– Suppose that

$$\Delta > (2C_\lambda)^{-1/2}\|S_N - \widehat{S}_N\|_p^{1/2},$$

we have then the following bound for the error between the expected reward $\overline{V}_{N-1}$ when stopping at time $\widehat{U}_1$ and the value function V_{N-1}

$$\|V_{N-1} - \overline{V}_{N-1}\|_p \leq \left\|V_{N-1} - \widehat{V}_{N-1}\right\|_p + \left\|V_N - \widehat{V}_N\right\|_p$$
$$+ d_N\|\Pi_{N-1} - \widehat{\Pi}_{N-1}\|_p + 2L_{v_N}\|\Pi_N - \widehat{\Pi}_N\|_p$$
$$+ b\|S_N - \widehat{S}_N\|_p^{1/2},$$

where $b = 2C_g\left(2C_\lambda\right)^{1/2}$ and $d_N = \sqrt{q}\left(7C_g + 4L_{v_N}\right)$.

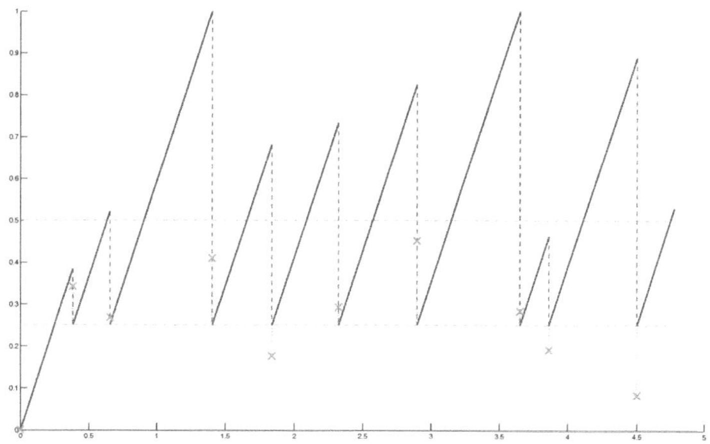

Figure 8.1. *A trajectory of the discretized TCP process drawn until the 10th jump time with $a = 3$, $v = 1$ and $E_0 = \{0; \frac{1}{4}; \frac{1}{2}\}$. The dotted lines represent the possible post-jump values*

8.6. Numerical example

We apply our procedure to a simple PDMP corresponding to a discretized version of the Transmission Control Protocol (TCP) process presented in section 1.8.2. Let $E = [0; 1)$. For $x \in E$ and $t \geq 0$, the flow is defined by $\Phi(x, t) = x + vt$ so that $t^*(x) = (1 - x)/v$. We set the jump rate to $\lambda(x) = ax$ for some $a > 0$ and the transition kernel $Q(x, \cdot)$ to the uniform distribution on a finite set $E_0 \subset E$. Thus, the process evolves toward 1 and the closer it gets to 1, the more likely it will jump back to some point of E_0. A trajectory is represented in Figure 8.1.

The observation process is $Y_n = \varphi(Z_n) + W_n$ where $\varphi(x) = x$ and $W_n \sim \mathcal{N}(0, \sigma^2)$ for some $\sigma^2 > 0$. The observations are given by the crosses in Figure 8.1. Finally, we choose the reward function $g(x) = x$. Our assumptions thus clearly hold.

Simulations are run with $a = 3, v = 1, E_0 = \{0, 1/4, 1/2\}, \sigma^2 = 0.25$ and $N = 10$. The numerical approximation is implemented as follows. First, we make an exact simulator for the sequence (Z_n, S_n). From the values of (Z_n), we build the observation sequence (Y_n) that allows for a recursive computation of the filter process due to proposition 8.1. The computed values corresponding to the trajectory in Figure 8.1 are given in Table 8.2.

Thus, we can simulate trajectories of the Markov chain (Π_n, S_n) that we feed into the competitive learning vector quantization (CLVQ) algorithm 3.3 to obtain quantization grids. By Monte Carlo simulations, we can also estimate the quantization errors. To run our numerical procedure, we then need to choose the parameter Δ satisfying conditions [8.13] and [8.14]. In this special case, they boil down to

$$6^{-1/2} \max_{0 \leq n \leq N-1} \|S_{n+1} - \widehat{S}_{n+1}\|_p^{1/2} < \Delta < \frac{1}{8}.$$

Quantization grids	Δ	$\overline{V}_0$	$\widehat{V}_0$	B_{em}	B_{th}
50 points	0.1179	0.7900	0.8135	0.181	683
100 points	0.0970	0.8031	0.8250	0.169	467
300 points	0.0731	0.8182	0.8407	0.154	271
500 points	0.0634	0.8250	0.8477	0.147	211
1,000 points	0.0535	0.8313	0.8545	0.140	152
2,000 points	0.0453	0.8361	0.8599	0.135	110
4,000 points	0.0381	0.8408	0.8643	0.130	80
6,000 points	0.0345	0.8430	0.8666	0.128	67
8,000 points	0.0321	0.8479	0.8725	0.122	58
10,000 points	0.0303	0.8497	0.8742	0.120	53
12,000 points	0.0290	0.8521	0.8771	0.117	49

Table 8.1. *Simulation results for the optimal stopping of the discretized TCP process. The terms B_{em} and B_{th}, respectively, denote an empirical bound and the theoretical bound provided by theorem 8.4 for the error $|V_0 - \widehat{V}_0|$*

Jump	0	1	2	3	4	5	6	7	8	9	10
Z_n	0	0.25	0.25	0.25	0.25	0.25	0.25	0.25	0.25	0.25	0.25
Y_n	0	0.34	0.27	0.41	0.18	0.29	0.45	0.28	0.19	0.08	0.26
$\mathbb{P}(\Pi_n = 0)$	1	0.00	0.03	0.00	0.21	0.01	0.00	0.02	0.16	0.74	0.03
$\mathbb{P}(\Pi_n = 0.25)$	0	0.67	0.91	0.29	0.78	0.87	0.13	0.89	0.83	0.26	0.92
$\mathbb{P}(\Pi_n = 0.5)$	0	0.31	0.06	0.71	0.01	0.11	0.87	0.09	0.01	0.00	0.05

Table 8.2. *Simulation of the observation and filter processes for the trajectory of the discretized TCP from Figure 8.1*

We have chosen Δ just above the Monte Carlo approximation of the lower bound. The values are given in the second column of Table 8.1 for different grid sizes.

Then, we recursively compute the approximated value functions $\widehat{v}_n$ on the quantization grids using algorithm 8.1. The conditional expectations are now merely weighted sums. The approximations we obtain for the value function of the partially observed optimal stopping problem are given in the fourth column of Table 8.1.

Finally, we implemented the construction of our ϵ-optimal stopping time according to algorithm 8.2 and ran 10^6 Monte Carlo simulations to compute its mean performance. The results are given in the third column of Table 8.1. Figure 8.2 shows the two examples of stopped trajectories. The first ones achieved a very high performance, close to 1, and the second one was stopped shortly after the 9-th jump and achieved a poorer performance for lack of enough time left to improve it. Recall that our stopping rule reaches optimality on average, and not almost surely.

As in section 7.7, the exact value of V_0 is unknown but we have

$$\overline{V}_0 = \mathbb{E}[g(X_{\widehat{U}_N})] \leq V_0 = \sup_{\sigma \in \Sigma_N^Y} \mathbb{E}[g(X_\sigma)] \leq \mathbb{E}\big[\sup_{0 \leq t \leq T_N} g(X_t)\big]. \quad [8.22]$$

Both the first and the last term may be estimated by Monte Carlo simulations. We have thus, with 10^6 trajectories, $\mathbb{E}[\sup_{0 \leq t \leq T_N} g(X_t)] = 0.9944$. The theoretical bound B_{th} of the error $|V_0 - \widehat{V}_0|$ provided by theorem 8.4 is computed using the approximated quantization errors. This bound decreases as the number of points in the quantization grids increases, as expected. Moreover, we computed the empirical bound given by equation [8.22] $B_{em} = \max\big\{|\overline{V}_0 - \widehat{V}_0|, |\mathbb{E}[\sup_{0 \leq t \leq T_N} g(X_t)] - \widehat{V}_0|\big\}$. As in Chapter 7, the generic bounds are not very sharp, but they do decrease as the number of points in the quantization grids increases.

We did not test this procedure on more involved examples. In practical cases, we must find a reasonable balance between a large set E_0 to reflect the variability of the underlying process and a small one to have a low-dimensional filter and accurate quantization grids. The question of optimal stopping when the jump times are not observed remains an important open problem.

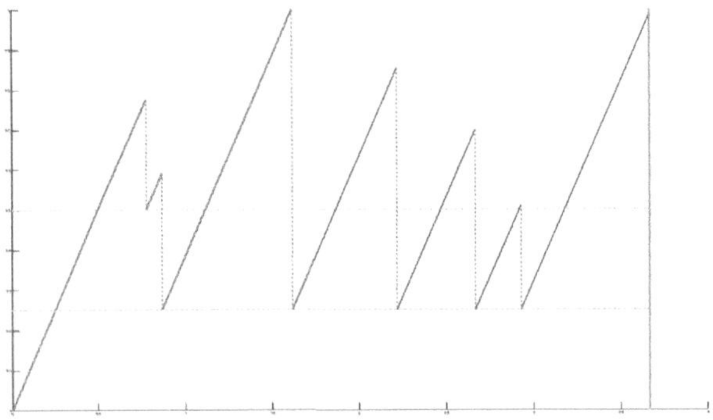

(a) Stop after jump 6

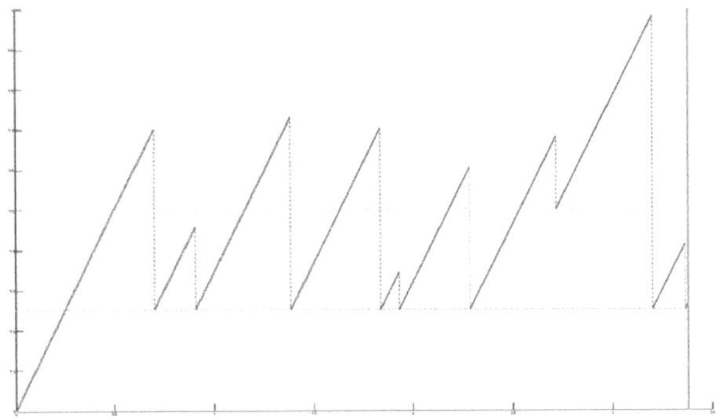

(b) Stop after jump 9

Figure 8.2. *Two optimally stopped trajectories of the discretized TCP process*

9

Example in Reliability: Maintenance Optimization

9.1. Introduction

A complex system is inherently sensitive to the failures of its components. We must therefore determine maintenance policies in order to maintain an acceptable operating condition. Optimizing the maintenance is a very important problem in the analysis of complex systems. It determines when maintenance tasks should be performed on the system in order to optimize a cost function: either maximizing a performance function or conversely minimizing a loss function. Moreover, this optimization must take into account the random nature of failures and random evolution and dynamics of the system.

The only maintenance operation considered here is the complete replacement of all the failed components and the system restarts in its initial equilibrium state. Partial repairs are not allowed; see section 10.6 for a discussion of partial repairs. Mathematically, this problem of preventive maintenance corresponds to a stochastic optimal stopping problem as explained, for example, in the book of Aven and Jensen [AVE 99]. A classical approach consists of using condition-based maintenance (CBM) to act on the system based on its current state and before its failure. We can, for example, calculate the remaining useful life (RUL) of the system and the preventive replacement is carried out when the deterioration level exceeds a certain threshold or enters in a certain state [VAN 09, GRA 02]. Our approach, based on the numerical method developed in Chapter 7, also takes into account the current state of the process, but our decision rule is neither based on damage accumulation nor does it correspond to hitting some threshold. Instead, it involves a performance function that reflects that the longer the system is in a functioning state the better.

Our aim is to demonstrate the high practical power of the theoretical methodology described in Chapter 7, by applying it to three different challenging examples. In section 9.2, we study the corrosion example first presented in section 2.2. In this special case, an optimal stopping time exists and we can precisely compare the performance of our approximation to the optimal one. It is no longer the case for remaining examples. In section 9.3, we study the maintenance of the air conditioning unit presented in section 1.8.3. Finally, section 9.4 is devoted to the maintenance optimization of the heated hold-up tank (see section 2.3). This latter case is a hard problem because of the closed loop interactions between the state of the components and the liquid level and temperature.

9.2. Corrosion process

Our first example concerns the maintenance of the metallic structure subject to corrosion presented in section 2.2. The main material of this section comes from [SAP 12c]. Note that this process was already studied regarding other reliability questions in Chapter 6, and that we use yet again the same quantization grids (see section 3.4.1) to solve the following optimal stopping problem

$$v(x_0) = \sup_{\tau \in \mathcal{M}_N} \mathbb{E}_{x_0}[g(X_\tau)],$$

where the horizon N is empirically set to $N = 25$, and the cost function g is given in Figure 9.1. It depends only on the loss of thickness, and reflects of the fact that on the one hand, we should not perform maintenance too early, when the thickness loss is low (low gain), but on the other hand we should not cross the security threshold of 0.2 mm (zero gain). It also satisfies our Lipschitz continuity requirements.

The reward function g is maximum (value 4) at abscissa 0.18 and the thickness loss is continuous and non-decreasing. Thus, it is easy to see that the true value function at our starting point is 4, which is the maximum of the reward function g, and an optimal stopping time is the first moment when the loss reaches 0.18 mm thick.

Our numerical procedure is valid for any sufficiently regular reward function, and the knowledge of the true value function or optimal stopping time will not be used in the numerical procedure. Recall that the thickness loss is not measured continuously. However, this exact optimal stopping time will serve us as a benchmark to evaluate the performance of our approximation. Figure 9.2 shows two examples of computation of the quasi optimal maintenance time on two specific simulated trajectories. The thick vertical line represents the moment provided by the algorithm to perform maintenance. The other vertical lines materialize the moments of change of environment, the horizontal dotted line the theoretical optimum. In both examples, we stop at a value very close to the optimum value. Moreover, the intervention did take place before the critical threshold of 0.2 mm.

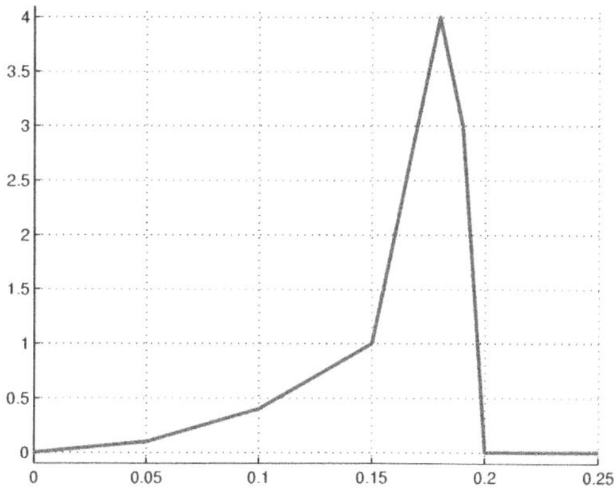

Figure 9.1. *Reward function for the maintenance of the corrosion process*

An approximate value function v is calculated in two ways. The first way is the direct method obtained by the algorithm described in section 7.3. The second way is obtained by Monte Carlo simulation using the quasi-optimal stopping time provided by our procedure. The numerical results obtained are summarized in Table 9.1. As expected, the greater the number of points in the quantization grid, the better our approximation becomes. Furthermore, the specific form of this cost function g indicates that at the threshold of 1, the intervention takes place between 0.15 and 0.21 mm and when the threshold increases, this range is narrowed. Our approximation is therefore good even for low numbers of grid points. The last column of the table also shows the validity of our stopping rule. It should be noted here that this rule does not use the optimal stopping time *stop at the first moment when the thickness loss reaches* 0.18 *mm*. The method used is general and implementable even when the optimal stopping time is unknown or does not exist, as will be illustrated in the following sections.

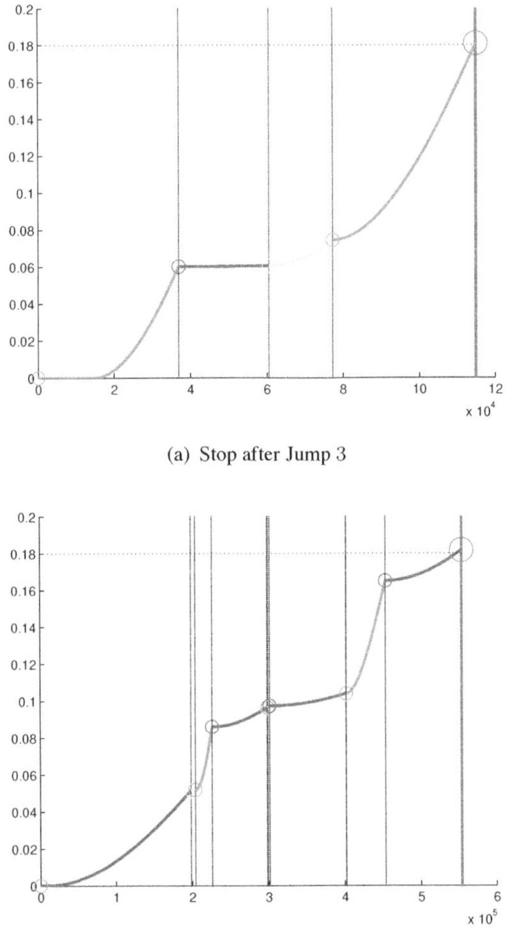

(a) Stop after Jump 3

(b) Stop after Jump 7

Figure 9.2. *Examples of stopped trajectories for the corrosion process*

Moreover, we can also construct a histogram of the values of our stopping time, that is to say, a histogram of the values of effective moments of maintenance and compare it with the exact optimal stopping time. Figure 9.3

shows the results for 10^5 Monte Carlo simulations. Image (a) is obtained by our approximation procedure and image (b) is the exact one. Both figures appear very similar, which is a strong point in favor of our approximation.

Number of points in the quantization grids	Approximation of the value function by the direct algorithm	Approximation of the value function by Monte Carlo with the quasi-optimal stopping time
10	2.48	0.94
50	2.70	1.84
100	2.94	2.10
200	3.09	2.63
500	3.39	3.15
1000	3.56	3.43
2000	3.70	3.60
5000	3.82	3.73
8000	3.86	3.75

Table 9.1. *Numerical results for the calculation of the value function for the corrosion process*

We can also estimate the probability that the maintenance date is below certain thresholds. These results are interesting for Astrium in the design phase of the structure to optimize margins from the specifications and to consolidate the design margins available. Thus, we can justify that with a given probability, no maintenance will be required before the termination date of the contract.

9.3. Air conditioning unit

We now turn to the maintenance of the air conditioning unit process, described in section 1.8.3. We choose the following reward function

$$g(m,t) = \frac{t}{p(m)}$$

where

- $p(1) = 6$ is the price of maintenance in stable mode;
- $p(2) = 6$ is the price of maintenance in degraded ball bearing mode;
- $p(3) = 5$ is the price of repair of electrovalve failure;
- $p(4) = 3.5$ is the price of repair of electronic failure;
- $p(5) = 12$ is the price of repair of ball bearing failure;

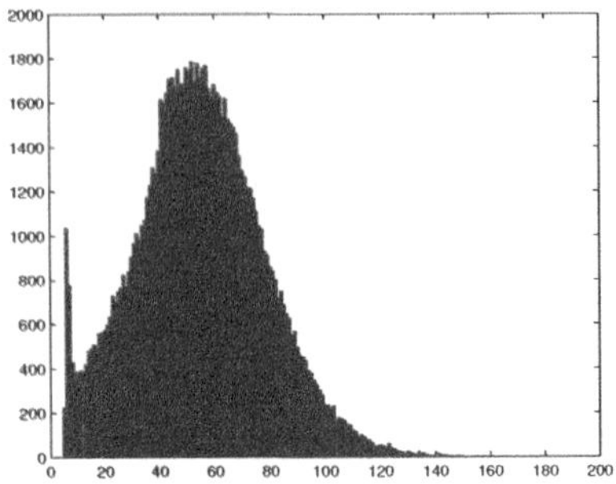

(a) Approximate stopping time

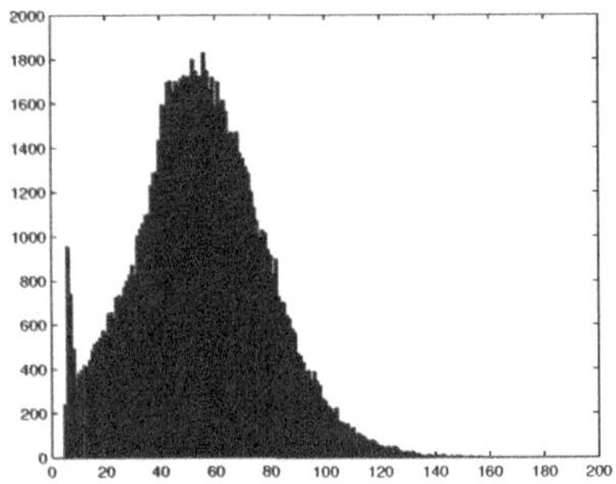

(b) Optimal stopping time

Figure 9.3. *Histograms of approximate and optimal stopping times for the corrosion process (10^5 Monte Carlo simulations)*

Thus, it is better to start a maintenance in mode 2 than wait for total failure mode 5, but the failure modes 3 and 4 cost less than maintenance. Therefore, intuitively we should let the electrovalve and electronic components fail, but perform a maintenance before the total failure of the ball bearing. The time at the numerator of g says that the longer the device is functioning, the higher the performance.

Although the underlying piecewise deterministic Markov process (PDMP) is much simpler than in the corrosion example, there does not exist an optimal stopping time in this case. The average performance of the system without maintenance is 342.72 and the average trajectories are given in Figure 9.4.

The average performance of the system with our maintenance rule is 592.47 and the average trajectories are given in Figure 9.5. Thus, our policy corresponds to an average gain multiplied by 1.73, compared to no maintenance, which is very significant. It also succeeds in behaving as expected: no early maintenance in mode 1 and a degraded ball bearing is almost always sent to maintenance.

Note that in this special case, the jump intensity λ does not satisfy our Lipschitz regularity assumption because of the exponents less than 2 in the Weibull distributions. Our procedure seems to work anyways, so probably local Lipschitz continuity are sufficient.

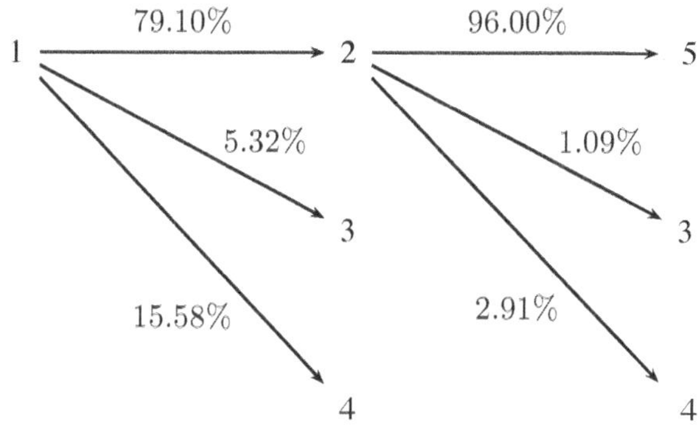

Figure 9.4. *Possible trajectories of the air conditioning unit without maintenance*

This example was provided by Thales Optronique SAS, that are currently using our procedures to optimize the maintenance of several of their equipments. Thus, our approach can also be conducted in an industrial framework.

9.4. The heated hold-up tank

We now turn to our third and final example, the maintenance of the heated hold-up tank first presented in section 2.3. See also section 3.4.2 for issues related to the quantization of this process.

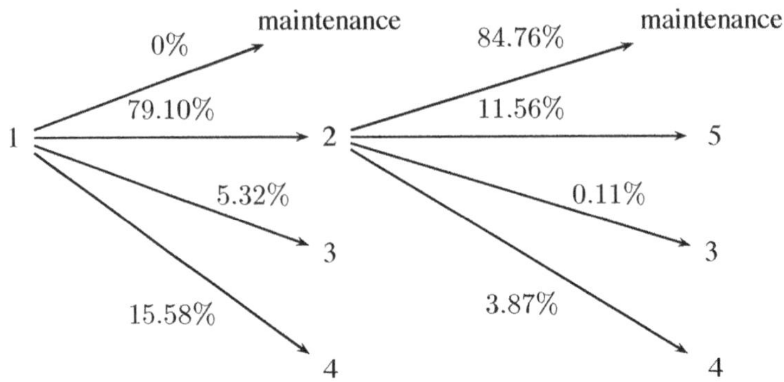

Figure 9.5. *Possible trajectories of the air conditioning unit without maintenance*

9.4.1. *Problem setting and simulation*

The simulation of the tank process and the computation of related reliability indices have long been a challenge for the reliability community (see [DEV 97, MAR 95, MAR 96, ZHA 08]). Here, we go one step further by optimizing its maintenance. To our best knowledge, we proposed the first results in the literature on the optimization of maintenance policies for the heated hold-up tank in [SAP 13].

Recall that the PDMP describing the behavior of the tank has two continuous variables, deterministic jumps when these variables hit some given boundaries and closed loop interactions between continuous and discrete variables. Furthermore, we consider a cost function that depends on both continuous variables as well as on the running time (due to the time-augmented process).

The reward function has two components $g(h, \theta, t) = f(h, \theta)t^{\alpha}$. The first component f depends on the liquid level and temperature, and reflects that the reward is maximal (set to 1) when h and θ are in the normal range (6 m $\leq h \leq 8$ m, $\theta \leq 50$°C), minimal (set to 0) when reaching the top events: dry out ($h < 4$ m), overflow ($h > 10$ m) or hot temperature ($\theta > 100$°C) and continuous in between, (see Figure 9.6 for an illustration). The second term t^{α} involves the time and reflects that the longer the system is functioning the higher the reward. The parameter α is set to 1.01 for smoothness.

Figure 9.6. *Reward function f as a function of h and θ for the heated hold-up tank*

The numerical procedure described in Chapter 7 is valid for a wide class of PDMPs. It has nice convergence properties. However, to implement it in practice, we must carefully choose the various parameters:

– the computation horizon N;

– the possibly state dependent time discretization steps $\Delta(z)$;

– the number of points in the quantization grids;

– the parameters required to build quantization grids.

We shall explain how these parameters can be suitably chosen to optimize the tank.

We used the exact C++ simulator of trajectories developed for [ZHA 08], suitably modified to take into account the possible failures of the command and interfaced with a matlab code for the optimization procedure. The jump horizon N was empirically set to 26 jumps, thus allowing all the trajectories to reach one of the top events $h < 4$ m, $h > 10$ m, $\theta > 100$°C or $t >$ 1000 h, (see Table 3.2). The construction of the quantization grids is explained in section 3.4.2.

9.4.2. *Numerical results and validation*

We ran our optimization procedure for a number of discretization points in the quantization grids. The results we obtained are given in Table 9.2. The optimal performance is our approximation of the value function v at the starting equilibrium point whereas the last column gives the mean performance achieved by our stopping rule (obtained by 10^5 Monte Carlo simulations). We can observe the convergence as the number of points in the quantization grids increases, and we can also see that our stopping rule is indeed close enough to optimality.

Number of points	Optimal performance	Stopping rule
200	334.34	305.55
300	333.04	319.45
400	332.95	322.20
800	330.43	323.63
1000	330.87	324.04

Table 9.2. *Optimal performance for the heated hold-up tank*

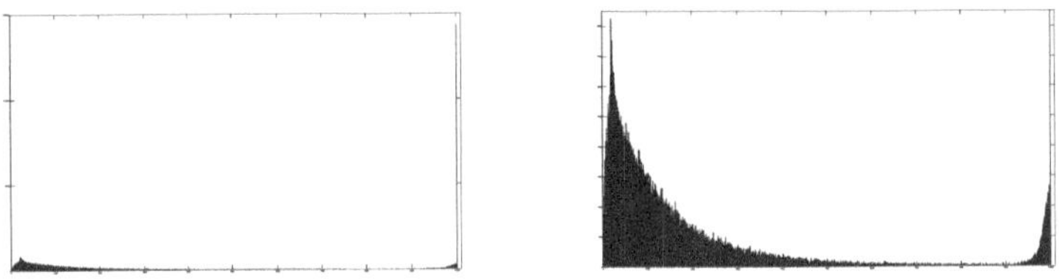

(a) distribution (b) zoom of the distribution

Figure 9.7. *Histogram of the distribution of the computed maintenance time (with and without the mode at 1,000) for the heated hold-up tank*

We can also obtain the distributions of the maintenance time by Monte Carlo simulations (10^5 simulations), given in Figure 9.7. The distribution is roughly bimodal, with a very high mode (14.16% of trajectories) at time 1000, which means that the tank remained in an acceptable state until the end of the experiment. For better readability of the histogram, only the value of $\tau^* < 1000$ is plotted on the right-hand side figure. This distribution illustrates that an average stopping rule would be far from optimal for the tank. The distribution of the liquid level and temperature at the maintenance time are given on Figure 9.8. The distribution of the liquid level is almost discrete with three main values at levels 6 m (15.16%), 7 m (12.68%) and 8 m (45.45%).

This is natural as the liquid level is often constant, and moves very fast between these 3 states. The distribution of the temperature is also almost discrete as it is a function of the liquid level for most modes. It has a maximum at the equilibrium temperature (15.60%). Note that the top events are never reached for the liquid level, and only 0.02% of trajectories ended at $\theta = 100°C$. This is a strong point in favor of our procedure.

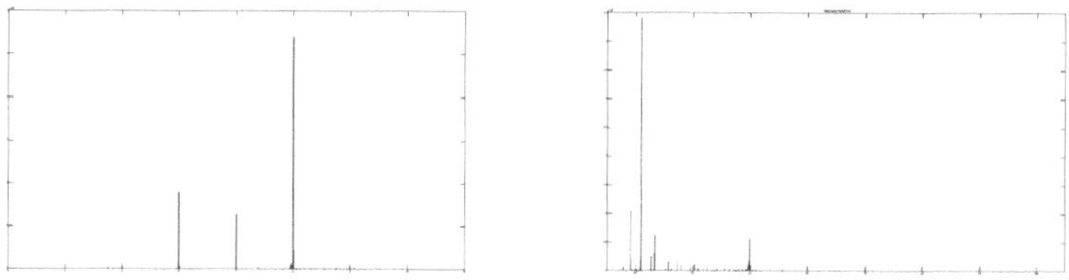

(a) Liquid level (b) Temperature

Figure 9.8. *Histogram of the liquid level (left) and temperature (right) at the maintenance time for the heated hold-up tank*

There is no analytical solution to our optimization problem, therefore it is impossible to compare our results with the true value function. However, it must be emphasized that there is a theoretical proof in Chapter 7 that this procedure converges to the true value as the number of points in the quantization grids goes to infinity. We can also conduct two kinds of simple experiments to validate our results. First, we can simulate new trajectories and check one by one if the intervention took place at a reasonable time. Figure 9.9 shows such an interesting example:

– the system starts in the equilibrium state, so that the reward function grows roughly linearly with time;

– at time 12.94 h, a jump occurs. The first unit is stuck OFF. The liquid level drops fast and the temperature rises slowly. As the liquid level and temperature are still within the acceptable bounds, the reward function is still growing roughly linearly with time;

– at time 13.61 h, the liquid level reaches the boundary 6 m and the controller switches the 3 units to stuck OFF, ON and OFF. The liquid level now rises whereas the temperature drops.

– at time 14.94 h, the liquid level reaches the boundary 8 m and the controller switches the 3 units back to stuck OFF, OFF and ON. The liquid level drops again and the temperature rises;

– at time 16.27 h, the liquid level reaches again the boundary 6 m and the controller switches the 3 units back to stuck OFF, ON and OFF. The liquid level rises and the temperature drops;

– at time 17.38 h, before the liquid level reaches again 8 m, the second unit fails and is now stuck ON. The liquid level still rises and the temperature drops;

– at time 17.60 h, the liquid level reaches the boundary 8 m and the controller switches the 3 units to stuck OFF, Stuck ON and ON. The liquid level now remains constant as the temperature drops down to its equilibrium state;

– at time 150.24 h, the third unit is stuck OFF. The liquid level rises again rapidly and goes above the acceptable threshold. The temperature remains constant. The algorithm decides to perform a maintenance at time 151.58. The final reward is 99.07.

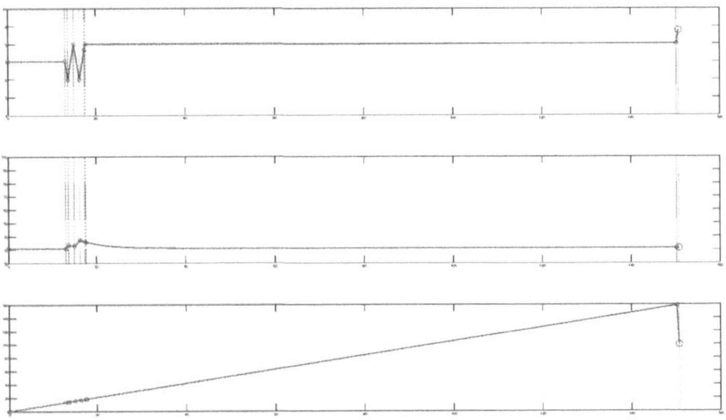

Figure 9.9. *An example of trajectory for the heated hold-up tank. The upper chart shows the liquid level, the middle chart the temperature, and the bottom chart the corresponding reward, as a function of time. The small circles represent the jumps of regime, and the large circle and red line the computed maintenance time*

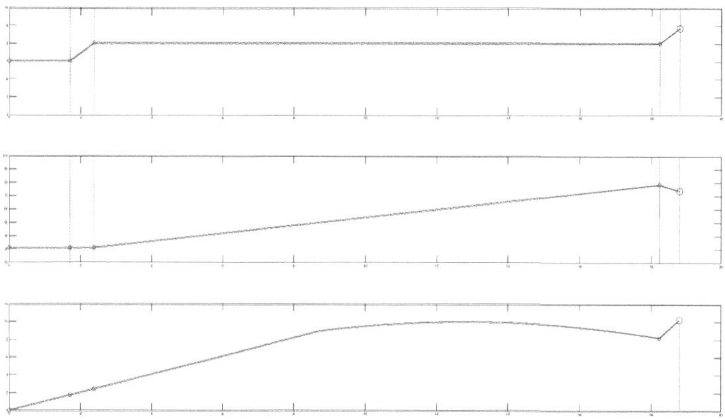

Figure 9.10. *An example of trajectory for the heated hold-up tank. The upper chart shows the fluid level, the middle chart the temperature, and the bottom chart the corresponding reward, as a function of time. The small circles represent the jumps of regime, and the large circle and red line the computed maintenance time*

Another example is given in Figure 9.10:

– the system starts in the equilibrium state, so that the reward function grows linearly with time;

– at time 1.71 h, a jump occurs. The third unit is stuck OFF. The liquid level rises and the temperature remains constant. As the liquid level and temperature are still within the acceptable bounds, the reward function is still growing linearly with time;

– at time 2.37 h, the liquid level reaches the threshold 8 m and triggers the command. The solicitation of the command is successful, and the first unit is turned OFF. The current state is now OFF for units 1 and 2, and stuck OFF for unit 3. The liquid level remains constant at 8 m, but the temperature rises. At about 9 h, the temperature crosses the threshold 50°C so that the reward function is now slowly decreasing. Note that this does not trigger an immediate maintenance;

– At time 18.22 h, another jump occurs and the second unit is now stuck ON, causing the liquid level to rise again, but the temperature to decrease. As a result, the reward function is now increasing again;

– the algorithm then selects the maintenance time to be 18.8 h, before the liquid level reaches the level 10 m causing a total failure of the system.

Note that at this intervention time, the first unit is not stuck, so that other jumps may happen in the future. In this special example, the intervention time occurs when the reward is maximal.

Second, we can simply compare the performances of the system when no maintenance is performed to those with our maintenance rule. The results are summarized in Table 9.3. These results are obtained with 10^5 Monte Carlo simulations and illustrate the power of our procedure as the mean performance is increased by 156% and the top events are almost always avoided.

	Without maintenance	With maintenance
Mean performance	211.80	330.87
Null gain	80.33%	0.02%
$6 \leq h \leq 8$	28.25%	90.02%
$\theta \leq 50°C$	80.33%	95.09%
$h = 4$	16.65%	0%
$h = 10$	54.55%	0%
$\theta = 100$	9.13%	0.02%

Table 9.3. *Comparison of the performances with and without maintenance*

9.5. Conclusion

The numerical method described in Chapter 7 has been applied to three practical industrial examples to approximate the value function of the optimal stopping problem and a quasi-optimal stopping time for a piecewise deterministic Markov process, that is the quasi optimal maintenance date for our processes. The quantization method proposed can sometimes be costly in computing time, but has a very interesting property: it can be calculated offline, and serve to compute all sorts of indices. Moreover, it depends only on the evolutionary characteristics of the model, and not on the cost function chosen, or the actual trajectory of the specific process we want to monitor. The calculation of the optimal maintenance time is done in real time. This method is especially attractive as its application requires knowledge of the system state only at moments of change of environment and not in continuous time. The optimal maintenance time is updated at the moments when the system switches to another environment or mode and has the form *intervene at such date if no change of mode takes place in the meantime*, which allows to schedule maintenance services in advance.

To apply this methodology in practice, we must observe the jump times and post-jump locations of the PDMPs at the jumps times. If the observation is not perfect, we discussed how to adapt our methodology in Chapter 8. If the jump times are not observed, we can adopt the following suboptimal strategy: first estimate the jump times, (see [BAY 13a]), then apply our optimization procedure with the estimated jump times.

The applications presented here are all examples of maintenance *as good as new* of the system. The next step will be to allow only partial repair of the system and study it for its whole lifecycle. The problem will then be to find simultaneously the optimal times of maintenance and optimal repair levels. Mathematically, it is an impulse control problem, which complexity exceeds widely that of the optimal stopping. See Chapter 10 for a first step in that direction.

10

Optimal Impulse Control

10.1. Introduction

In this chapter we present a numerical method to compute the value function of an impulse control problem for a piecewise deterministic Markov process (PDMP). An impulse control strategy consists of a sequence of single interventions introducing a jump of the process at some controller-specified stopping time and moving the process at that time to some new point in the state space. The selection of a new starting point after each intervention is the main difference with the optimal stopping problems studied in Chapters 7 and 8 and yields significant additional technical difficulties. Our impulse control problem consists of choosing a strategy (if it exists) that minimizes the expected sum of discounted running and intervention costs up to infinity, and computing the optimal cost thus achieved. Many applied problems fall into this class, such as inventory problems in which a sequence of restocking decisions is made, or optimal maintenance of complex systems with components subject to failure and repair.

Impulse control problems of PDMPs in the context of an expected discounted cost have been considered in [COS 89, DEM 95, GAT 91, GAT 92, LEN 89]. Generally speaking, in [COS 89], the authors study this impulse control problem by using the value improvement approach, while in [DEM 95, GAT 91, GAT 92, LEN 89] the authors choose to analyze it by using the variational inequality approach. In [COS 89], the authors also consider a numerical procedure. By showing that iteration of the single-jump-or-intervention operator generates a sequence of functions converging to the value function of the problem, they derive an algorithm to compute an approximation of that value function. Their approach is also based on a uniform discretization of the state space similar to the one proposed by Kushner in [KUS 77]. In particular, they derive a convergence result for the approximation scheme but no estimation of the rate of convergence is given. To the best of our knowledge, it is the only paper presenting a computational method for solving the impulse control problem for a PDMP in the context of discounted cost. Note that a similar procedure has been applied by Costa in [COS 93] to derive a numerical scheme for the impulse control problem with a long run average cost.

Our approach is also based on the iteration of the single-jump-or-intervention operator, but we want to derive a convergence rate for our approximation. Our method does not rely on a *blind* discretization of the state space, but on a discretization that depends on time and takes into account the random nature of the process and involves the quantization procedure described in Chapter 3.

Although the value function of the impulse control problem can be computed by iterating implicit optimal stopping problems, see [COS 89, Proposition 2] or [DAV 93, Proposition 54.18], from a numerical point of view the impulse control is much more difficult to handle than the optimal stopping problem described in Chapter 7. Indeed, for the optimal stopping problem, the value function is computed as the limit of a sequence (v_n) constructed by iterating the operator L defined by equation [7.6]. This iteration procedure yields an iterative construction of a sequence of random variables $V_n = v_n(Z_n)$. This was the keystone of our approximation procedure. As regards impulse control, the iterative construction for the corresponding random variables does not hold anymore, see section 10.2 for more details. This is mostly due to the fact that not only does the controller choose times to stop the process, but it also chooses a new starting point for the process to restart from after each intervention. This makes the single-jump-or-intervention operator significantly more complicated to iterate that the single-jump-or-stop operator used for optimal stopping. We manage to overcome this extra difficulty by using two series of quantization grids instead of just one we used for optimal stopping.

The main material of this chapter comes from [SAP 12a]. The chapter is organized as follows. In section 10.2, we present the impulse control problem and recall the iterative construction of the value function presented in [COS 89]. Technical results related to Lipschitz continuity issues are presented in section 10.3. In section 10.4, we explain our approximation procedure and prove its convergence with error bounds. In section 10.5, we present a numerical application to the Transmission Control Protocol (TCP) process. Finally, in section 10.6, we present some concluding remarks and remaining open questions.

10.2. Impulse control problem

The formal probabilistic apparatus necessary to precisely define the impulse control problem is rather cumbersome, and will not be used in the sequel, therefore, for the sake of simplicity, we only present an informal description of the problem. The interested readers are referred to [COS 89] for a rigorous definition.

A strategy $\mathcal{S} = (\tau_n, R_n)_{n\geq 1}$ is a sequence of non-anticipative intervention times $(\tau_n)_{n\geq 1}$ and non-anticipative E-valued random variables $(R_n)_{n\geq 1}$ on a measurable space $(\overline{\Omega}, \overline{\mathcal{F}})$. Between the intervention times τ_i and τ_{i+1}, the motion of the system is determined by the PDMP (X_t) starting from R_i. If an intervention takes place at $x \in E$, then the set of admissible points where the decision maker can send the system to is denoted by $\mathbb{U} \subset E$. The following assumption on the control set $\mathbb{U}$ will be in force throughout this chapter.

ASSUMPTION 10.1.– The control set $\mathbb{U}$ is finite and does not depend on x. The cardinal of the set $\mathbb{U}$ is denoted by u

$$\mathbb{U} = \{y^i : 1 \leq i \leq u\}.$$

The strategy $\mathcal{S}$ induces a family of probability measures $P_x^{\mathcal{S}}$, $x \in E$, on $(\overline{\Omega}, \overline{\mathcal{F}})$. We define the class $\mathbb{S}$ of admissible strategies as the strategies $\mathcal{S}$ which satisfy $\tau_\infty = \infty$ $P_x^{\mathcal{S}}$-a.s. for all $x \in E$. Associated with the strategy $\mathcal{S}$, the following discounted cost for a process starting at $x \in E$ is given by

$$\mathcal{J}^{\mathcal{S}}(x) = E_x^{\mathcal{S}}\left[\int_0^\infty e^{-\alpha s} f(Y_s)ds + \sum_{i=1}^{\infty} e^{-\alpha\tau_i} c(Y_{\tau_i}, Y_{\tau_i^+})\right],$$

where $E_x^{\mathcal{S}}$ is the expectation with respect to $P_x^{\mathcal{S}}$ and (Y_t) is the process with interventions. The function f then corresponds to the running cost and $c(x, y)$ corresponds to the intervention cost of moving the process from x to y, α is a positive discount factor.

We suppose in this chapter that assumptions 1.3, 1.4, 1.5, 1.6 and 10.1 hold. Moreover, we make the following hypothesis on the cost functions.

ASSUMPTION 10.2.– The running cost f is a positive function in $\mathbf{L}_{\Phi}(E)$.

ASSUMPTION 10.3.– The intervention cost function c is in $\mathbf{B}(\overline{E} \times \mathbb{U})$ and there exist $[c]_1 \in \mathbb{R}^+$, $[c]_2 \in \mathbb{R}^+$ and $[c]_* \in \mathbb{R}^+$ such that

1) for any $(x, y) \in E^2$, $u \in [0, t^*(x) \wedge t^*(y)]$,

$$\max_{z \in \mathbb{U}} \left| c(\Phi(x, u), z) - c(\Phi(y, u), z) \right| \leq [c]_1 |x - y|,$$

2) for any $x \in E$, and $(t, s) \in [0, t^*(x)]^2$,

$$\max_{z \in \mathbb{U}} \left| c(\Phi(x, t), z) - c(\Phi(x, s), z) \right| \leq [c]_2 |t - s|,$$

3) for any $(x, y) \in E^2$,

$$\max_{z \in \mathbb{U}} \left| c(\Phi(x, t^*(x)), z) - c(\Phi(y, t^*(y)), z) \right| \leq [c]_* |x - y|,$$

4) for any $(x, y) \in \overline{E} \times \mathbb{U}, 0 < c_0 \leq c(x, y) \leq C_c$,

5) for any $(x, y, z) \in \overline{E} \times \mathbb{U} \times \mathbb{U}$,

$$c(x, y) + c(y, z) \geq c(x, z).$$

The last item of the previous assumption implies that the cost of taking two or more interventions instantaneously will not be lower than taking a single intervention. Finally, the value function for the discounted infinite horizon impulse control problem is defined for all x in E by

$$\mathcal{V}(x) = \inf_{\mathcal{S} \in \mathbb{S}} \mathcal{J}^{\mathcal{S}}(x).$$

Note that unlike the optimal stopping problems studied in Chapters 7 and 8, we are now interested in a minimization problem instead of a maximization problem, hence the term *cost* functions instead of *reward* functions. Nevertheless, the approach to solve this optimization problem follows the same lines as in the previous chapters with suitable adjustments.

Associated with this impulse control problem, we define the following operators. For $x \in E, t \geq 0, v, w \in \mathbf{L}_\Phi(E)$, set

$$\begin{aligned}
F(x,t) &= \int_0^{t\wedge t^*(x)} \mathrm{e}^{-\alpha s-\Lambda(x,s)} f\big(\Phi(x,s)\big)ds, \\
&= \mathbb{E}_x\Big(\int_0^{T_1\wedge t} e^{-\alpha s} f(\Phi(x,s))ds\Big), \\
Hv(x,t) &= \mathrm{e}^{-\alpha(t\wedge t^*(x))-\Lambda(x,(t\wedge t^*(x)))} v\big(\Phi(x,t\wedge t^*(x))\big) \\
&= \mathbb{E}_x\Big(e^{-\alpha(t\wedge t^*(Z_0))} v\big(\Phi(Z_0,t\wedge t^*(Z_0))\big)\mathbb{1}_{\{S_1\geq t\wedge t^*(Z_0)\}}\Big), \\
Iw(x,t) &= \int_0^{t\wedge t^*(x)} \mathrm{e}^{-\alpha s-\Lambda(x,s)} \lambda Qw\big(\Phi(x,s)\big)ds, \\
&= \mathbb{E}_x\Big(e^{-\alpha S_1} w(Z_1)\mathbb{1}_{\{S_1< t\wedge t^*(Z_0)\}}\Big).
\end{aligned}$$

Finally, for notational convenience, let us introduce for $v, w \in \mathbf{L}_\Phi(E), x \in E$ and $t \geq 0$.

$$J(v,w)(x,t) = F(x,t) + Hv(x,t) + Iw(x,t), \qquad [10.1]$$

$$Kw(x) = F(x,t^*(x)) + HQw(x,t^*(x)) + Iw(x,t^*(x)). \qquad [10.2]$$

Note that these operators involve the original non-controlled process (X_t) and only depend on the embedded Markov chain $(\Theta_n) = (Z_n, S_n)$. The above equalities are valid for all n because (Θ_n) is a homogeneous Markov chain. Finally, for $v, w \in \mathbf{L}_\Phi(E)$, φ defined on $\mathbb{U}$ and $x \in E$, set

$$M\varphi(x) = \inf_{y\in\mathbb{U}} \big(c(x,y) + \varphi(y)\big),$$

$$L(v,w)(x) = \inf_{t\in\mathbb{R}+} J(v,w)(x,t) \wedge Kw(x), \qquad [10.3]$$

$$\mathcal{L}w(x) = L(Mw,w)(x).$$

As explained in [COS 89], operator $\mathcal{L}$ applied to w is the value function of the single-jump-or-intervention problem with cost function w and the value function $\mathcal{V}$ can be computed by iterating $\mathcal{L}$. More precisely, let h be the cost associated with the no-impulse strategy

$$h(x) = \mathbb{E}_x\left[\int_0^\infty e^{-\alpha s} f(X_s)ds\right],$$

for all $x \in E$. Then, we recall proposition 4 of [COS 89].

PROPOSITION 10.1.– Assume that g is in $\mathbf{L}_\Phi(E)$ and $g \geq h$. Define $\mathcal{V}_0^g = g$ and $\mathcal{V}_{n+1}^g = \mathcal{L}\mathcal{V}_n^g$, for all $n \geq 0$. Then, for all $x \in E$

$$\mathcal{V}(x) = \lim_{n \to \infty} \mathcal{V}_n^g(x).$$

As pointed out in [COS 89], if we choose exactly $g = h$, then $\mathcal{V}_n^h$ corresponds to the value function of the impulse problem where only n jumps plus interventions are allowed, and after that, there are no further interventions.

Note that operator L is quite similar to the operator used in Chapter 7. However, the iteration procedure here does not rely on L but on $\mathcal{L}$. The difference between operators L and $\mathcal{L}$ comes from the operator M that chooses optimally the next starting point after an intervention. This is one of the main technical differences between approximating the value functions of an optimal stopping and impulse problems, and it makes the approximation scheme significantly more difficult, as explained in section 10.4.

10.3. Lipschitz-continuity properties

This section gathers technical results related to the Lipschitz-continuity properties of operators and functions involved in the recursive computation of the value function.

10.3.1. *Lipschitz properties of the operators*

We start with preliminary results on operators M, H, F and I.

LEMMA 10.1.– For any g defined on $\mathbb{U}$, $Mg \in \mathbf{L}_\Phi(E)$. Moreover,

$$[Mg]_1 \leq [c]_1, \quad [Mg]_2 \leq [c]_2, \quad [Mg]_* \leq [c]_*, \quad C_{Mg} \leq C_c + C_g.$$

PROOF.– By using the fact that $\left|Mg(x) - Mg(y)\right| \leq \sup_{z \in \mathbb{U}} \left|c(x,z) - c(y,z)\right|$ and the assumption on the cost function c, the result follows easily. □

LEMMA 10.2.– Let $v \in \mathbf{L}_\Phi(E)$. Then, for all $(x, y) \in E^2$ and $(t, u) \in (\mathbb{R}^+)^2$, we have

$$\left|Hv(x,t) - Hv(y,u)\right| \leq D_1(v)|x - y| + D_2(v)|t - u|,$$

where

– if $t < t^*(x)$ and $u < t^*(y)$,

$$D_1(v) = [v]_1 + C_v C_{t^*}[\lambda]_1, \qquad D_2(v) = [v]_2 + C_v(C_\lambda + \alpha),$$

– if $t = t^*(x)$ and $u = t^*(y)$,

$$D_1(v) = [v]_* + C_v\big(C_{t^*}[\lambda]_1 + (C_\lambda + \alpha)L_{t^*}\big), \qquad D_2(v) = 0,$$

– otherwise,

$$D_1(v) = [v]_1 + [v]_2 L_{t^*} + C_v\big(C_{t^*}[\lambda]_1 + (C_\lambda + \alpha)L_{t^*}\big),$$
$$D_2(v) = [v]_2 + C_v(C_\lambda + \alpha).$$

PROOF.– This is a straightforward consequence of lemma 7.2. □

LEMMA 10.3.– For all $w \in \mathbf{L}_\Phi(E)$, $(x, y) \in E^2$ and $(t, u) \in (\mathbb{R}^+)^2$, we have

$$\Big|F(x,t) - F(y,u)\Big| \leq \frac{1}{\alpha}\Big([f]_1 + C_f C_{t^*}[\lambda]_1\Big)|x-y| + C_f\big(|t-u| \vee L_{t^*}|x-y|\big),$$

$$\Big|Iw(x,t) - Iw(y,u)\Big| \leq \frac{1}{\alpha}\Big([Q][w]_1 C_\lambda + C_w[\lambda]_1\big(1 + C_\lambda C_{t^*}\big)\Big)|x-y| + C_w C_\lambda\Big(|t-u| \vee L_{t^*}|x-y|\Big).$$

PROOF.– Suppose, without loss of generality, that $t \wedge t^*(x) \leq u \wedge t^*(y)$. Then, we have

$$\begin{aligned}
&\Big|F(x,t) - F(y,u)\Big| \\
&\leq \int_0^{t\wedge t^*(x)} e^{-\alpha s}\Big|f\big(\Phi(x,s)\big)e^{-\Lambda(x,s)} - f\big(\Phi(y,s)\big)e^{-\Lambda(y,s)}\Big| ds \\
&\quad + \int_{t\wedge t^*(x)}^{u\wedge t^*(y)} \Big|f\big(\Phi(y,s)\big)e^{-\alpha s - \Lambda(y,s)}\Big| ds \\
&\leq \Big([f]_1 + C_f C_{t^*}[\lambda]_1\Big)\int_0^\infty e^{-\alpha s} ds |x-y| \\
&\quad + C_f\Big|u \wedge t^*(y) - t \wedge t^*(x)\Big|.
\end{aligned}$$

From the fact that $\left|u \wedge t^*(y) - t \wedge t^*(x)\right| \leq |t-u| \vee L_{t^*}|x-y|$, we get the first inequality. By using similar arguments, it is easy to obtain the last result.

□

Now, we turn to the Lipschitz property of operators K and J. First, introduce the following constants

$$E_1 = C_{t^*}[\lambda]_1 + (C_\lambda + \alpha)L_{t^*}, \qquad E_2 = C_\lambda L_{t^*} + [\lambda]_1 \frac{1 + C_\lambda C_{t^*}}{\alpha},$$

$$E_3 = [f]_1 \frac{1}{\alpha} + C_f\Big(\frac{C_{t^*}[\lambda]_1}{\alpha} + L_{t^*}\Big).$$

Definitions [10.2] and [10.1] together with lemmas 10.2 and 10.3 readily yield the following lemmas 10.4, 10.5 and 10.6.

LEMMA 10.4.– For $w \in \mathbf{L}_\Phi(E)$ and $(x,y) \in E^2$, we have

$$\left|Kw(x) - Kw(y)\right| \leq \Big([Q][w]_1 \frac{C_\lambda}{\alpha} + [Q][w]_* + C_w(E_1 + E_2) + E_3\Big)|x-y|.$$

LEMMA 10.5.– For all $(v,w) \in \mathbf{L}_\Phi(E)^2, x \in E$ and $(t,u) \in (\mathbb{R}^+)^2$, we have

$$\left|J(v,w)(x,t) - J(v,w)(x,u)\right| \leq \big(C_f + C_w C_\lambda + [v]_2 + C_v(C_\lambda + \alpha)\big)|t-u|.$$

LEMMA 10.6.– For all $(v,w) \in \mathbf{L}_\Phi(E)^2$, $(x,y) \in E^2$ and $t \geq 0$, we have

$$\left|J(v,w)(x,t) - J(v,w)(y,t)\right| \leq \Big([v]_1 + [v]_2 L_{t^*} + C_v E_1 + [Q][w]_1 \frac{C_\lambda}{\alpha} + C_w E_2 + E_3\Big)|x-y|.$$

It is now easy to obtain that for $(v,w) \in \mathbf{L}_\Phi(E)^2, s \in \mathbb{R}^+$ and $(x,y) \in E^2$,

$$\left|\inf_{t \geq s} J(v,w)(x,t) - \inf_{t \geq s} J(v,w)(y,t)\right| \leq \sup_{t \geq 0}\left|J(v,w)(x,t) - J(v,w)(y,t)\right|. \qquad [10.4]$$

LEMMA 10.7.– Let $(v,w) \in \mathbf{L}_\Phi(E)^2$. Then, for all $x \in E$ and $(s,t) \in (\mathbb{R}^+)^2$,

$$\left|\inf_{u \geq t} J(v,w)(x,u) - \inf_{u \geq s} J(v,w)(x,u)\right| \leq \Big(C_f + C_w C_\lambda + [v]_2 + C_v(C_\lambda + \alpha)\Big)|t-s|.$$

PROOF.– Without loss of generality, it can be assumed that $s \leq t$. Therefore, we have

$$\left| \inf_{u \geq t} J(v,w)(x,u) - \inf_{u \geq s} J(v,w)(x,u) \right|$$

$$= \inf_{u \geq t} J(v,w)(x,u) - \inf_{u \geq s} J(v,w)(x,u). \quad [10.5]$$

Note that $\inf_{u \geq s} J(w,g)(x,u) = J(w,g)(x,\overline{s})$ for some $\overline{s} \in [s \wedge t^*(x), t^*(x)]$. Consequently, if $\overline{s} \geq t \wedge t^*(x)$ then we have $\left| \inf_{u \geq t} J(v,w)(x,u) - \inf_{u \geq s} J(v,w)(x,u) \right| = 0$. Now, if $\overline{s} \in [s \wedge t^*(x), t \wedge t^*(x)[$, then we have

$$\inf_{u \geq t} J(v,w)(x,u) - \inf_{u \geq s} J(v,w)(x,u) \leq J(v,w)(x,t) - J(v,w)(x,\overline{s}).$$

From lemma 10.5, we obtain the following inequality

$$\inf_{u \geq t} J(v,w)(x,u) - \inf_{u \geq s} J(v,w)(x,u)$$

$$\leq \Big(C_f + C_w C_\lambda + [v]_2 + C_v(C_\lambda + \alpha) \Big) |t - \overline{s}|. \quad [10.6]$$

Combining [10.5] and [10.6] and the fact that $|t - \overline{s}| \leq |t - s|$, the result follows. □

10.3.2. *Lipschitz properties of the operator $\mathcal{L}$*

Now, we study the Lipschitz continuity of our main operator.

LEMMA 10.8.– For all $(v,w) \in \mathbf{L}_\Phi(E)^2$, $x \in E$ and $t \in [0, t^*(x))$ and $u \in \mathbb{R}^+$, we have

$$F(\Phi(x,t),u) = e^{\alpha t + \Lambda(x,t)} \Big(F(x,t+u) - F(x,t) \Big),$$

$$Iw(\Phi(x,t),u) = e^{\alpha t + \Lambda(x,t)} \Big(Iw(x,t+u) - Iw(x,t) \Big),$$

$$Hv(\Phi(x,t),u) = e^{\alpha t + \Lambda(x,t)} Hv(x,t+u).$$

PROOF.– By using the semi-group property of Φ, we have $\Lambda\big(\Phi(x,t),u\big) = \Lambda(x,t+u) - \Lambda(x,t)$ for $t+u < t^*(x)$ and noting that $t^*(\Phi(x,t)) = t^*(x) - t$ for $t < t^*(x)$, a simple change of variable yields

$$F\big(\Phi(x,t),u\big) = e^{\alpha t+\Lambda(x,t)} \int_t^{(t+u)\wedge t^*(x)} \mathrm{e}^{-\alpha s-\Lambda(x,s)} f\big(\Phi(x,s)\big)ds,$$

and we get the first equation. The other equalities can be obtained by using similar arguments. □

LEMMA 10.9.– For all $(v,w) \in \mathbf{L}_\Phi(E)^2, x \in E$ and $t \in [0,t^*(x))$,

$$L(v,w)\big(\Phi(x,t)\big) = e^{\alpha t+\Lambda(x,t)}\bigg(\Big(\inf_{s\geq t} J(v,w)(x,s) \wedge Kw(x)\Big) - F(x,t) - Iw(x,t)\bigg).$$

PROOF.– For $t \in [0,t^*(x))$ and $u \in \mathbb{R}^+$, we have from lemma 10.8, [10.1] and [10.2]

$$J(v,w)\big(\Phi(x,t),u\big) = e^{\alpha t+\Lambda(x,t)}\big(J(v,w)(x,t+u) - F(x,t) - Iw(x,t)\big),$$
$$Kw\big(\Phi(x,t)\big) = e^{\alpha t+\Lambda(x,t)}\big(Kw(x) - F(x,t) - Iw(x,t)\big).$$

Consequently, from equation [10.3], it follows

$$L(v,w)\big(\Phi(x,t)\big) = e^{\alpha t+\Lambda(x,t)}\bigg(\Big(\inf_{u\geq 0} J(v,w)(x,t+u) \wedge Kw(x)\Big) - F(x,t) - Iw(x,t)\bigg),$$

showing the result. □

PROPOSITION 10.2.– For all $w \in \mathbf{L}_\Phi(E)$, $\mathcal{L}w \in \mathbf{L}_\Phi(E)$, and

$$
\begin{aligned}
[\mathcal{L}w]_1 \leq e^{(\alpha+C_\lambda)C_{t^*}} \bigg(& [\lambda]_1 C_{t^*}(C_c + \frac{C_f}{\alpha}) \\
& + \Big([c]_1 + [c]_2 L_{t^*} + C_c E_1\Big) \vee \Big([Q][w]_*\Big) \\
& + 2E_3 + \frac{2[Q]C_\lambda}{\alpha}[w]_1 + \Big(E_1 + 2E_2 + [\lambda]_1 C_{t^*}(1 + C_\lambda/\alpha)\Big)C_w\bigg),
\end{aligned}
$$

$$
\begin{aligned}
[\mathcal{L}w]_2 \leq e^{(\alpha+C_\lambda)C_{t^*}} \Big(& 3C_f + [c]_2 + 2C_c(C_\lambda + \alpha) + \frac{C_f C_\lambda}{\alpha} \\
& + C_w\big[4C_\lambda + \frac{C_\lambda^2}{\alpha} + \alpha\big]\Big),
\end{aligned}
$$

$$
[\mathcal{L}w]_* \leq [\mathcal{L}w]_1 + [\mathcal{L}w]_2 L_{t^*},
$$

$$
\begin{aligned}
L_{\mathcal{L}w} \leq & (E_1 + E_2)C_w + \frac{[Q]C_\lambda}{\alpha}[w]_1 + E_3 \\
& + \Big([c]_1 + [c]_2 L_{t^*} + C_c E_1\Big) \vee \Big([Q][w]_*\Big).
\end{aligned}
$$

PROOF.– Let us denote $\mathcal{L}w$ by g. We have for $(x, y) \in E^2$ and $t \in [0, t^*(x) \wedge t^*(y)]$

$$
\begin{aligned}
& \Big|\mathcal{L}w(\Phi(x,t)) - \mathcal{L}w(\Phi(y,t))\Big| \\
& \quad \leq e^{\alpha t + \Lambda(y,t)}\Big(\big|F(x,t) - F(y,t)\big| + \big|Iw(x,t) - Iw(y,t)\big| \\
& \qquad + \big|\inf_{s \geq t} J(Mw,w)(x,s) - \inf_{s \geq t} J(Mw,w)(y,s)\big| \\
& \qquad \vee \big|Kw(x) - Kw(y)\big|\Big) \\
& \qquad + \bigg|\Big(\inf_{s \geq t} J(Mw,w)(x,s) \wedge Kw(x)\Big) - F(x,t) - Iw(x,t)\bigg| \\
& \qquad \times \Big|e^{\alpha t + \Lambda(x,t)} - e^{\alpha t + \Lambda(y,t)}\Big|
\end{aligned}
$$

It is easy to show that for $x \in E, t \in [0, t^*(x)]$ and $w \in \mathbf{L}_\Phi(E)$, we have

$$
e^{\alpha t + \Lambda(x,t)} \leq e^{(\alpha + C_\lambda)C_{t^*}},
$$

$$\left|\left(\inf_{s\geq t} J(Mw,w)(x,s)\wedge Kw(x)\right) - F(x,t) - Iw(x,t)\right|$$
$$\leq \frac{1}{\alpha}(C_f + C_\lambda C_w) + C_c + C_w$$

and for $(x,y)\in E^2$ and $t\in[0,t^*(x)\wedge t^*(y)]$

$$\left|e^{\alpha t+\Lambda(x,t)} - e^{\alpha t+\Lambda(y,t)}\right| \leq e^{(\alpha+C_\lambda)C_{t^*}}[\lambda]_1 C_{t^*}|x-y|.$$

Consequently, by using lemmas 10.3 and 10.4 and equation [10.4], we get the first equation. For $x\in E$ and $(s,t)\in[0,t^*(x)]^2$, we have

$$\left|\mathcal{L}w(\Phi(x,s)) - \mathcal{L}w(\Phi(x,t))\right| \leq e^{\alpha t+\Lambda(x,t)}\Big(\left|Iw(x,s)-Iw(x,t)\right|$$
$$+\left|\inf_{u\geq s} J(Mw,w)(x,u) - \inf_{u\geq t} J(Mw,w)(x,u)\right| + \left|F(x,s)-F(x,t)\right|\Big)$$
$$+\left|e^{\alpha s+\Lambda(x,s)} - e^{\alpha t+\Lambda(x,t)}\right|$$
$$\times\left|\left(\inf_{u\geq s} J(Mw,w)(x,u)\wedge Kw(x)\right) - F(x,s) - Iw(x,s)\right|.$$

Note that for $x\in E$, $(s,t)\in[0,t^*(x)]^2$

$$\left|e^{\alpha s+\Lambda(x,s)} - e^{\alpha t+\Lambda(x,t)}\right| \leq e^{(\alpha+C_\lambda)C_{t^*}}(C_\lambda+\alpha)|t-s|.$$

Consequently, by using lemmas 10.3 and 10.7, we obtain the second equation. The third inequality is straightforward and finally, for $(x,y)\in E^2$ we have

$$\left|\mathcal{L}w(x) - \mathcal{L}w(y)\right| \leq \left|\inf_{s\geq 0} J(Mw,w)(x,s) - \inf_{s\geq 0} J(Mw,w)(y,s)\right|$$
$$\vee\left|Kw(x)-Kw(y)\right|.$$

By using equation [10.4] and lemma 10.4, we get the last equation. □

10.4. Approximation of the value function

From now on, we assume that the distribution of X_0 is given by δ_{x_0} for some fixed point x_0 in the state space E. We also choose a function g in $\mathbf{L}_\Phi(E)$ satisfying $g\geq h$. Our approximation of the value function at x_0 is based on proposition 10.1. Following the approach proposed by Davis and

Costa in [COS 89], we now suppose that we have selected a suitable index N such that $\mathcal{V}(x_0) - \mathcal{V}_N^g(x_0)$ is small enough, see the example in section 10.5 for a practical determination of N. We turn to the approximation of $\mathcal{V}_N^g(x_0)$ which is the main objective of this chapter. In all generality, finding an index N such that $\mathcal{V}(x_0) - \mathcal{V}_N^g(x_0)$ is below a prescribed level is a very difficult problem to solve. However, in particular cases we can hope to be able to evaluate the distance between $\mathcal{V}(x_0)$ and $\mathcal{V}_N^g(x_0)$. As suggested by Davis and Costa in [COS 89], a value of N can be chosen by calculating $\mathcal{V}_n^g(x_0)$ for different values of n and stopping when the difference between two consecutive values is small enough. Our results of convergence are derived for a fixed but arbitrary N.

Recall that if $\mathcal{V}_0 = h$, then $\mathcal{V}_N^h$ corresponds to the value function of the impulse problem where only N jumps plus interventions are allowed. This is an interesting problem to be solved in itself. For notational convenience, we will change our notation in the sequel and reverse the indices for the sequence $(\mathcal{V}_n^g)_{0 \leq n \leq N}$.

DEFINITION 10.1.– *Set*

$$\begin{cases} v_N = g = \mathcal{V}_0^g, \\ v_n = \mathcal{L}v_{n+1} = \mathcal{V}_{N-n}^g, \quad \text{for all } 0 \leq n < N. \end{cases}$$

As explained in the introduction, the keystone of the approximation procedure for optimal stopping in Chapter 7 is that the analogue of proposition 10.1 yields a recursive construction of the random variables $v_n(Z_n)$. Unfortunately, this key and important property does not hold anymore here. Indeed, we have

$$\begin{aligned} v_n(Z_n) &= \mathcal{L}v_{n+1}(Z_n) \\ &= \bigg(\inf_{t \in \mathbb{R}+} \mathbb{E}_{x_0} \Big(F(Z_n, t) + e^{-\alpha S_{n+1}} v_{n+1}(Z_{n+1}) \mathbb{1}_{\{S_{n+1} < t \wedge t^*(Z_n)\}} \\ &\quad + e^{-\alpha t \wedge t^*(Z_n)} M v_{n+1}(\Phi(Z_n, t \wedge t^*(Z_n)) \mathbb{1}_{\{S_{n+1} \geq t \wedge t^*(Z_n)\}} \big| Z_n \Big) \bigg) \\ &\quad \wedge \mathbb{E}_{x_0} \Big(F(Z_n, t^*(Z_n)) + e^{-\alpha S_{n+1}} v_{n+1}(Z_{n+1}) \big| Z_n \Big). \end{aligned}$$

Here, $Mv_{n+1}(\Phi(Z_n, t \wedge t^*(Z_n))$ cannot be written as a function of $v_{n+1}(Z_{n+1})$. Hence, we have no recursive construction of the random variables $v_n(Z_n)$ and we cannot apply the same procedure that we used for optimal stopping in section 7.2. Thus, we propose a new self-contained procedure to evaluate $Mv_{n+1}(\Phi(Z_n, t \wedge t^*(Z_n))$ separately from the main computation of the value function.

Note that for all $0 \leq n < N$, to compute Mv_{n+1} at any point, we actually only need to evaluate the value functions v_{n+1} at the points of the control grid $\mathbb{U}$. We propose again a recursive computation based on the embedded Markov chain (Z_n, S_n) but with a different starting point. Set $Z_0^y = y \in \mathbb{U}$ and $S_0^y = 0$. We denote by (Z_n^y, S_n^y) the Markov chain starting from this point $(y, 0)$. We clearly know $v_N = g$ on $\mathbb{U}$. Now, suppose we have computed all the v_n on $\mathbb{U}$ for $k+1 \leq n \leq N$. Therefore, all functions Mv_n are known everywhere. We can then propose the following recursive computation to evaluate v_k at $y \in \mathbb{U}$

$$\begin{cases} v_N(Z_{N-k}^y) = g(Z_{N-k}^y) \\ v_{k+n}(Z_n^y) = L(Mv_{k+n+1}, v_{k+n+1})(Z_n^y), \\ \qquad \text{for all } 0 \leq n \leq N-k-1. \end{cases} \quad [10.7]$$

In this way, we obtain $v_k(Z_0^y)$ that exactly equals $v_k(y)$. Note that, since the functions Mv_{k+n} are known, this provides a tractable recurrence relation on the random variables $v_{k+n}(Z_k^y)$.

REMARK 10.1.– Note that this procedure requires the knowledge of function g for all the random variables $(Z_n^y)_{n \leq N-1}$ defined for the different starting points $y \in \mathbb{U}$. This is why, in general, we are not able to use the no-impulse cost function h. Indeed, it is hard to compute this function, especially if we need to know it everywhere on the state space. The most practical solution is to take g equal to a upper bound of h, and therefore constant.

There is yet another new difficulty hidden in the recurrence relation [10.7] above as regards its discretization. Indeed, to compute $v_k(y)$, we need first to compute all the $v_{k+n}(Z_n^y)$ with $1 \leq n \leq N-k$, and to compute $v_{k+1}(y)$, for instance, we have already computed all the $v_{k+n}(Z_{n-1}^y)$ for $2 \leq n \leq N-k$. Unfortunately, we cannot reuse the values of $v_{k+n}(Z_{n-1}^y)$ to compute those of $v_{k+n}(Z_n^y)$, so the computation has to be started all over again each time, and we have to be very careful in the design of the approximation scheme. However, all these computations can be done with the same discretization grids for (Z_n^y, S_n^y), so that our procedure is still reasonably fast, see section 10.4.2 for details, and Figure 10.1 for a graphical illustration of our procedure.

The recursive procedure [10.7] is triangular in the sense that we need to compute all the $v_{k+n}(Z_n^y)$ for $0 \leq k \leq N$ and $0 \leq n \leq N-k$, see also section 10.4.4 for practical implementation details.

Our approximation procedure is in three steps, as explained in the following sections. The first step consists of replacing the continuous minimization in the definition of operator L by a discrete-time minimization, on path adapted grids. The second step is specific to the impulse problem, and is due to the operator M as explained in details above. The second step hence consists of carefully approximating the value functions v_n on the control grid $\mathbb{U}$. The last step will then be similar to the approximation of the optimal stopping problem 7.2 and will consist of approximating the value functions at the points of the quantization grids of the *no impulse* process.

Finally, we end this section by showing the Lipschitz continuity of the functions (v_n) introduced in definition 10.1.

COROLLARY 10.1.– For all $0 \leq n \leq N$, the value functions v_n are in $\mathbf{L}_\Phi(E)$.

PROOF.– As $v_N = g$ is in $\mathbf{L}_\Phi(E)$ by assumption, a recursive application of proposition 10.2 yields the result. □

10.4.1. *Time discretization*

We define the path-adapted discretization grids as follows.

DEFINITION 10.2.– *For $z \in E$, set $\Delta(z) \in (0, t^*(z))$. Define $n(z) = \left\lfloor \frac{t^*(z)}{\Delta(z)} \right\rfloor - 1$, where $\lfloor x \rfloor$ denotes the greatest integer smaller than or equal to x. The set of points $(t_i)_{i \in \{0,\ldots,n(z)\}}$ with $t_i = i\Delta(z)$ is denoted by $G(z)$. This is the grid associated with the time interval $[0, t^*(z)]$.*

REMARK 10.2.– It is important to note that, for all $z \in E$, we have $t^*(z) \notin G(z)$, but also $\max G(z) = t_{n(z)} \leq t^*(z) - \Delta(z)$. This property is crucial for the sequel, as in the previous chapters.

We propose the following approximation of operator L, where the continuous minimization is replaced by a discrete-time minimization on the path-adapted grids.

DEFINITION 10.3.– *For $(v, w) \in \mathbf{L}_\Phi(E)^2$ and $x \in E$, set*

$$L^d(v, w)(x) = \min_{t \in G(x)} J(v, w)(x, t) \wedge Kw(x).$$

Now, we compute the error induced by the replacement of the continuous minimization by the discrete one.

LEMMA 10.10.– Let $(v, w) \in \mathbf{L}_\Phi(E)^2$. Then for all $x \in E$,

$$\Big| \inf_{t \le t^*(x)} J(v,w)(x,t) - \min_{s \in G(x)} J(v,w)(x,s) \Big|$$
$$\le 2\Big(C_f + C_w C_\lambda + [v]_2 + C_v(C_\lambda + \alpha)\Big)\Delta(x).$$

PROOF.– We have

$$\Big| \inf_{t \le t^*(x)} J(v,w)(x,t) - \min_{s \in G(x)} J(v,w)(x,s) \Big|$$
$$= \min_{s \in G(x)} J(v,w)(x,s) - \inf_{t \le t^*(x)} J(v,w)(x,t).$$

Clearly, there exists $\bar{t} \in [0, t^*(x)]$ such that $\inf_{t \le t^*(x)} J(v,w)(x,t) = J(v,w)(x,\bar{t})$. Moreover, there exists $0 \le i \le n(x)$ such that $\bar{t} \in [t_i, t_{i+1}]$ (with $t_{n(x)+1} = t^*(x)$). Consequently, lemma 10.5 yields

$$\Big| \inf_{t \le t^*(x)} J(v,w)(x,t) - \min_{s \in G(x)} J(v,w)(x,s) \Big| \le J(v,w)(x,t_i) - J(v,w)(x,\bar{t})$$
$$\le \Big(C_f + C_w C_\lambda + [v]_2 + C_v(C_\lambda + \alpha)\Big)|\bar{t} - t_i|.$$

implying the result. □

LEMMA 10.11.– Let $(v, w) \in \mathbf{L}_\Phi(E)^2$ be non-negative functions. Then for all $x \in E$, we have

$$\Big| L(v,w)(x) - L^d(v,w)(x) \Big| \le 2\Big(C_f + C_w C_\lambda + [v]_2 + C_v(C_\lambda + \alpha)\Big)\Delta(x).$$

PROOF.– Since the functions v and w are non-negative, it follows from the definition of L and L^d that

$$\Big| L(v,w)(x) - L^d(v,w)(x) \Big| \le \Big| \inf_{t \le t^*(x)} J(v,w)(x,t) - \min_{s \in G(x)} J(v,w)(x,s) \Big|.$$

Now in view of the previous lemma, we obtain the result. □

10.4.2. *Approximation of the value functions on the control grid* $\mathbb{U}$

We now need to introduce the quantized approximations of the embedded Markov chains $\Theta_n^y = (Z_n^y, S_n^y)$. More precisely, we need several approximations at this stage, one for each starting point y in the control set $\mathbb{U}$. Recall that $\mathbb{U} = \{y^i,\ 1 \leq i \leq u\}$. For all $1 \leq i \leq u$, let $(Z_n^i, S_n^i)_{0 \leq n \leq N-1}$ be the Markov chain $(Z_n, S_n)_{0 \leq n \leq N-1}$ with starting point $Z_0 = y^i$, $S_0 = 0$, and let $(\widehat{Z}_n^i, \widehat{S}_n^i)_{0 \leq n \leq N-1}$ be the quantized approximation of the sequence $(Z_n^i, S_n^i)_{0 \leq n \leq N-1}$. The quantization algorithm 3.3 provides us with a finite grid $\Gamma_n^{i,\Theta} \subset E \times \mathbb{R}^+$ at each time $0 \leq n \leq N-1$ as well as weights for each point of the grid and transition probabilities from one grid to the next grid. Set $p \geq 1$ such that Θ_n^i has finite moments at least up to the order $p + \epsilon$ for some positive ϵ and let $p_n^i = proj_{\Gamma_n^{i,\Theta}}$ be the closest neighbor projection from $E \times \mathbb{R}^+$ onto $\Gamma_n^{i,\Theta}$. Then, the quantization of Θ_n^i conditionally to $Z_0 = y^i, S_0 = 0$ is defined by

$$\widehat{\Theta}_n^i = (\widehat{Z}_n^i, \widehat{S}_n^i) = p_n^i(Z_n^i, S_n^i).$$

We will also denote by $\Gamma_n^{i,Z}$ the projection of $\Gamma_n^{i,\Theta}$ on E and by $\Gamma_n^{i,S}$ the projection of $\Gamma_n^{i,\Theta}$ on $\mathbb{R}^+$.

Although (Z_n^i, S_n^i) is a Markov chain, its quantized approximation is usually not a Markov chain. Therefore, the following quantized approximations of operators H, I, K, J and L^d depend on both indices n and i.

DEFINITION 10.4.– *For $v \in \mathbf{L}_\Phi(E)$, w in $\mathbf{B}(\Gamma_{n+1}^{i,Z})$, $t \in \mathbb{R}^+$, $0 \leq n \leq N-1$, $1 \leq i \leq u$ and $z \in \Gamma_n^{i,Z}$, consider*

$$\begin{aligned}
\widehat{H}_{n+1}^i v(z,t) &= \mathbb{E}_{y^i}\big(e^{-\alpha(t \wedge t^*(\widehat{Z}_n^i))} v\big(\Phi(\widehat{Z}_n^i, t \wedge t^*(\widehat{Z}_n^i))\big) \\
&\quad \times \mathbb{1}_{\{\widehat{S}_{n+1}^i \geq t \wedge t^*(\widehat{Z}_n^i)\}} \big| \widehat{Z}_n^i = z\big), \\
\widehat{I}_{n+1}^i w(z,t) &= \mathbb{E}_{y^i}\Big(e^{-\alpha \widehat{S}_{n+1}^i} w(\widehat{Z}_{n+1}^i) \mathbb{1}_{\{\widehat{S}_{n+1}^i < t \wedge t^*(\widehat{Z}_n^i)\}} \Big| \widehat{Z}_n^i = z\Big), \\
\widehat{K}_{n+1}^i w(z) &= \mathbb{E}_{y^i}\Big(F(\widehat{Z}_n^i, t^*(\widehat{Z}_n^i)) + e^{-\alpha \widehat{S}_{n+1}^i} w(\widehat{Z}_{n+1}^i) \big| \widehat{Z}_n^i = z\Big), \\
\widehat{J}_{n+1}^i(v,w)(z,t) &= \mathbb{E}_{y^i}\Big(F(\widehat{Z}_n^i, t) + e^{-\alpha \widehat{S}_{n+1}^i} w(\widehat{Z}_{n+1}^i) \\
&\quad \times \mathbb{1}_{\{\widehat{S}_{n+1}^i < t \wedge t^*(\widehat{Z}_n^i)\}} \big| \widehat{Z}_n^i = z\Big) \\
&\quad + \mathbb{E}_{y^i}\Big(e^{-\alpha(t \wedge t^*(\widehat{Z}_n^i))} v(\Phi(\widehat{Z}_n^i, t \wedge t^*(\widehat{Z}_n^i))) \mathbb{1}_{\{\widehat{S}_{n+1}^i \geq t \wedge t^*(\widehat{Z}_n^i)\}} \Big| \widehat{Z}_n^i = z\Big), \\
\widehat{L}_{n+1}^{i,d}(v,w)(z) &= \min_{t \in G(z)} \widehat{J}_{n+1}^i(v,w)(z,t) \wedge \widehat{K}_{n+1}^i w(z).
\end{aligned}$$

Our approximation scheme goes backward in time, in as much as it is initialized with computing v_N at the points of the last quantization grids $\Gamma_N^{i,Z}$, then v_{N-1} is computed on $\Gamma_{N-1}^{i,Z}$ and so on.

DEFINITION 10.5.– *Set* $\widetilde{v}_N(y^i) = g(y^i)$ *for* $1 \leq i \leq u$. *Then, for* $1 \leq k \leq N-1$ *and* $1 \leq i \leq u$, *set* $\widetilde{v}_k(y^i) = \widehat{v}_k^{i,k}(y^i)$, *where*

$$\widehat{v}_N^{i,k}(z) = g(z), \qquad z \in \Gamma_{N-k}^{i,Z},$$

$$\widehat{v}_{k+n-1}^{i,k}(z) = \widehat{L}_n^{i,d}(M\widetilde{v}_{k+n}, \widehat{v}_{k+n}^k)(z),$$

where $z \in \Gamma_{n-1}^{i,Z}$, *and* $n \in \{1, \ldots, N-k\}$.

See Figure 10.1 for a graphical illustration of this numerical procedure. Note the use of both $\widetilde{v}_{k+n}$ and $\widehat{v}_{k+n}^k$ in the scheme above. This is due to the fact that we have to reset all our calculations for each value function $\widetilde{v}_k$ and cannot use the calculations made, for example, $\widetilde{v}_{k+1}$ because the value functions are evaluated at different points, and are approximated with different discrete operators. This is mostly because the quantized process $(\widehat{Z}_n^i, \widehat{S}_n^i)$ is not a homogeneous Markov chain.

We can now state our first result on the convergence rate of this approximation.

THEOREM 10.1.– For all $1 \leq k \leq N-1, 0 \leq n \leq N-k-1$ and $1 \leq i \leq u$, suppose that $\Delta(z)$ for $z \in \Gamma_n^{i,Z}$ is such that

$$\left(\frac{2}{C_\lambda}\Big(\big\|S_{n+1}^i - \widehat{S}_{n+1}^i\big\|_p + L_{t^*}\big\|Z_n^i - \widehat{Z}_n^i\big\|_p\Big)\right)^{1/2} < \min_{z \in \Gamma_n^{i,Z}} \{\Delta(z)\}.$$

Then, we have

$$\begin{aligned}
&\|v_{k+n}(Z_n^i) - \widehat{v}_{k+n}^{i,k}(\widehat{Z}_n^i)\|_p \\
&\leq \big\|v_{k+n+1}(Z_{n+1}^i) - \widehat{v}_{k+n+1}^{i,k}(\widehat{Z}_{n+1}^i)\big\|_p + \max_{y \in \mathbb{U}} \big|v_{k+n+1}(y) - \widetilde{v}_{k+n+1}(y)\big| \\
&\quad + d_{k,n}^1\|Z_n^i - \widehat{Z}_n^i\|_p + d_{k,n}^2\big\|\Delta(\widehat{Z}_n^i)\big\|_p \\
&\quad + 3L_{v_{k+n+1}}\big\|Z_{n+1}^i - \widehat{Z}_{n+1}^i\big\|_p + 2\alpha C_{v_{n+k+1}}\big\|S_{n+1}^i - \widehat{S}_{n+1}^i\big\|_p \\
&\quad + 2(2C_{v_{k+n+1}} + C_c)\Big(2C_\lambda\Big(\big\|S_{n+1}^i - \widehat{S}_{n+1}^i\big\|_p + L_{t^*}\big\|Z_n^i - \widehat{Z}_n^i\big\|_p\Big)\Big)^{1/2},
\end{aligned}$$

with

$$d^1_{k,n} = \Big(2[c]_1 + 2[c]_2 L_{t^*} + C_c E_1 + \alpha L_{t^*}(C_{v_{k+n+1}} + C_c)\Big)$$
$$\vee \Big([Q][v_{k+n+1}]_* + 2E_3\Big)$$
$$+ \Big(\frac{C_\lambda}{\alpha}[Q][v_{k+n+1}]_1 + C_{v_{n+k+1}}(E_1 + E_2) + L_{v_{k+n}}\Big),$$
$$d^2_{k,n} = 2\Big(C_f + C_{v_{k+n+1}} C_\lambda + [c]_2 + (C_c + C_{v_{k+n+1}})(C_\lambda + \alpha)\Big).$$

REMARK 10.3.– Recall that $v_N = \widehat{v}^{i,k}_N = \widetilde{v}_N = g$. Hence, we have

$$\|v_N(Z^i_{N-k}) - \widehat{v}^{i,k}_N(\widehat{Z}^i_{N-k})\|_p \le L_g \big\|\widehat{Z}^i_N - Z^i_N\big\|_p$$

and

$$\max_{y \in \mathbb{U}} \big|v_N(y) - \widetilde{v}_N(y)\big| = 0.$$

In addition, the quantization error $\|\Theta^i_n - \widehat{\Theta}^i_n\|_p$ goes to zero as the number of points in the grids goes to infinity, see theorem 3.1. Therefore, according to definition 10.5 and by using an induction procedure $\max_{y\in\mathbb{U}} \big|v_k(y) - \widetilde{v}_k(y)\big|$ can be made arbitrarily small by an adequate choice of the discretization parameters. From a theoretical point of view, the error can be calculated by iterating the result of theorem 10.1. However, this result is not presented here because it would lead to an intricate expression. From a numerical point of view, a computer can easily estimate this error as shown in the example of section 10.5.

The proof is going to be detailed in the following lemmas. As in the preceding chapters, we first split the error into four terms. For all $1 \le k \le N-1, 0 \le n \le N-k-1$ and $1 \le i \le u$, we have

$$\|v_{k+n}(Z^i_n) - \widehat{v}^{i,k}_{k+n}(\widehat{Z}^i_n)\|_p \le \sum_{j=1}^{4} \Upsilon^i_j,$$

where

$$\Upsilon^i_1 = \|v_{k+n}(Z^i_n) - v_{k+n}(\widehat{Z}^i_n)\|_p,$$
$$\Upsilon^i_2 = \|L(Mv_{k+n+1}, v_{k+n+1})(\widehat{Z}^i_n) - L^d(Mv_{k+n+1}, v_{k+n+1})(\widehat{Z}^i_n)\|_p,$$
$$\Upsilon^i_3 = \|L^d(Mv_{k+n+1}, v_{k+n+1})(\widehat{Z}^i_n) - \widehat{L}^{i,d}_{n+1}(Mv_{k+n+1}, v_{k+n+1})(\widehat{Z}^i_n)\|_p,$$
$$\Upsilon^i_4 = \|\widehat{L}^{i,d}_{n+1}(Mv_{k+n+1}, v_{k+n+1})(\widehat{Z}^i_n) - \widehat{L}^{i,d}_{n+1}(M\widetilde{v}_{k+n+1}, \widehat{v}^{i,k}_{k+n+1})(\widehat{Z}^i_n)\|_p.$$

The first two terms are easy to handle due to corollary 10.1 and lemma 10.11.

LEMMA 10.12.– An upper bound for Υ_1^i is

$$\|v_{k+n}(Z_n^i) - v_{k+n}(\widehat{Z}_n^i)\|_p \leq L_{v_{k+n}} \|Z_n^i - \widehat{Z}_n^i\|_p.$$

LEMMA 10.13.– An upper bound for Υ_2^i is

$$\|L(Mv_{k+n+1}, v_{k+n+1})(\widehat{Z}_n^i) - L^d(Mv_{k+n+1}, v_{k+n+1})(\widehat{Z}_n^i)\|_p$$
$$\leq 2\Big(C_f + C_{v_{k+n+1}} C_\lambda + [c]_2 + (C_c + C_{v_{k+n+1}})(C_\lambda + \alpha)\Big) \big\|\Delta(\widehat{Z}_n^i)\big\|_p.$$

The fourth term is also easy enough to deal with as it is a mere comparison of two finite weighted sums.

LEMMA 10.14.– An upper bound for Υ_4^i is

$$\|\widehat{L}_{n+1}^{i,d}(Mv_{k+n+1}, v_{k+n+1})(\widehat{Z}_n^i) - \widehat{L}_{n+1}^{i,d}(M\widetilde{v}_{k+n+1}, \widehat{v}_{k+n+1}^{i,k})(\widehat{Z}_n^i)\|_p$$
$$\leq L_{v_{k+n+1}} \big\|Z_{n+1}^i - \widehat{Z}_{n+1}^i\big\|_p + \big\|v_{k+n+1}(Z_{n+1}^i) - \widehat{v}_{k+n+1}^{i,k}(\widehat{Z}_{n+1}^i)\big\|_p$$
$$+ \max_{y \in \mathbb{U}} \big|v_{k+n+1}(y) - \widetilde{v}_{k+n+1}(y)\big|.$$

PROOF.– We clearly have

$$\big\|\widehat{L}_{n+1}^{i,d}(Mv_{k+n+1}, v_{k+n+1})(\widehat{Z}_n^i) - \widehat{L}_{n+1}^{i,d}(M\widetilde{v}_{k+n+1}, \widehat{v}_{k+n+1}^{i,k})(\widehat{Z}_n^i)\big\|_p$$
$$\leq \Big\| \max_{t \in G(\widehat{Z}_n^i)} \big|\widehat{J}_{n+1}^i(Mv_{k+n+1}, v_{k+n+1})(\widehat{Z}_n^i, t)\big|$$
$$- \widehat{J}_{n+1}^i(M\widetilde{v}_{k+n+1}, \widehat{v}_{k+n+1}^{i,k})(\widehat{Z}_n^i, t)\big|\Big\|_p$$
$$\vee \Big\|\widehat{K}_{n+1}^i v_{k+n+1}(\widehat{Z}_n^i) - \widehat{K}_{n+1}^i \widehat{v}_{k+n+1}^{i,k}(\widehat{Z}_n^i)\Big\|_p$$
$$\leq \big\|\mathbb{E}[v_{k+n+1}(\widehat{Z}_{n+1}^i) - \widehat{v}_{k+n+1}^{i,k}(\widehat{Z}_{n+1}^i) | \widehat{Z}_n^i]\big\|_p$$
$$+ \Big\| \max_{t \in G(\widehat{Z}_n^i)} \Big|\mathbb{E}\Big(Mv_{k+n+1}(\Phi(\widehat{Z}_n^i, t \wedge t^*(\widehat{Z}_n^i)))$$
$$- M\widetilde{v}_{k+n+1}(\Phi(\widehat{Z}_n^i, t \wedge t^*(\widehat{Z}_n^i))) \Big| \widehat{Z}_n^i\Big)\Big|\Big\|_p$$
$$\leq \big\|v_{k+n+1}(\widehat{Z}_{n+1}^i) - v_{k+n+1}(Z_{n+1}^i)\big\|_p$$

$$+ \left\| v_{k+n+1}(Z^i_{n+1}) - \widehat{v}^{i,k}_{k+n+1}(\widehat{Z}^i_{n+1}) \right\|_p$$
$$+ \max_{y \in \mathbb{U}} \left| v_{k+n+1}(y) - \widetilde{v}_{k+n+1}(y) \right|,$$

showing the result. □

We now turn to the third term. This is the key step of the error evaluation, because on the one hand, this is where we deal with the indicator functions. The main idea is that although they are not continuous, we prove in the following two lemmas that the set where the discontinuity actually occurs is of small enough probability. This is also where our special choice of time discretization grids is crucial. On the other hand, we use here the specific properties of quantization.

We now look up the error made in replacing K by $\widehat{K}^i_{n+1}$. This is where we use the specific properties of quantization.

LEMMA 10.15.– For all $1 \leq i \leq u, k \in \{1, \ldots, N-1\}$ and $n \in \{1, \ldots, N-k\}$, we have

$$\left\| Kv_{k+n+1}(\widehat{Z}^i_n) - \widehat{K}^i_{n+1} v_{k+n+1}(\widehat{Z}^i_n) \right\|_p \leq \alpha C_{v_{n+k+1}} \| S^i_{n+1} - \widehat{S}^i_{n+1} \|_p$$
$$+ L_{v_{k+n+1}} \| Z^i_{n+1} - \widehat{Z}^i_{n+1} \|_p + \left([Q][v_{k+n+1}]_1 \frac{C_\lambda}{\alpha} + [Q][v_{k+n+1}]_* \right.$$
$$\left. + C_{v_{n+k+1}} \Big(E_1 + E_2 \Big) + 2E_3 \right) \| Z^i_n - \widehat{Z}^i_n \|_p.$$

PROOF.– We have

$$\begin{aligned}
\Big| Kv_{k+n+1}(\widehat{Z}^i_n) - &\widehat{K}^i_{n+1} v_{k+n+1}(\widehat{Z}^i_n) \Big| \\
&\leq \left| Kv_{k+n+1}(\widehat{Z}^i_n) - \mathbb{E}_{y_i}\left[Kv_{k+n+1}(Z^i_n) \middle| \widehat{Z}^i_n \right] \right| \\
&\quad + \left| \mathbb{E}_{y_i}\left[Kv_{k+n+1}(Z^i_n) \middle| \widehat{Z}^i_n \right] - \widehat{K}^i_{n+1} v_{k+n+1}(\widehat{Z}^i_n) \right| \\
&\leq \mathbb{E}_{y_i}\left(\left| Kv_{k+n+1}(\widehat{Z}^i_n) - Kv_{k+n+1}(Z^i_n) \right| \middle| \widehat{Z}^i_n \right) \\
&\quad + \left| \mathbb{E}_{y_i}\left[Kv_{k+n+1}(Z^i_n) \middle| \widehat{Z}^i_n \right] - \widehat{K}^i_{n+1} v_{k+n+1}(\widehat{Z}^i_n) \right|.
\end{aligned} \tag{10.8}$$

By using the Lipschitz property of K stated in lemma 10.4, we obtain

$$\Big\|\mathbb{E}_{y_i}\Big(\big|Kv_{k+n+1}(\widehat{Z}_n^i) - Kv_{k+n+1}(Z_n^i)\big|\Big|\widehat{Z}_n^i\Big)\Big\|_p \leq \Big([Q][v_{k+n+1}]_1\frac{C_\lambda}{\alpha} + [Q][v_{k+n+1}]_* + C_{v_{k+n+1}}\Big(E_1 + E_2\Big) + E_3\Big)\|Z_n^i - \widehat{Z}_n^i\|_p. \quad [10.9]$$

It now follows from the definition of K given in equation [10.2] that

$$\begin{aligned}
&\Big|\mathbb{E}_{y_i}\big[Kv_{k+n+1}(Z_n^i)\big|\widehat{Z}_n^i\big] - \widehat{K}_{n+1}^i v_{k+n+1}(\widehat{Z}_n^i)\Big| \\
&\quad\leq \mathbb{E}_{y_i}\Big(\big|F(Z_n^i, t^*(Z_n^i)) - F(\widehat{Z}_n^i, t^*(\widehat{Z}_n^i))\big|\Big|\widehat{Z}_n^i\Big) \\
&\qquad + \Big|\mathbb{E}_{y_i}\big[\mathbb{E}_{y_i}\big[e^{-\alpha S_{n+1}^i}v_{k+n+1}(Z_{n+1}^i)\big|\sigma\{Z_n^i, S_n^i\}\big]\big|\widehat{Z}_n^i\big] \\
&\qquad - \mathbb{E}_{y_i}\big[e^{-\alpha \widehat{S}_{n+1}^i}v_{k+n+1}(\widehat{Z}_{n+1}^i)\big|\widehat{Z}_n^i\big]\Big|.
\end{aligned}$$

Then, recalling that $(\widehat{Z}_n^i, \widehat{S}_n^i) = p_n^i(Z_n^i, S_n^i)$. Hence, we have the following crucial property: $\sigma\{\widehat{Z}_n^i\} \subset \sigma\{Z_n^i, S_n^i\}$ and so,

$$\begin{aligned}
&\Big\|\mathbb{E}_{y_i}\big[Kv_{k+n+1}(Z_n^i)\big|\widehat{Z}_n^i\big] - \widehat{K}_{n+1}^i v_{k+n+1}(\widehat{Z}_n^i)\Big\|_p \\
&\quad\leq \Big\|F(Z_n^i, t^*(Z_n^i)) - F(\widehat{Z}_n^i, t^*(\widehat{Z}_n^i))\Big\|_p \\
&\qquad + \Big\|e^{-\alpha S_{n+1}^i}v_{k+n+1}(Z_{n+1}^i) - e^{-\alpha \widehat{S}_{n+1}^i}v_{k+n+1}(\widehat{Z}_{n+1}^i)\Big\|_p.
\end{aligned} \quad [10.10]$$

From lemma 10.3, we readily obtain

$$\Big\|F(Z_n^i, t^*(Z_n^i)) - F(\widehat{Z}_n^i, t^*(\widehat{Z}_n^i))\Big\|_p \leq E_3\|Z_n^i - \widehat{Z}_n^i\|_p, \quad [10.11]$$

and it is easy to show that

$$\begin{aligned}
&\Big\|e^{-\alpha S_{n+1}^i}v_{k+n+1}(Z_{n+1}^i) - e^{-\alpha \widehat{S}_{n+1}^i}v_{k+n+1}(\widehat{Z}_{n+1}^i)\Big\|_p \\
&\qquad\leq L_{v_{k+n+1}}\|Z_{n+1}^i - \widehat{Z}_{n+1}^i\|_p + \alpha C_{v_{k+n+1}}\|S_{n+1}^i - \widehat{S}_{n+1}^i\|_p.
\end{aligned} \quad [10.12]$$

Finally, by combining equations [10.8]–[10.12], we obtain the expected result.

□

We turn to the error made in replacing J by $\widehat{J}^i_{n+1}$. Here, we use the specific properties of quantization again, and lemma 7.12 on indicator functions for (Z^i_n, S^i_n) and $(\widehat{Z}^i_n, \widehat{S}^i_n)$.

LEMMA 10.16.– For all $1 \leq i \leq u, k \in \{1, \ldots, N-1\}, n \in \{1, \ldots, N-k\}$, and $0 < \eta < \min_{z \in \Gamma^{i,Z}_n} \{\Delta(z)\}$, we have

$$
\begin{aligned}
&\Big\| \max_{t \in G(\widehat{Z}^i_n)} \big| J(Mv_{k+n+1}, v_{k+n+1})(\widehat{Z}^i_n, t) - \widehat{J}^i_{n+1}(Mv_{k+n+1}, v_{k+n+1})(\widehat{Z}^i_n, t) \big| \Big\|_p \\
&\leq L_{v_{k+n+1}} \big\| Z^i_{n+1} - \widehat{Z}^i_{n+1} \big\|_p + \alpha C_{v_{k+n+1}} \big\| S^i_{n+1} - \widehat{S}^i_{n+1} \big\|_p \\
&\quad + \Big(\frac{1}{\alpha} [Q][v_{k+n+1}]_1 C_\lambda + 2[c]_1 + 2[c]_2 L_{t^*} + (C_{v_{k+n+1}} + C_c) E_1 \\
&\qquad + C_{v_{k+n+1}} E_2 + \alpha L_{t^*} (C_{v_{k+n+1}} + C_c) \Big) \| Z^i_n - \widehat{Z}^i_n \|_p \\
&\quad + (2C_{v_{k+n+1}} + C_c) \Big(C_\lambda \eta + \frac{2}{\eta} \Big(\big\| S^i_{n+1} - \widehat{S}^i_{n+1} \big\|_p + L_{t^*} \big\| Z^i_n - \widehat{Z}^i_n \big\|_p \Big) \Big).
\end{aligned}
$$

PROOF.– By definition of J, we have

$$
\begin{aligned}
&\Big\| \max_{t \in G(\widehat{Z}^i_n)} \big| J(Mv_{k+n+1}, v_{k+n+1})(\widehat{Z}^i_n, t) - \widehat{J}^i_{n+1}(Mv_{k+n+1}, v_{k+n+1})(\widehat{Z}^i_n, t) \big| \Big\|_p \\
&\qquad \leq \Big\| \max_{t \in G(\widehat{Z}^i_n)} \big| Iv_{k+n+1}(\widehat{Z}^i_n, t) - \widehat{I}^i_{n+1} v_{k+n+1}(\widehat{Z}^i_n, t) \big| \Big\|_p \\
&\qquad\quad + \Big\| \max_{t \in G(\widehat{Z}^i_n)} \Big| HMv_{k+n+1}(\widehat{Z}^i_n, t) - \widehat{H}^i_{n+1} Mv_{k+n+1}(\widehat{Z}^i_n, t) \Big| \Big\|_p .
\end{aligned}
\tag{10.13}
$$

For the first term on the right-hand side of equation [10.13], we proceed as for K in the preceding lemma

$$
\begin{aligned}
&\Big\| \max_{t \in G(\widehat{Z}^i_n)} \big| Iv_{k+n+1}(\widehat{Z}^i_n, t) - \widehat{I}^i_{n+1} v_{k+n+1}(\widehat{Z}^i_n, t) \big| \Big\|_p \\
&\leq \Big\| \max_{t \in G(\widehat{Z}^i_n)} \mathbb{E}_{y_i} \Big(\big| Iv_{k+n+1}(\widehat{Z}^i_n, t) - Iv_{k+n+1}(Z^i_n, t) \big| \Big| \widehat{Z}^i_n \Big) \Big\|_p \\
&\quad + \Big\| \max_{t \in G(\widehat{Z}^i_n)} \Big| \mathbb{E}_{y_i} \big[Iv_{k+n+1}(Z^i_n, t) \big| \widehat{Z}^i_n \big] - \widehat{I}^i_{n+1} v_{k+n+1}(\widehat{Z}^i_n, t) \Big| \Big\|_p .
\end{aligned}
\tag{10.14}
$$

On the one hand, it follows from lemma 10.3 that

$$\Big\| \max_{t\in G(\widehat{Z}^i_n)} \mathbb{E}_{y_i}\Big(\big|Iv_{k+n+1}(\widehat{Z}^i_n,t) - Iv_{k+n+1}(Z^i_n,t)\big|\Big|\widehat{Z}^i_n\Big)\Big\|_p$$
$$\leq \Big(\frac{1}{\alpha}[Q][v_{k+n+1}]_1 C_\lambda + C_{v_{k+n+1}}E_2\Big)\|Z^i_n - \widehat{Z}^i_n\|_p. \qquad [10.15]$$

On the other hand, we use again the fact that $\sigma\{\widehat{Z}^i_n\} \subset \sigma\{Z^i_n, S^i_n\}$ to obtain

$$\Big|\mathbb{E}_{y_i}\big[Iv_{k+n+1}(Z^i_n,t)\big|\widehat{Z}^i_n\big] - \widehat{I}^i_{n+1}v_{k+n+1}(\widehat{Z}^i_n,t)\Big|$$
$$\leq \mathbb{E}_{y_i}\Big(\big|e^{-\alpha S^i_{n+1}}v_{k+n+1}(Z^i_{n+1}) - e^{-\alpha\widehat{S}^i_{n+1}}v_{k+n+1}(\widehat{Z}^i_{n+1})\big|\Big|\widehat{Z}^i_n\Big)$$
$$+ C_{v_{k+n+1}}\mathbb{E}_{y_i}\Big(\big|\mathbb{1}_{\{S^i_{n+1}<t\wedge t^*(Z^i_n)\}} - \mathbb{1}_{\{\widehat{S}^i_{n+1}<t\wedge t^*(\widehat{Z}^i_n)\}}\big|\Big|\widehat{Z}^i_n\Big).$$

Recalling that v_{k+n+1} is Lipschitz, lemma 7.12 yields

$$\Big\| \max_{t\in G(\widehat{Z}^i_n)} \Big|\mathbb{E}_{y_i}\big[Iv_{k+n+1}(Z^i_n,t)\big|\widehat{Z}^i_n\big] - \widehat{I}^i_{n+1}v_{k+n+1}(\widehat{Z}^i_n,t)\Big|\Big\|_p$$
$$\leq C_{v_{k+n+1}}C_\lambda\eta + (\alpha + \frac{2}{\eta})C_{v_{k+n+1}}\big\|S^i_{n+1} - \widehat{S}^i_{n+1}\big\|_p$$
$$+ L_{v_{k+n+1}}\big\|Z^i_{n+1} - \widehat{Z}^i_{n+1}\big\|_p + \frac{2C_{v_{k+n+1}}L_{t^*}}{\eta}\big\|Z^i_n - \widehat{Z}^i_n\big\|_p. \qquad [10.16]$$

Regarding the second term on the right-hand side of equation [10.13], we have

$$\Big\| \max_{t\in G(\widehat{Z}^i_n)} \Big|HMv_{k+n+1}(\widehat{Z}^i_n,t) - \widehat{H}^i_{n+1}Mv_{k+n+1}(\widehat{Z}^i_n,t)\Big|\Big\|_p$$
$$\leq \Big\| \max_{t\in G(\widehat{Z}^i_n)} \mathbb{E}_{y_i}\Big(\big|HMv_{k+n+1}(\widehat{Z}^i_n,t) - HMv_{k+n+1}(Z^i_n,t)\big|\Big|\widehat{Z}^i_n\Big)\Big\|_p$$
$$+\Big\| \max_{t\in G(\widehat{Z}^i_n)} \Big|\mathbb{E}_{y_i}\big[HMv_{k+n+1}(Z^i_n,t)\big|\widehat{Z}^i_n\big] - \widehat{H}^i_{n+1}Mv_{k+n+1}(\widehat{Z}^i_n,t)\Big|\Big\|_p. \qquad [10.17]$$

By using the same arguments as above and lemmas 10.1 and 10.2, we obtain easily

$$\Big\| \max_{t\in G(\widehat{Z}^i_n)} \mathbb{E}_{y_i}\Big(\big|HMv_{k+n+1}(\widehat{Z}^i_n,t) - HMv_{k+n+1}(Z^i_n,t)\big|\Big|\widehat{Z}^i_n\Big)\Big\|_p$$
$$\leq \Big([c]_1 + [c]_2L_{t^*} + (C_{v_{k+n+1}} + C_c)E_1\Big)\|Z^i_n - \widehat{Z}^i_n\|_p, \qquad [10.18]$$

and

$$\left\| \max_{t \in G(\widehat{Z}_n^i)} \left| \mathbb{E}_{y_i}\left[HMv_{k+n+1}(Z_n^i, t) \middle| \widehat{Z}_n^i\right] - \widehat{H}_{n+1}^i M v_{k+n+1}(\widehat{Z}_n^i, t) \right| \right\|_p$$

$$\begin{aligned}
&\leq \frac{2}{\eta}(C_{v_{k+n+1}} + C_c)\|S_{n+1}^i - \widehat{S}_{n+1}^i\|_p + C_\lambda(C_{v_{k+n+1}} + C_c)\eta \\
&\quad + \big(\alpha L_{t^*}(C_{v_{k+n+1}} + C_c) + [c]_1 + [c]_2 L_{t^*} \\
&\quad + \frac{2L_{t^*}}{\eta}(C_{v_{k+n+1}} + C_c)\big)\|Z_n^i - \widehat{Z}_n^i\|_p.
\end{aligned} \qquad [10.19]$$

Combining equations [10.13]–[10.19], we get the result. □

We now add up the preceding results to obtain the following upper bound for the third error term Υ_3^i.

LEMMA 10.17.– An upper bound for Υ_3^i is

$$\left\| L^d(Mv_{k+n+1}, v_{k+n+1})(\widehat{Z}_n^i) - \widehat{L}_{n+1}^{i,d}(Mv_{k+n+1}, v_{k+n+1})(\widehat{Z}_n^i) \right\|_p$$

$$\begin{aligned}
&\leq 2\alpha C_{v_{n+k+1}}\|S_{n+1}^i - \widehat{S}_{n+1}^i\|_p + 2L_{v_{k+n+1}}\|Z_{n+1}^i - \widehat{Z}_{n+1}^i\|_p \\
&\quad + \Big(\frac{C_\lambda}{\alpha}[Q][v_{k+n+1}]_1 + C_{v_{n+k+1}}(E_1 + E_2)\Big)\|Z_n^i - \widehat{Z}_n^i\|_p \\
&\quad + \Big(2[c]_1 + 2[c]_2 L_{t^*} + C_c E_1 + \alpha L_{t^*}(C_{v_{k+n+1}} + C_c)\Big) \\
&\qquad \vee \Big([Q][v_{k+n+1}]_* + 2E_3\Big)\|Z_n^i - \widehat{Z}_n^i\|_p \\
&\quad + (2C_{v_{k+n+1}} + C_c)\big(C_\lambda \eta + \frac{2}{\eta}(\big\|S_{n+1}^i - \widehat{S}_{n+1}^i\big\|_p + L_{t^*}\big\|Z_n^i - \widehat{Z}_n^i\big\|_p)\big).
\end{aligned}$$

10.4.3. *Approximation of the value function*

Now, we have computed the value functions on the control grid, we turn to the actual approximation of v_0. As in the preceding section, we define the quantized approximation of the embedded Markov chain (Θ_n) starting from $(x_0, 0)$, the actual starting point of the PDMP. Let $(\widehat{Z}_n, \widehat{S}_n)_{0 \leq n \leq N-1}$ be the quantized approximation of the sequence $(Z_n, S_n)_{0 \leq n \leq N-1}$. The quantization algorithm provides us with another series of finite grids $\Gamma_n^\Theta \subset E \times \mathbb{R}^+$ for all $0 \leq n \leq N-1$ as well as weights for each point of the grids and transition probabilities from one grid to the next grid. Let p_n be the closest neighbor projection from $E \times \mathbb{R}^+$ onto Γ_n^Θ. Then, the quantization of Θ_n conditionally to $Z_0 = x_0$ is defined by

$$\widehat{\Theta}_n = (\widehat{Z}_n, \widehat{S}_n) = p_n(Z_n, S_n).$$

We will also denote by Γ_n^Z the projection of Γ_n^Θ on E and by Γ_n^S the projection of Γ_n^Θ on $\mathbb{R}^+$. We use yet again new quantized approximations of operators H, I, K, J and L^d.

DEFINITION 10.6.– *For* $v \in \mathbf{L}_\Phi(E)^2$, $w \in \mathbf{B}(\Gamma_{n+1}^Z)$, $t \in \mathbb{R}^+$, $n \in \{0, \ldots, N-1\}$ *and* $z \in \Gamma_n^Z$, *consider*

$$\begin{aligned}
\widehat{H}_{n+1}v(z,t) &= \mathbb{E}_{x_0}\big(e^{-\alpha(t \wedge t^*(\widehat{Z}_n))}v(\Phi(\widehat{Z}_n, t \wedge t^*(\widehat{Z}_n)))\\
&\qquad \times \mathbb{1}_{\{\widehat{S}_{n+1} \geq t \wedge t^*(\widehat{Z}_n)\}} \big| \widehat{Z}_n = z\big),\\
\widehat{I}_{n+1}w(z,t) &= \mathbb{E}_{x_0}\Big(e^{-\alpha \widehat{S}_{n+1}} w(\widehat{Z}_{n+1}) \mathbb{1}_{\{\widehat{S}_{n+1} < t \wedge t^*(\widehat{Z}_n)\}} \Big| \widehat{Z}_n = z\Big),\\
\widehat{K}_{n+1}v(z) &= \mathbb{E}_{x_0}\Big(F(\widehat{Z}_n, t^*(\widehat{Z}_n)) + e^{-\alpha \widehat{S}_{n+1}} v(\widehat{Z}_{n+1}) \big| \widehat{Z}_n = z\Big),\\
\widehat{J}_{n+1}(v,w)(z,t) &= \mathbb{E}_{x_0}\Big(F(\widehat{Z}_n, t) + e^{-\alpha \widehat{S}_{n+1}} w(\widehat{Z}_{n+1})\\
&\qquad \times \mathbb{1}_{\{\widehat{S}_{n+1} < t \wedge t^*(\widehat{Z}_n)\}} \big| \widehat{Z}_n = z\Big)\\
&\qquad + \mathbb{E}_{x_0}\Big(e^{-\alpha(t \wedge t^*(\widehat{Z}_n))} v(\Phi(\widehat{Z}_n, t \wedge t^*(\widehat{Z}_n)))\\
&\qquad \times \mathbb{1}_{\{\widehat{S}_{n+1} \geq t \wedge t^*(\widehat{Z}_n)\}} \big| \widehat{Z}_n = z\Big),\\
\widehat{L}_{n+1}^d(v,w)(z) &= \min_{t \in G(z)} \widehat{J}_{n+1}(v,w)(z,t) \wedge \widehat{K}_{n+1}w(z).
\end{aligned}$$

With these discretized operators and the previous evaluation of the $\widetilde{v}_k$, we propose the following approximation scheme.

DEFINITION 10.7.– *Consider* $\widehat{v}_N(z) = g(z)$ *for all* $z \in \Gamma_N^Z$, *and for* $k \in \{1, \ldots, N\}$

$$\widehat{v}_{k-1}(z) = \widehat{L}_k^d(M\widetilde{v}_k, \widehat{v}_k)(z), \qquad [10.20]$$

where $z \in \Gamma_{k-1}^Z$.

See Figure 10.1 for a graphical illustration of this numerical procedure. Therefore, $\widehat{v}_0(\widehat{Z}_0)$ will be an approximation of $v_0(Z_0) = v_0(x_0)$. The derivation of the error bound for this scheme follows exactly the same lines as in the preceding section. Therefore, we omit it and only state our main result.

THEOREM 10.2.– For any $0 \leq n \leq N-1$, suppose that $\Delta(z)$ for $z \in \Gamma_n^Z$ is such that

$$\Big(\frac{2}{C_\lambda}\Big(\big\|S_{n+1} - \widehat{S}_{n+1}\big\|_p + L_{t^*}\big\|Z_n - \widehat{Z}_n\big\|_p\Big)\Big)^{1/2} < \min_{z\in\Gamma_n^Z}\{\Delta(z)\}.$$

Then, we have

$$\begin{aligned}
&\|v_n(Z_n) - \widehat{v}_n(\widehat{Z}_n)\|_p \\
&\leq \big\|v_{n+1}(Z_{n+1}) - \widehat{v}_{n+1}(\widehat{Z}_{n+1})\big\|_p + \max_{y\in\mathbb{U}}\big|v_{n+1}(y) - \widetilde{v}_{n+1}(y)\big| \\
&\quad + D_n^1\|Z_n - \widehat{Z}_n\|_p + D_n^2\big\|\Delta(\widehat{Z}_n^i)\big\|_p \\
&\quad + 3L_{v_{n+1}}\big\|Z_{n+1} - \widehat{Z}_{n+1}\big\|_p + 2\alpha C_{v_{n+1}}\big\|S_{n+1} - \widehat{S}_{n+1}\big\|_p \\
&\quad + 2(2C_{v_{n+1}} + C_c)\Big(2C_\lambda\Big(\big\|S_{n+1} - \widehat{S}_{n+1}\big\|_p + L_{t^*}\big\|Z_n - \widehat{Z}_n\big\|_p\Big)\Big)^{1/2},
\end{aligned}$$

with

$$\begin{aligned}
D_n^1 &= \Big(2[c]_1 + 2[c]_2L_{t^*} + C_cE_1 + \alpha L_{t^*}(C_{v_{n+1}} + C_c)\Big) \\
&\quad \vee \Big([Q][v_{n+1}]_* + 2E_3\Big) \\
&\quad + \Big(\frac{C_\lambda}{\alpha}[Q][v_{n+1}]_1 + C_{v_{n+1}}(E_1 + E_2) + L_{v_n}\Big), \\
D_n^2 &= 2\Big(C_f + C_{v_{n+1}}C_\lambda + [c]_2 + (C_c + C_{v_{n+1}})(C_\lambda + \alpha)\Big).
\end{aligned}$$

PROOF.– The proof of this result is similar to theorem 10.1. We just present the main steps. We first split the error into four terms. For any $0 \leq n \leq N-1$, we have

$$\|v_n(Z_n) - \widehat{v}_n(\widehat{Z}_n)\|_p \leq \sum_{j=1}^{4}\Upsilon_j,$$

where

$$\Upsilon_1 = \|v_n(Z_n) - v_n(\widehat{Z}_n)\|_p,$$
$$\Upsilon_2 = \|L(Mv_{n+1}, v_{n+1})(\widehat{Z}_n) - L^d(Mv_{n+1}, v_{n+1})(\widehat{Z}_n)\|_p,$$
$$\Upsilon_3 = \|L^d(Mv_{n+1}, v_{n+1})(\widehat{Z}_n) - \widehat{L}^d_{n+1}(Mv_{n+1}, v_{n+1})(\widehat{Z}_n)\|_p,$$
$$\Upsilon_4 = \|\widehat{L}^d_{n+1}(Mv_{n+1}, v_{n+1})(\widehat{Z}_n) - \widehat{L}^d_{n+1}(M\widetilde{v}_{n+1}, \widehat{v}_{n+1})(\widehat{Z}_n)\|_p.$$

Similarly to lemmas 10.12, 10.13, 10.14 and 10.17, it is easy to obtain the following inequalities. An upper bound for Υ_1^i is given by

$$\|v_n(Z_n) - v_n(\widehat{Z}_n)\|_p \leq L_{v_n}\|Z_n - \widehat{Z}_n\|_p.$$

An upper bound for Υ_2^i is

$$\|L(Mv_{n+1}, v_{n+1})(\widehat{Z}_n) - L^d(Mv_{n+1}, v_{n+1})(\widehat{Z}_n)\|_p$$
$$\leq 2\Big(C_f + C_{v_{n+1}}C_\lambda + [c]_2 + (C_c + C_{v_{n+1}})(C_\lambda + \alpha)\Big)\big\|\Delta(\widehat{Z}_n)\big\|_p.$$

For the third term, we have

$$\Big\|L^d(Mv_{n+1}, v_{n+1})(\widehat{Z}_n) - \widehat{L}^d_{n+1}(Mv_{n+1}, v_{n+1})(\widehat{Z}_n)\Big\|_p$$
$$\leq 2\alpha C_{v_{n+1}}\|S_{n+1} - \widehat{S}_{n+1}\|_p + 2L_{v_{n+1}}\|Z_{n+1} - \widehat{Z}_{n+1}\|_p$$
$$+ \Big(\frac{C_\lambda}{\alpha}[Q][v_{n+1}]_1 + C_{v_{n+1}}(E_1 + E_2)\Big)\|Z_n - \widehat{Z}_n\|_p$$
$$+ \Big(2[c]_1 + 2[c]_2 L_{t^*} + C_c E_1 + \alpha L_{t^*}(C_{v_{n+1}} + C_c)\Big)$$
$$\vee \Big([Q][v_{n+1}]_* + 2E_3\Big)\|Z_n - \widehat{Z}_n\|_p$$
$$+ (2C_{v_{n+1}} + C_c)\Big(C_\lambda\eta + \frac{2}{\eta}\Big(\big\|S_{n+1} - \widehat{S}_{n+1}\big\|_p + L_{t^*}\big\|Z_n - \widehat{Z}_n\big\|_p\Big)\Big).$$

The last term is bounded by

$$\|\widehat{L}^d_{n+1}(Mv_{n+1}, v_{n+1})(\widehat{Z}_n) - \widehat{L}^d_{n+1}(M\widetilde{v}_{n+1}, \widehat{v}_{n+1})(\widehat{Z}_n)\|_p$$
$$\leq L_{v_{n+1}}\big\|Z_{n+1} - \widehat{Z}_{n+1}\big\|_p + \big\|v_{n+1}(Z_{n+1}) - \widehat{v}_{n+1}(\widehat{Z}_{n+1})\big\|_p$$
$$+ \max_{y\in\mathbb{U}}\big|v_{n+1}(y) - \widetilde{v}_{n+1}(y)\big|.$$

Combining these four inequalities, we get the result. □

By using the same arguments as in remark 10.3, $\|v_n(Z_n) - \widehat{v}_n(\widehat{Z}_n)\|_p$ can be made arbitrarily small by an adequate choice of the discretization parameters. From a theoretical point of view, the error can be calculated by iterating the result of theorem 10.2. However, this result is not presented here because it would lead to an intricate expression. From a numerical point of view, a computer can easily estimate this error as shown in the example of section 10.5.

10.4.4. *Step-by-step description of the algorithm*

Recall that the main objective of our algorithm is to compute the approximation $\widehat{v}_0(x_0)$ of the value function of the impulse control problem $v_0(x_0)$. The global recursive procedure is described in Figure 10.1.

The calculation of $\widehat{v}_0(x_0)$ is based on the backward recursion given in definition 10.7 and described in the first line of Figure 10.1. It involves the operators $\widehat{L}_j^d$ constructed with the quantized process $\widehat{\Theta}_n$ starting from x_0. This recursion is not self-contained and requires previous evaluation of the functions $\widetilde{v}_j$ on the control set $\mathbb{U}$.

The lower part of Figure 10.1 shows how to compute these functions $\widetilde{v}_j$ at each point of the control grid $\mathbb{U}$. This is the triangular backward recursion given in definition 10.5. More precisely, define $\widetilde{v}_N = g$ and set $j < N$ and suppose that all the $\widetilde{v}_l$ for all $l > j$ have already been computed everywhere on the control set $\mathbb{U}$. We then compute $\widetilde{v}_j$ in the following way, following the j-th line of Figure 10.1 counting from the bottom. We first iterate the operators $\widehat{L}_k^{1,d}$ and use the quantized process $\widehat{\Theta}_n^1$ to obtain $\widetilde{v}_j(y^1)$. Then, we iterate the operators $\widehat{L}_k^{2,d}$ and use the quantized process $\widehat{\Theta}_n^2$ to obtain $\widetilde{v}_j(y^2)$ and so on until the last point $\widetilde{v}_j(y^u)$. Thus, we obtain $\widetilde{v}_j$ at all points of the control set $\mathbb{U}$.

The complete procedure is summed up in algorithms 10.1 and 10.2.

10.4.5. *Practical implementation*

The procedure defined above is the natural one to obtain convergence rates for our approximations. However, in practice, we proceed in a different order.

The first step is to fix the computational horizon N. This point was discussed in section 10.4. The second step is not the time discretization, but

Algorithm 10.1: Approximation of the value functions on the control grid $\mathbb{U}$

input : Quantization grids $(\Gamma_n^{i,\Theta})_{0\leq n\leq N}$, Transition matrices $(\widehat{P}_n^{i,\Theta})_{1\leq n\leq N}$, $1 \leq i \leq u$, initial function g, time discretization step Δ

output : $\widetilde{v}_n(y^i), 1 \leq n \leq N, 1 \leq i \leq u$

begin
  for $1 \leq i \leq u$ **do**
    $\widetilde{v}_N(y^i) \leftarrow g(y^i)$
  end
  for $n \leftarrow N-1$ **to** 1 **by step**: -1
  do
    for $1 \leq i \leq u$ **do**
      for $z \in \Gamma_{N-n}^{i,Z}$ **do**
        $\widehat{v}_N^n(z) \leftarrow g(z)$
      end
      for $k \leftarrow N-n-1$ **to** 0 **by step**: -1
      do
        for $z \in \Gamma_k^{i,Z}$ **do**
          $\widehat{K\widehat{v}}_{n+k+1}^n(z) \leftarrow \sum_{(z',s')\in\Gamma_{k+1}^{i,\Theta}} \widehat{P}_{k+1}^{i,\Theta}(z,(z',s'))$
          $\big(e^{-\alpha s'}\widehat{v}_{n+k+1}^n(z') + F(z,t^*(z))\big)$
          for $t \leftarrow 0$ **to** $t^*(z) - \Delta$ **by step**: Δ
          do
            $w(z,t) = \min_{1\leq j\leq u} \{c(\Phi(z,t),y^j) + \widetilde{v}_{n+1+k}(y^j)\}$
            $\widehat{J\widehat{v}}_{n+k+1}^n(z,t) \leftarrow \sum_{(z',s')\in\Gamma_{k+1}^{i,\Theta}} \widehat{P}_{k+1}^{i,\Theta}(z,(z',s'))\big(F(z,t)$
            $+e^{-\alpha s'}\widehat{v}_{n+k+1}^n(z')\mathbb{1}_{\{s'<t\}} + e^{-\alpha t}w(z,t)\mathbb{1}_{\{s'\geq t\}}\big)$
          end
          $\widehat{v}_{n+k}^n(z) \leftarrow \max_t\{\widehat{J\widehat{v}}_{n+k+1}^n(z,t)\} \vee \widehat{K\widehat{v}}_{n+k+1}^n(z)$
        end
      end
      $\widetilde{v}_n(y^i) \leftarrow \widehat{v}_n^n(y^i)$
      return : $\widetilde{v}_n(y^i)$
    end
  end
end

Algorithm 10.2: Approximation of the value function $v(x_0)$

input : Quantization grids $(\Gamma_n^{\Theta})_{0 \leq n \leq N}$, Transition matrices $(\widehat{P}_n^{\Theta})_{1 \leq n \leq N}$, initial function g, time discretization step Δ, output of algorithm 10.1

output : $\widehat{v}_0$

begin
  for $z \in \Gamma_N^Z$ **do**
    $\widehat{v}_N(z) \leftarrow g(z)$
  end
  for $k \leftarrow N-1$ **to** 0 **by step**: -1
  do
    for $z \in \Gamma_k^Z$ **do**
      $\widehat{K}\widehat{v}_{k+1}(z) \leftarrow \sum_{(z',s') \in \Gamma_{k+1}^{\Theta}} \widehat{P}_{k+1}^{\Theta}(z,(z',s'))$
        $\left(e^{-\alpha s'}\widehat{v}_{k+1}(z') + F(z,t^*(z))\right)$
      for $t \leftarrow 0$ **to** $t^*(z) - \Delta$ **by step**: Δ
      do
        $w(z,t) = \min_{1 \leq j \leq u} \{c(\Phi(z,t), y^j) + \widetilde{v}_{k+1}(y^j)\}$
        $\widehat{J}\widehat{v}_{k+1}(z,t) \leftarrow \sum_{(z',s') \in \Gamma_{k+1}^{\Theta}} \widehat{P}_{k+1}^{\Theta}(z,(z',s'))\big(F(z,t)$
          $+e^{-\alpha s'}\widehat{v}_{k+1}(z')\mathbb{1}_{\{s'<t\}} + e^{-\alpha t}w(z,t)\mathbb{1}_{\{s' \geq t\}}\big)$
      end
      $\widehat{v}_k(z) \leftarrow \max_t\{\widehat{J}\widehat{v}_{k+1}(z,t)\} \vee \widehat{K}\widehat{v}_{k+1}(z)$
    end
  end
  return : $\widetilde{v}_n(y^i)$
end

the computation of the quantized approximations of the sequences (Θ_n) and (Θ_n^i). The quantization algorithm may be quite long to run. However, it must be pointed out that this quantization step only depends on the optimization procedure through the control set $\mathbb{U}$ but it does not depend on the cost functions f and c. The sequence $(\widehat{\Theta}_n)$ is obtained in a straightforward way. As for the $(\widehat{\Theta}_n^i)$, if the control set is very small, it is possible to run as many sequences of grids as there are points in the control set. Otherwise, we can do with only one sequence of grids computed with the Markov chain (Θ_n^{μ}) with a random starting point $Z_0 = Z_0^{\mu}$ uniformly distributed on the control set $\mathbb{U}$. To derive the point-wise approximation error, we simply use the finiteness of $\mathbb{U}$ and the definition of the L_p norm.

$$\begin{aligned}|v_k(y^i) - \widetilde{v}_k(y^i)| &\leq u \sum_{i=1}^{u} |v_k(y^i) - \widetilde{v}_k(y^i)| \frac{1}{u} \\ &\leq u^{1/p} \|v_k(Z_0^\mu) - \widetilde{v}_k(\widehat{Z}_0^\mu)\|_p\end{aligned}$$

where u is the cardinal of $\mathbb{U}$. Notice that the last term is bounded in theorem 10.1. Hence, we really only need two series of quantization grids.

Once the quantization grids are computed and stored, we compute the path-adapted time grids $G(z)$ for all z in all the quantization grids, which is only a finite number of z. The step $\Delta(z)$ can usually be chosen constant equal to Δ, so that either we can store the whole time grids, or we only need to store the values of Δ and $t^*(z)$ for all z in the quantization grids.

Once these preliminary computations are done, we can finally compute the value function. This last step is comparatively faster. The only point left to discuss is how to choose the initializing function g. The most interesting starting function is the cost h of the no impulse strategy, because then the value function $\mathcal{V}_N$ has a natural interpretation. However, in general, we need additional assumptions on Q to ensure that h is in $\mathbf{L}_\Phi(E)$. Another problem is that in general computing h is a difficult problem, especially as we need to know its value at many different points, as explained in remark 10.1. To overcome these difficulties, we can choose g to be an upper bound of h, for instance, $g = \alpha^{-1}C_f$. In the special cases where h can be explicitly computed, we advise to use h.

10.5. Example

Now, we apply our procedure to the transmission control protocol (TCP) process presented in section 1.8.2. The running cost is defined on E by $f(x) = 1 - x$ and the intervention cost is a constant c_0. Therefore, the best performance is obtained when the process is close to the boundary 1. But, then the probability to jump backward is high. The control set $\mathbb{U}$ is the set of $\frac{k}{u}$, $0 \leq k \leq u - 1$ for some fixed integer u. In this special case, the control grid is already a discretization of the whole state space of the process. Therefore, we need only one series of grids starting from the control points to obtain an approximation of the value function at each point of the control grid.

We ran our algorithm for the parameters $x_0 = 0$, $v = 1$, $\beta = 3$, $c_0 = 0.08$, the discount factor $\alpha = 2$ and $u = 50$ points in the control grid and several values of the horizon N. For a horizon $N = 5$ (respectively, $N = 10, N = 15$) interventions or jumps Figure 10.2(a) (respectively, (b) and (c)) gives the approximated value function we obtained (computed at the 50 points of the control grid) for 50, 100 and 500 discretization points in each quantization grid. As expected, the approximation gets smoother and lower as the number of points in the quantization grids increases.

The theoretical errors corresponding to the horizon $N = 5$ (respectively, $N = 10, N = 15$) are given in Table 10.1 (respectively, Tables 10.2 and 10.3). The values of the error are fairly high and conservative, but it must be pointed out that on the one hand, they do decrease as the number of points in the quantization grids increases, as expected; on the other hand their expressions are calculated and valid for a very wide and general class of PDMPs, hence when applied to a specific example, they cannot be very sharp.

Number of points in the quantization grids	$\|\|v_0(Z_0) - \widehat{v}_0(Z_0)\|\|_2$
50	4,636
100	3,700
500	2,141

Table 10.1. *Theoretical errors for the value function for the impulse control of the TCP process for* $N = 5$

Number of points in the quantization grids	$\|\|v_0(Z_0) - \widehat{v}_0(Z_0)\|\|_2$
50	$5.341 \cdot 10^{11}$
100	$4.501 \cdot 10^{11}$
500	$2.567 \cdot 10^{11}$

Table 10.2. *Theoretical errors for the value function for the impulse control of the TCP process for* $N = 10$

Also note that the approximated value function obtained for the horizon of $N = 10$ jumps or interventions is much lower than that obtained for the horizon $N = 5$ jumps or interventions. This is natural as it is a minimization problem, and the more there are possible interventions the lower the value function is. This also suggests that the horizon $N = 5$ is not large enough to approximate the infinite horizon problem. Figure 10.3 gives the approximated value function we obtained (computed at the 50 points of the control grid) for, respectively, 500 points in the quantization grids and respective horizons of $N = 5$, $N = 10$ and $N = 15$ jumps or interventions. There is very little difference between the results for $N = 10$ and $N = 15$, suggesting that it is enough to take a horizon of 10 jumps or interventions to approximate the infinite time optimization problem.

Number of points in the quantization grids	$\|v_0(Z_0) - \widehat{v}_0(Z_0)\|_2$
50	$1.460 \cdot 10^{21}$
100	$1.288 \cdot 10^{21}$
500	$0.750 \cdot 10^{21}$

Table 10.3. *Theoretical errors for the value function for the impulse control of the TCP process for $N = 15$*

10.6. Conclusion

The impulse control of PDMPs corresponds to a wide range of possible applications, including, for instance, maintenance optimization. As explained in Chapter 9, in the context of a system subject to failure of its components, and intervention may correspond to a partial repair of the structure, for instance the change of a component. Thus, the system does not restart as good as new, unlike the optimal stopping control case. We can thus optimize the maintenance operations during all the lifecycle of the structure. Unfortunately, the important question of finding a computable strategy reaching a performance close to optimality is still an open question on which the authors are now currently working.

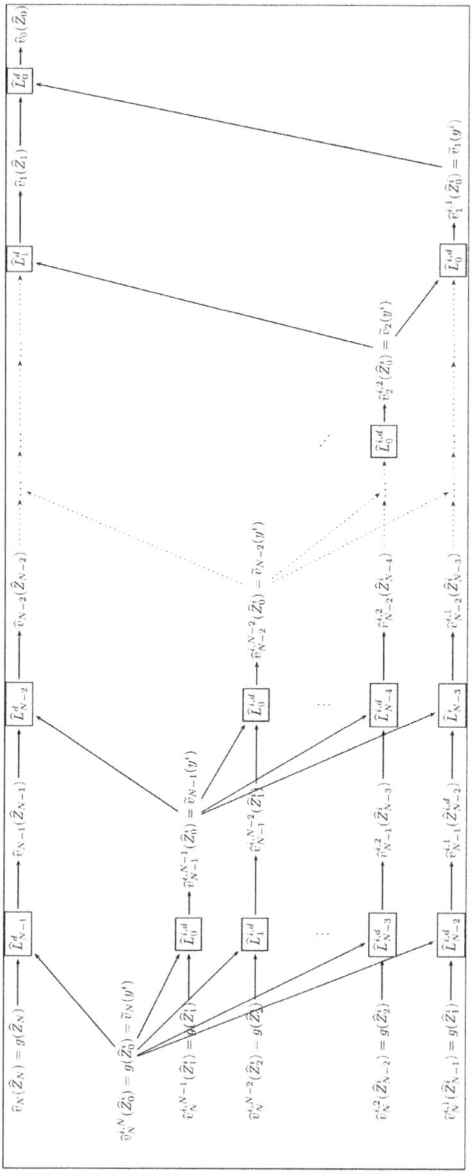

Figure 10.1. *Step-by-step procedure for approximating the value function of the impulse control problem*

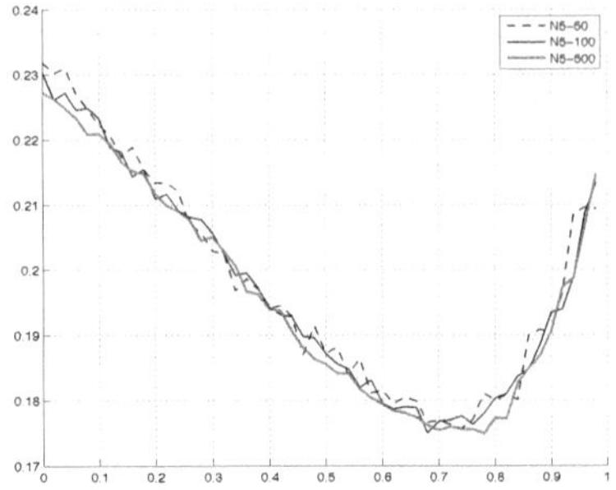

(a) Horizon $N = 5$

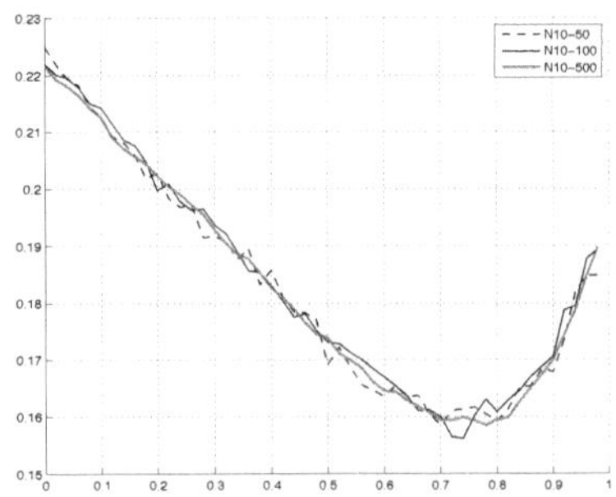

(b) Horizon $N = 10$

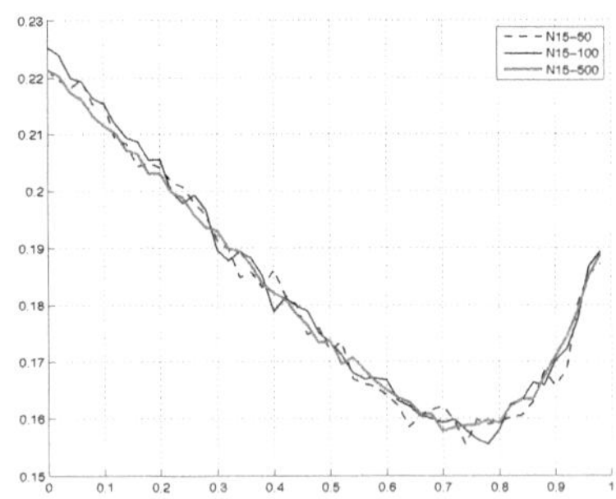

(c) Horizon $N = 15$

Figure 10.2. *Approximated value function for the impulse control of the TCP process*

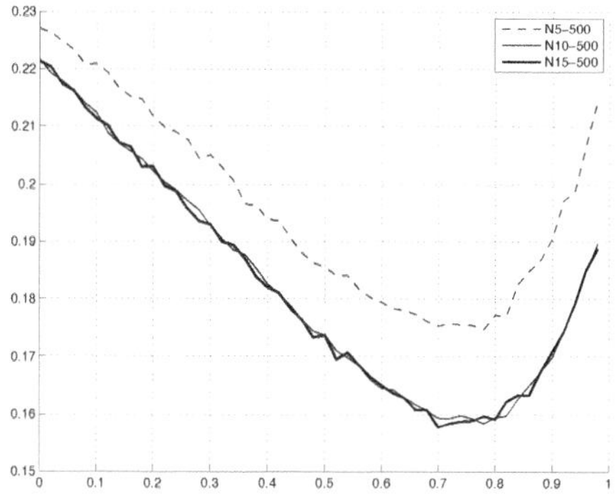

Figure 10.3. *Approximated value function for* $N = 5$, $N = 10$ *and* $N = 15$ *for* 500 *points in the quantization grids for the TCP process*

Bibliography

[ALD 87] ALDEMIR T., "Computer-assisted Markov failure modeling of process control systems", *IEEE Transactions on Reliability*, vol. 36, no. 4, pp. 133–144, 1987.

[ARJ 92] ARJAS E., HAARA P., NORROS I., "Filtering the histories of a partially observed marked point process", *Stochastic Processes and their Applicationss*, vol. 40, no. 2, pp. 225–250, 1992.

[AVE 99] AVEN T., JENSEN U., *Stochastic Models in Reliability*, Springer, New York, 1999.

[AZA 14] AZAÏS R., BARDET J.-B., GÉNADOT A. *et al.*, "Piecewise deterministic Markov process, recent results", *ESAIM: Proceedings*, vol. 44, pp. 276–290, 2014.

[BAL 03] BALLY V., PAGÈS G., "A quantization algorithm for solving multi-dimensional discrete-time optimal stopping problems", *Bernoulli*, vol. 9, no. 6, pp. 1003–1049, 2003.

[BAL 05] BALLY V., PAGÈS G., PRINTEMS J., "A quantization tree method for pricing and hedging multidimensional American options", *Mathematical Finance*, vol. 15, no. 1, pp. 119–168, 2005.

[BAY 13a] BAYSSE C., Analyse et optimisation de la fiabilité d'un équipement opto-électronique équipé de HUMS, PhD Thesis, Université Sciences et Technologies Bordeaux I, 2013.

[BAY 13b] BAYSSE C., BIHANNIC D., DE SAPORTA B. *et al.*, "Optimisation de la maintenance d'un équipement optronique", *45émes Journées de Statistique*, pp. 1–6, Toulouse, 2013.

[BAY 13c] BAYSSE C., BIHANNIC D., GÉGOUT PETIT A. *et al.*, "Maintenance optimization of optronic equipment", *Prognostics and System Health Management Conference*, Milano, vol. 33, pp. 709–714, Italy, 2013.

[BIH 13] BIHANNIC D., BAYSSE C., DE SAPORTA B. *et al.*, Procédé de maintenance d'un équipement, Patent: France N° 1303026 no. N/Ref.: 068689 FR MPH/MAG, Thales, Inria, 2013.

[BOY 10] BOYER F., COLIN M., SANTUGINI K., Evaluation des performances, rapport partiel, ANR report, available at http://fautocoes.bordeaux.inria.fr/files/PartialReport2-1.pdf, 2010.

[BOY 12] BOYER F., COLIN M., SANTUGINI K., Evaluation des performances, rapport final, RANR report, available at http://fautocoes.bordeaux.inria.fr/files/FinalReport2-1.pdf, 2012.

[BRA 11] BRANDEJKY A., DE SAPORTA B., DUFOUR F. *et al.*, "Numerical method for the distribution of a service time", *ESREL 11*, Troyes, pp. 1–7, France, 2011.

[BRA 12a] BRANDEJSKY A., Méthodes numériques pour les processus markoviens déterministes par morceaux, PhD Thesis, University of Bordeaux 1, 2012.

[BRA 12b] BRANDEJSKY A., DE SAPORTA B., DUFOUR F., "Numerical method for expectations of piecewise deterministic Markov processes", *Communications in Applied Mathematics and Computational Science*, vol. 7, no. 1, pp. 63–104, 2012.

[BRA 12c] BRANDEJSKY A., DE SAPORTA B., DUFOUR F., "Numerical methods for the exit time of a piecewise-deterministic Markov process", *Advances in Applied Probability*, vol. 44, no. 1, pp. 196–225, 2012.

[BRA 13] BRANDEJSKY A., DE SAPORTA B., DUFOUR F., "Optimal stopping for partially observed piecewise-deterministic Markov processes", *Stochastic Processes and their Applications*, vol. 123, no. 8, pp. 3201–3238, 2013.

[BRÉ 81] BRÉMAUD P., *Point Processes and Queues – Martingale Dynamics*, Springer Series in Statistics, Springer-Verlag, New York, 1981.

[CHI 08] CHIQUET J., LIMNIOS N., "A method to compute the transition function of a piecewise deterministic Markov process with application to reliability", *Statistics & Probability Letters*, vol. 78, no. 12, pp. 1397–1403, 2008.

[COC 97] COCOZZA-THIVENT C., *Processus stochastiques et fiabilité des systèmes*, Springer-Verlag, Berlin, 1997.

[COC 06] COCOZZA-THIVENT C., EYMARD R., MERCIER S., "A finite-volume scheme for dynamic reliability models", *IMA Journal of Numerical Analysis*, vol. 26, no. 3, pp. 446–471, 2006.

[COJ 96] COJAZZI G., "The DYLAM approach for dynamic reliability analysis of systems", *Reliability Engineering and System Safety*, vol. 52, pp. 279–296, 1996.

[COS 88] COSTA O.L.V., DAVIS M.H.A., "Approximations for optimal stopping of a piecewise-deterministic process", *Mathematics of Control, Signals, and Systems*, vol. 1, no. 2, pp. 123–146, 1988.

[COS 89] COSTA O.L.V., DAVIS M.H.A., "Impulse control of piecewise-deterministic processes", *Mathematics of Control, Signals, and Systems*, vol. 2, no. 3, pp. 187–206, 1989.

[COS 93] COSTA O.L.V., "Discretizations for the average impulse control of piecewise deterministic processes", *Journal of Applied Probability*, vol. 30, no. 2, pp. 405–420, 1993.

[COS 00] COSTA O.L.V., RAYMUNDO C.A.B., DUFOUR F., "Optimal stopping with continuous control of piecewise deterministic Markov processes", *Stochastics and Stochastics Reports*, vol. 70, nos. 1–2, pp. 41–73, 2000.

[COS 08] COSTA O.L.V., DUFOUR F., "Stability and ergodicity of piecewise deterministic Markov processes", *SIAM Journal on Control and Optimization*, vol. 47, no. 2, pp. 1053–1077, 2008.

[DAV 84] DAVIS M.H.A., "Piecewise deterministic Markov processes: a general class of non-distribution stochastic models", *Journal of the Royal Statistical Society*, vol. 46, no. 3, pp. 353–388, 1984.

[DAV 93] DAVIS M.H.A., *Markov Models and Optimization*, Chapman & Hall, London 1993.

[DEM 95] DEMPSTER M.A.H., YE J.J., "Impulse control of piecewise deterministic Markov processes", *The Annals of Applied Probability*, vol. 5, no. 2, pp. 399–423, 1995.

[DEV 97] DEVOOGHT J., *Dynamic Reliability*, Chapman and Hall, Berlin, 1997.

[DOU 15] DOUMIC M., HOFFMANN M., KRELL N. *et al.*, "Statistical estimation of a growth-fragmentation model observed on a genealogical tree", *Bernoulli*, vol. 21, no. 3, pp. 1760–1799, 2015.

[DUF 99] DUFOUR F., COSTA O.L.V., "Stability of piecewise-deterministic Markov processes", *SIAM Journal on Control and Optimization*, vol. 37, no. 5, pp. 1483–1502, 1999.

[DUF 02] DUFOUR F., DUTUIT Y., "Dynamic reliability: a new model", *Proceedings of ESREL 2002 Lambda-Mu 13 Conference*, pp. 350–353, 2002.

[DUT 97] DUTUIT Y., CHÂTELET E., SIGNORET J. *et al.*, "Dependability modelling and evaluation by using stochastic Petri nets: application to two test cases", *Reliability Engineering and System Safety*, vol. 55, pp. 117–124, 1997.

[EDO 10] EDOLI E., RUNGGALDIER W.J., "On optimal investment in a reinsurance context with a point process market model", *Insurance: Mathematics & Economics*, vol. 47, no. 3, pp. 315–326, 2010.

[ELL 82] ELLIOTT R.J., *Stochastic Calculus and Applications*, Springer-Verlag, New York, 1982.

[EYM 08] EYMARD R., MERCIER S., PRIGNET A., "An implicit finite volume scheme for a scalar hyperbolic problem with measure data related to piecewise deterministic Markov processes", *Journal of Computational and Applied Mathematics*, vol. 222, no. 2, pp. 293–323, 2008.

[FRI 14] FRITSCH C., Probabilistic and numerical approaches of individual-based chemostat model, PhD thesis, University of Montpellier 2, December 2014.

[GAT 91] GATAREK D., "On first-order quasi-variational inequalities with integral terms", *Applied Mathematics and Optimization*, vol. 24, no. 1, pp. 85–98, 1991.

[GAT 92] GATAREK D., "Optimality conditions for impulsive control of piecewise-deterministic processes", *Mathematics of Control, Signals, and Systems*, vol. 5, no. 2, pp. 217–232, 1992.

[GIL 77] GILLESPIE D.T., "Exact stochastic simulation of coupled chemical reactions", *The Journal of Physical Chemistry*, vol. 81, no. 25, pp. 2340–2361, 1977.

[GIL 01] GILLESPIE D.T., "Approximate accelerated stochastic simulation of chemically reacting systems", *The Journal of Chemical Physics*, vol. 115, no. 4, pp. 1716–1733, 2001.

[GON 10] GONZALEZ K., Contribution à l'étude des processus markoviens déterministes par morceaux, PhD Thesis, University of Bordeaux 1, 2010.

[GRA 98] GRAY R.M., NEUHOFF D.L., "Quantization", *IEEE Transactions on Information Theory*, vol. 44, no. 6, pp. 2325–2383, 1998.

[GRA 00] GRAF S., LUSCHGY H., *Foundations of Quantization for Probability Distributions*, Springer-Verlag, Berlin, 2000.

[GRA 02] GRALL A., BÉRENGUER C., DIEULLE L., "A conditional-based maintenance policy for a stochastically deteriorating system", *Reliability Engineering and System Safety*, vol. 76, no. 2, pp. 167–180, 2002.

[GUG 86] GUGERLI U.S., "Optimal stopping of a piecewise-deterministic Markov process", *Stochastics*, vol. 19, no. 4, pp. 221–236, 1986.

[HEL 01] HELMES K., RÖHL S., STOCKBRIDGE R.H., "Computing moments of the exit time distribution for Markov processes by linear programming", *Operations Research*, vol. 49, no. 4, pp. 516–530, 2001.

[KUS 77] KUSHNER H.J., *Probability Methods for Approximations in Stochastic Control and for Elliptic Equations*, Mathematics in Science and Engineering, Academic Press, New York, vol. 129, 1977.

[LAS 04] LASSERRE J.-B., PRIETO-RUMEAU T., "SDP vs. LP relaxations for the moment approach in some performance evaluation problems", *Stochastic Models*, vol. 20, no. 4, pp. 439–456, 2004.

[LEN 85] LENHART S., LIAO Y.C., "Integro-differential equations associated with optimal stopping time of a piecewise-deterministic process", *Stochastics*, vol. 15, no. 3, pp. 183–207, 1985.

[LEN 89] LENHART S.M., "Viscosity solutions associated with impulse control problems for piecewise-deterministic processes", *International Journal of Mathematics and Mathematical Sciences*, vol. 12, no. 1, pp. 145–157, 1989.

[LI 11] LI J., MOSLEHB A., KANGA R., "Likelihood ratio gradient estimation for dynamic reliability applications", *Reliability Engineering and System Safety*, vol. 96, pp. 1667–1679, 2011.

[LUD 12] LUDKOVSKI M., SEZER S.O., "Finite horizon decision timing with partially observable Poisson processes", *Stochastic Models*, vol. 28, no. 2, pp. 207–247, 2012.

[MAR 94] MARSEGUERRA M., "Approximated physical modelling in dynamic PSA using artificial neural networks", *Reliability Engineering & System Safety*, vol. 45, pp. 47–56, 1994.

[MAR 95] MARSEGUERRA M., ZIO E., "The cell-to-boundary method in Monte Carlo-based dynamic PSA", *Reliability Engineering and System Safety*, vol. 48, no. 3, pp. 199–204, 1995.

[MAR 96] MARSEGUERRA M., ZIO E., "Monte Carlo approach to PSA for dynamic process systems", *Reliability Engineering and System Safety*, vol. 52, no. 3 pp. 227–241, 1996.

[VAN 09] VAN NOORTWIJK J.M., "A survey of the application of gamma processes in maintenance", *Reliability Engineering and System Safety*, vol. 94, no. 1, pp. 2–21, 2009.

[OUV 04] OUVRARD J.-Y., *Probabilites 2*, Cassini, 2004.

[PAG 98] PAGÈS G., "A space quantization method for numerical integration", *Journal of Computational and Applied Mathematics*, vol. 89, no. 1, pp. 1–38, 1998.

[PAG 04a] PAGÈS G., PHAM H., PRINTEMS J., "An optimal Markovian quantization algorithm for multi-dimensional stochastic control problems", *Stochastics and Dynamics*, vol. 4, no. 4, pp. 501–545, 2004.

[PAG 04b] PAGÈS G., PHAM H., PRINTEMS J., "Optimal quantization methods and applications to numerical problems in finance", *Handbook of Computational and Numerical Methods in Finance*, Birkhäuser, Boston, pp. 253–297, 2004.

[PAG 15] PAGÈS G., "Introduction to optimal vector quantization and its applications to numerics", *ESAIM: Proceedings and Surveys*, vol. 48, pp. 29–79, 2015.

[PAK 10] PAKDAMAN K., THIEULLEN M., WAINRIB G., "Fluid limit theorems for stochastic hybrid systems with application to neuron models", *Advances in Applied Probability*, vol. 42, no. 3, pp. 761–794, 2010.

[PHA 05] PHAM H., RUNGGALDIER W., SELLAMI A., "Approximation by quantization of the filter process and applications to optimal stopping problems under partial observation", *Monte Carlo Methods and Applications*, vol. 11, no. 1, pp. 57–81, 2005.

[RIE 12] RIEDLER M.G., THIEULLEN M., WAINRIB G., "Limit theorems for infinite-dimensional piecewise deterministic Markov processes. Applications to stochastic excitable membrane models", *Electronic Journal of Probability*, vol. 17, no. 55, pp. 48, 2012.

[RIE 13] RIEDLER M.G., "Almost sure convergence of numerical approximations for piecewise deterministic Markov processes", *Journal of Computational and Applied Mathematics*, vol. 239, pp. 50–71, 2013.

[ROB 14] ROBERT L., HOFFMANN M., KRELL N. *et al.*, "Division in Escherichia coli is triggered by a size-sensing rather than a timing mechanism", *BMC Biology*, vol. 12, no. 17, pp. 1–10, 2014.

[SAP 09] DE SAPORTA B., DUFOUR F., GONZALEZ K., "Numerical method for optimal stopping of hybrid processes", *3rd IFAC International Conference on Analysis and Design of Hybrid Systems*, Zaragoza Spain, pp. 114–119, 2009.

[SAP 10a] DE SAPORTA B., DUFOUR F., "Numerical method for optimal stopping of piecewise deterministic Markov processes", in PIUNOVSKIY A. (ed.), *Modern Trends in Controlled Stochastic Processes*, Luniver Press, pp. 44–64, 2010.

[SAP 10b] DE SAPORTA B., DUFOUR F., GONZALEZ K., "Numerical method for optimal stopping of piecewise deterministic Markov processes", *The Annals of Applied Probability*, vol. 20, no. 5, pp. 1607–1637, 2010.

[SAP 10c] DE SAPORTA B., DUFOUR F., ZHANG H. *et al.*, "Arrêt optimal pour la maintenance prédictive", *Lambda-Mu 17*, La Rochelle France, pp. 1–8, 2010.

[SAP 11] DE SAPORTA B., DUFOUR F., ZHANG H., "Approximation of the value function of an impulse control problem", *IFAC 18th World Congress*, Milan, Italy, pp. 3474–3479, 2011.

[SAP 12a] DE SAPORTA B., DUFOUR F., "Numerical method for impulse control of piecewise deterministic Markov processes", *Automatica*, vol. 48, no. 5, pp. 779–793, 2012.

[SAP 12b] DE SAPORTA B., DUFOUR F., ZHANG H., "Predictive maintenance for the heated hold-up tank", *PSAM11-ESREL12*, Helsinki, Finland, pp. 1–10, 2012.

[SAP 12c] DE SAPORTA B., DUFOUR F., ZHANG H. *et al.*, "Optimal stopping for the predictive maintenance of a structure subject to corrosion", *Proceedings of the Institution of Mechanical Engineers, Part O: Journal of Risk and Reliability*, vol. 226, no. 2, pp. 169–181, 2012.

[SAP 13] DE SAPORTA B., ZHANG H., "Predictive maintenance for the heated hold-up tank", *Reliability Engineering & System Safety*, vol. 115, pp. 82–90, 2013.

[SCH 06] SCHOENIG R., AUBRY J., CAMBOIS T. *et al.*, "An aggregation method of Markov graphs for the reliability analysis of hybrid systems", *Reliability Engineering and System Safety*, vol. 91, pp. 137–148, 2006.

[SIU 94] SIU N., "Risk assessment for dynamic systems: an overview", *Reliability Engineering and System Safety*, vol. 43, pp. 43–73, 1994.

[TOM 96] TOMBUYSES B., ALDEMIR T., "Continuous cell-to-cell mapping and dynamic PSA", *Proceedings of ICONE 4 Conference*, pp. 431–438, 1996.

[VEL 15] VELTZ R., A new twist for the simulation of hybrid systems using the true jump method, available at http://arxiv.org/abs/1504.06873, 2015.

[ZHA 08] ZHANG H., DUFOUR F., DUTUIT Y. *et al.*, "Piecewise deterministic Markov processes and dynamic reliability", *Proceedings of the Institution of Mechanical Engineers, Part O: Journal of Risk and Reliability*, vol. 222, no. 4, pp. 545–551, 2008.

Index

A, B

C, D, E

T, V, Z

Other titles from

iSTE

in

Mathematics and Statistics

2015

DEVOLDER Pierre, JANSSEN Jacques, MANCA Raimondo
Basic Stochastic Processes

2014

COOKE Roger M., NIEBOER Daan, MISIEWICZ Jolanta
Fat-tailed Distributions: Data, Diagnostics and Dependence

MACKEVIČIUS Vigirdas
Integral and Measure: From Rather Simple to Rather Complex

PASCHOS Vangelis Th
Combinatorial Optimization – 3-volume series – 2nd edition
Concepts of Combinatorial Optimization / Concepts and Fundamentals – volume 1
Paradigms of Combinatorial Optimization – volume 2
Applications of Combinatorial Optimization – volume 3

2013

COUALLIER Vincent, GERVILLE-RÉACHE Léo, HUBER Catherine, LIMNIOS Nikolaos, MESBAH Mounir
Statistical Models and Methods for Reliability and Survival Analysis

JANSSEN Jacques, MANCA Oronzio, MANCA Raimondo
Applied Diffusion Processes from Engineering to Finance

SERICOLA Bruno
Markov Chains: Theory, Algorithms and Applications

2012

BOSQ Denis
Mathematical Statistics and Stochastic Processes

CHRISTENSEN Karl Bang, KREINER Svend, MESBAH Mounir
Rasch Models in Health

DEVOLDER Pierre, JANSSEN Jacques, MANCA Raimondo
Stochastic Methods for Pension Funds

2011

MACKEVIČIUS Vigirdas
Introduction to Stochastic Analysis: Integrals and Differential Equations

MAHJOUB Ridha
Recent Progress in SCombinatorial Optimization – ISCO2010

RAYNAUD Hervé, ARROW Kenneth
Managerial Logic

2010

BAGDONAVIČIUS Vilijandas, KRUOPIS Julius, NIKULIN Mikhail
Nonparametric Tests for Censored Data

BAGDONAVIČIUS Vilijandas, KRUOPIS Julius, NIKULIN Mikhail
Nonparametric Tests for Complete Data

IOSIFESCU Marius *et al.*
Introduction to Stochastic Models

VASSILIOU PCG
Discrete-time Asset Pricing Models in Applied Stochastic Finance

2008

ANISIMOV Vladimir
Switching Processes in Queuing Models

FICHE Georges, HÉBUTERNE Gérard
Mathematics for Engineers

HUBER Catherine, LIMNIOS Nikolaos *et al.*
Mathematical Methods in Survival Analysis, Reliability and Quality of Life

JANSSEN Jacques, MANCA Raimondo, VOLPE Ernesto
Mathematical Finance

2007

HARLAMOV Boris
Continuous Semi-Markov Processes

2006

CLERC Maurice
Particle Swarm Optimization

Printed and bound by CPI Group (UK) Ltd, Croydon, CR0 4YY

13/09/2026

14961017-0002